Pectins and pectinases

Pectins and pectinases

edited by

H.A. Schols, R.G.F. Visser and A.G.J. Voragen

Wageningen Academic Publishers

ISBN 978-90-8686-108-8

Cover image: Marie-Christine Ralet

First published, 2009

Preface

Pectin is one of the major components of the cell wall of dicotyledonous plants and probably one of the most complex macromolecules in nature. It is present in the middle lamella, primary and secondary walls and is deposited in the early stages of growth during cell expansion. It provides quite divers functionalities to plants ranging from rigidity of plant tissues to resistance to plant pathogens.

Pectic substances also have a great impact on the quality of fresh and processed foods particularly fruits and vegetables. Pectin extracted from suitable plant sources is used as food ingredient for its gelling, stabilizing and thickening functionalities. Plant products, fresh, extracted or processed, constitute a large part of the human diet. As a fibre, naturally present in these food products, pectic substances fulfil a nutritional function and gain increasingly interest as a health promoting polysaccharide.

To understand all these functions of pectin we need more knowledge about its biosynthesis and deposition in the plant cell wall, the architecture of the wall, the chemical fine structure and physical behaviour.

This book compiles chapters which were submitted to a review process. It provides an update about the current state of the art and research and clearly demonstrates that significant progress has been made during the past years particularly in the fields of biosynthesis, health modulating activities of pectin fractions, chemical structure analysis of pectin fractions of various sources and cell wall regions using advanced spectroscopic techniques and immunodetection combined with microscopy and chromatography, genomics of pectic enzymes of *Aspergillus niger*, and interaction of pectins with proteins. This progress allows us more and more to identify (and influence) the functionality of pectins and pectic enzymes, both *in vitro* after isolation as well as in plants themselves (*in planta*). This progress is also reflected in new applications of pectin and pectin degrading enzymes.

'Pectins and pectinases' provides a state-of-the-art account for both beginners and experienced researchers in almost all disciplines of pectin research. We hope that it will satisfy your interest and inspire you in your contributions to this important and fast developing research field.

Henk Schols, Richard Visser and Fons Voragen, editors

Table of contents

Part 2 – Pectinases

Part 3 – Pectins in plant cell walls

Part 4 – Functionality and applications of pectins

Part 5 – Health aspects of pectins

Part 1

Structure of pectins

Revealing pectin's structure

H.A. Schols, G.-J. Coenen and A.G.J. Voragen
Wageningen University, Laboratory of Food Chemistry, Bomenweg 2, 6703 HD Wageningen, the Netherlands

Abstract

Research towards revealing the structure of the plant cell wall polymer pectin which is being considered as the most complex molecule in nature, lasts already for many decades. Nature's ability to tailor the chemical structure of pectin molecule according to the plant cell wall and plant tissue needs at that particular moment really complicates such efforts to reveal the structure enormously. As already stated by the past professor Pilnik, the pre-eminent pectin expert in his time, it is more and more realised that 'no pectin molecule is alike'. As a result of the ongoing search, pectin models proposed in the early nineties are considered not to cover all phenomena observed and being not able to explain all new findings. A new model has been proposed some years ago. This new model further initiated a renewed discussion on the precise chemical structure of the various pectin structural elements, their interconnection and their assembly within the cell wall and has initiated quite some new research. Part of the efforts made in the Laboratory of Food Chemistry in Wageningen within the PhD research of Coenen (2007) in order to reveal the connection points between the different pectic subunits and the establishment/development of novel analytical techniques needed for such an investigation are summarised in this paper. Oligosaccharides, obtained by controlled acid hydrolysis of apple pectin, were characterized using optimized LC-MS and NMR approaches. The structures established for these oligosaccharides indicated that in apple pectin both homogalacturonan (HG) and xylogalacturonan (XGA) were covalently linked to rhamnogalacturonan I (RGI). A newly developed method coupled capillary electrophoresis (CE) on-line to mass spectrometry (MS) and allowed identification of compounds present in CE electropherograms of complex apple XGA and RGI digests. By combining β-eliminative breakdown with CE-LIF and CE-UV-MS^n analysis, detailed information on the neutral sugar side chain structure and distribution was obtained of some enzyme resistant structural elements of soy pectin. Using selective degradation by β-elimination, it was demonstrated that arabinans containing a rudimentary RGI backbone fragment play an important role within the complement fixing activity of white cabbage (Brassica oleracea) pectin. Based on the results obtained in this research and data reported in literature, a pectin model is proposed which positions HG both in-line and as a side chain of RGI. In addition, new insights in the position of XGA with respect to HG and RGI were obtained. A further increase of our knowledge on pectin and polysaccharide structures can be achieved by combining the 'combined approaches' described in this contribution with other chemical and novel enzymatic degradation methods and by implementing these methods more quantitatively.

Keywords: homogalacturonan, rhamnogalacturonan I, pectin, hypothetical model

1. Introduction

Pectin is an important component of the cell wall of many plant tissues. As such it plays an important role during ripening, storage and processing of plant raw materials for foods. It is also extracted from suitable plant sources and as such widely used in food industry because of its gelling, stabilizing, and thickening properties (Laurent and Boulenguer, 2003). Recently, pectin has gained interest as a functional food, because of possible health promoting effects (Yamada *et al.*, 2003, Yamada and Kiyohara, 2007). In all of these applications, the chemical fine structure of pectin strongly determines its functionality.

Even after more than 180 years of pectin research, the structure of pectin is still under debate (Braconnot, 1825; Vincken *et al.*, 2003a; Vincken *et al.*, 2003b). At this moment, there are two favorite models for describing the pectin structure: the smooth and hairy region model (Schols and Voragen, 1996), and the rhamnogalacturonan (RG) backbone model (Figure 1).

In the first model, pectin is composed of hairy regions, consisting of rhamnogalacturonan I (RGI) decorated with neutral sugar side-chains, interspersed with smooth regions of (methyl esterified) homogalacturonan (HG) (Schols and Voragen, 1996). The second model positions HG as a side chain of RGI, similar to the neutral sugar side chains (Vincken *et al.*, 2003a). The placement of xylogalacturonan (XGA) in respect to RGI is indistinct, but regarded exchangeable with HG (Huisman *et al.*, 2001; Vincken *et al.*, 2003b).

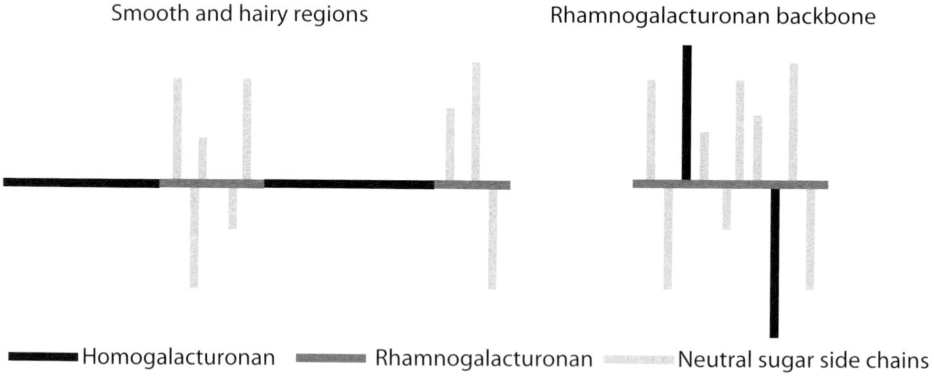

Figure 1. Schematic representation of two different models describing the hypothetical pectin structure, adapted from Vincken et al. *(2003a).*

As the main difference between these models is formed by the connection of HG to RGI, our research focused on the identification and characterization of oligosaccharides, which constitute these connection points and contain information about the linkages between different structure elements. In order to be able to analyze these components the analytical toolbox had to be extended. By combining different analytical techniques we were able to develop new methodology and to identify some of these connecting oligosaccharides, which provided us with new insight in the macromolecular build up of the pectin polymer.

2. Research strategy

To identify the connecting fragments in an oligosaccharide mixture the off-line HPAEC-Maldi-tof MS method (Kabel *et al.*, 2001) was further improved, in order to enable MS analysis of galacturonic acid (GalA) containing oligosaccharides. For the first time, oligosaccharides of hybrid nature, consisting out of both HG and RGI sequences together with an oligosaccharide composed out of both XGA and RGI sequences were isolated out of an apple modified hairy regions (MHR) hydrolysate and characterized using chromatography and mass spectrometry (Coenen *et al.*, 2007). For identification of the linkages between HG/XGA and RGI the enzymes exo polygalacturonase (ExoPG) and rhamnogalacturonan galacturono hydrolase (RGGH) were used as diagnostic tools. This allowed refinement of the currently used pectin models, with emphasis on the linkages between the different structural elements.

As part of our research strategy, methods were improved to be used in polysaccharide characterization. An effective separation of differently substituted GalA residues was obtained by low pH capillary electrophoresis-laser induced fluorescence (CE-LIF) analysis (Coenen *et al.*, 2008). By adapting the buffer pH and capillary used, on-line MS^n was enabled. With this combined CE-MS approach peaks present in the LIF electropherograms of XGA and RGI digests could be annotated.

CE-LIF-MS^n method was also used to identify products obtained after chemical degradation of pectins. The application of mild chemical conditions promoting β-eliminative breakdown (Deng *et al.*, 2006) allowed specific hydrolysis of the (enzyme resistant) RGI backbone while preserving the various neutral sugar side chains.

Another study was conducted on white cabbage pectin, which was found to activate the human complement (Westereng, 2007; Westereng *et al.*, 2009). The complement system plays an important role in the human body as a primary defense system against bacterial and viral infections (Helmy *et al.*, 2006). After β-eliminative degradation, purification and identification of pectin fragments, it was found that short pectin side chains had no bioactivity, while larger side chain fragments contained a (much) lower activity than polymers containing intact RGI with more arabinan side chains, which implied that structural elements containing multiple side chains expressed a higher complement fixing activity.

The newly obtained information about pectin structure is summarized and the implications of these new findings for the pectin models are discussed. The importance of having detailed structure information for the study of pectin in biological, physiological, pharmaceutical, medical and health applications is mentioned briefly. In addition, the possible applications of CE-MS within carbohydrate research are discussed.

3. Elucidation of the chemical structure of pectin

Over the years numerous publications dealing with the structure of pectin have been published. The findings recently published by Yapo *et al.* (2007), were of specific importance for the interpretation of the results obtained in our study. It was found that citrus peel pectin consisted of HG segments with a DP in the range of 81-117 GalA units, irrespective of their extraction method. Pectins were hydrolysed in such way that they were able to recover around 90% of all RGI segments as chains having an average length of 30-40 residues.

3.1 Identification of homo- or xylogalacturonan segments attached to RGI

In a structure characterization study using controlled acid hydrolysis, three different hybrid oligosaccharides were identified in apple MHR, composed out of part of a HG segment and part of a RGI segment. For one hybrid oligosaccharide the connection between the different structural elements was confirmed to be an GalA*p*α-(1,2) Rha*p* linkage (Coenen *et al.*, 2007). In addition, a $GalA_6Rha_3Xyl_1$ oligosaccharide was identified, which consisted out of both a XGA and a RGI sequence, connected via an GalA*p*α-(1,2)Rha*p* linkage. The xylose substitution was putatively located on the second GalA moiety on the non-reducing end. These oligosaccharides indicated that in apple pectin both HG and XGA were covalently linked to RGI. The length of the GalA sequences, within the different HG-RGI hybrid oligosaccharides, was three GalA residues. Possibly, this reoccurring length is the result of Endo-polygalacturonase (EndoPG) and Exo-Polygalacturonase (ExoPG) enzyme activity during the MHR preparation due to sterical hindrance of neighboring sugar moieties or the requirement for at least three unsubstituted GalA residues on the reducing end of the enzyme' active site. On the other hand, the controlled acid hydrolysis treatment used to generate these elements out of apple MHR is expected to cleave some of the backbone linkages, hereby directly influencing the remaining pectin structure. Nevertheless, the identification of three different oligosaccharides, containing 3 galacturonic acid residues on the non-reducing end was remarkable.

The limited HG chain length of the hybrid HG-RGI oligosaccharides gives room for the hypotheses that these fragments are resulting from a linkage between XGA and RGI, as demonstrated in the fourth characterized oligosaccharide $GalA_6Rha_3Xyl_1$. It is possible that the GalA sequences (which consist of three GalA sugar moieties) are originating from XGA as well. It is not unconceivable that a plant has a fixed building pattern for the arrangement of the different pectin structure elements. Therefore, it could be possible that all of the characterized hybrid oligosaccharides are resulting from a XGA-RGI linkage (Figure 2). Hypothetically, HG would than be connected on the terminal end of the rather homogeneous XGA (Figure 2a), or the degree of xylosidation would decrease when coming closer to the pure HG segment at the non-reducing end (Figure 2b).

Such a substitution pattern could be caused by a xylosidase, releasing xylose residues from XGA, which would be (sterically) hindered when approaching the RGI backbone. Xylogalacturonan is considered to be synthesized analogous to apiogalacturonan, where the apiose residue is likely to be incorporated during the synthesis of the galacturonan backbone (Mohnen, 2002). This could imply that highly substituted galacturonans are incorporated into pectin and are further modified by putative xylosidases.

This would explain the release of XGA when hairy regions polysaccharides were treated by RGH (Mutter *et al.*, 1996), which indicates a covalent connection of XGA and RGI in the cell wall (Albersheim *et al.*, 1996). Schols *et al.* (1995) observed that after removal of HG from MHR by EndoPG, XGA was retained in the high M_w region, indicating as well that XGA is located close, or is part of, the RGI backbone. In addition, such observations suggest that if there is a direct linkage between XGA and HG, this linkage will only be located at one side, since no XGA segments are liberated by this EndoPG treatment. Furthermore, the increased degradability of XGA by XGH after removal of HG by EndoPG, indicates that these elements are located in each other vicinity within the pectin molecule (Zandleven, 2006).

The absence of xylose substitutions on the hybrid HG-RGI oligosaccharides, could indicate that also HG structural elements are attached to the non-reducing end of RGI. It has been hypothized that the presence of XGA could be restricted to certain reproductive tissues (Albersheim *et al.*, 1996). The low level of xylose (XGA) in

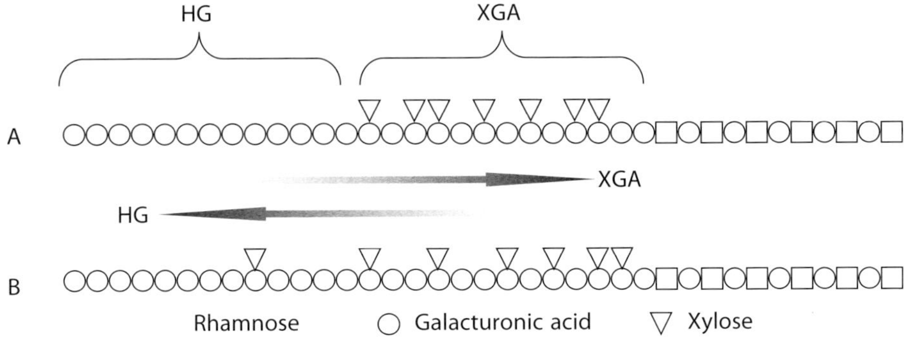

Figure 2. Two possibilities of linking HG via XGA structure to RGI. A) a clear and distinctive border between HG and XGA; B) a gradual increase in xylosidation of the galacturonan backbone, when approaching the RGI segment.

citrus pectin (Ros *et al.*, 1998) indicates that the statement that XGA is present in reproductive tissues is not valid. On the other hand, Zandleven *et al.* (2007) reported the presence of XGA in different plant tissues of Arabidopsis thaliana, which indicates that XGA is ubiquitous present in other dicotyledonous plant cell walls in contrast to what was believed previously (Albersheim *et al.*, 1996), albeit sometimes in small amounts. The suggestion that XGA is present on a wider scale in plant cell walls is further substantiated for e.g. white cabbage (Westereng *et al.*, 2009).

3.2 Positioning of structure elements in relation to RGI

In our structure characterization studies (Coenen, 2007, Coenen *et al.*, 2007) no sugar fragment was identified that contained a (xylo)galacturonan segment located on the reducing end of the RGI polymer. This could indicate that there are no alternating HG-RGI-HG-RGI segments present in the pectin backbone, as proposed by De Vries *et al.* (1982) and Schols and Voragen (1996). During controlled acid hydrolysis the more stable linkages were partly degraded, which resulted in the release of XGA, HG and RGI type oligosaccharides. The presence of RGI oligosaccharides demonstrates that controlled acid hydrolysis was mild enough to retain a part of the Rha-1,4-GalA in the backbone linkages, which are more susceptible to acid hydrolysis. Therefore it is expected that putative hybrid RGI-HG oligosaccharide structures would be liberated containing the HG/XGA also at the 'reducing end' of RGI, if such structures would exist. However, the absence of such hybrid oligosaccharides could be caused by the different susceptibilities of the structure elements for acid hydrolysis. The weaker linkage stability of rhamnose (Rha)-1,4-GalA compared to GalA-1,4-GalA and GalA-1,2- Rha, would result in a lower abundance of oligosaccharides with the RGI segment located on the non-reducing side, which could inhibit detection.

It has been reported that a RGI element is approximately 12 kDa, which is a chain of about 40 Rha-GalA moieties. In addition, a HG element equals about 17.5 kDa, which approximates 100 GalA units (Prade *et al.*, 1999; Yapo *et al.*, 2007; Zhan *et al.*, 1998). Consequently, a RGI-HG element has a Rha:GalA ratio of 1:3.7 However, the published Rha:GalA ratios of pectin from different fruit and vegetable sources are rather different: citrus peel (1:31), apple (1:21), bilberries (1:21), black currant (1:20), sugar beet (1:8), and soy (1:3.5) (Hilz *et al.*, 2005; Voragen *et al.*, 2001; Yapo *et al.*, 2007). Comparing both the reported ratios with the calculated ratio, it becomes clear that in general an excess of GalA exists (Coenen *et al.*, 2007; (Yapo *et al.*, 2007)). As a consequence, more HG chains per RGI unit should be present in pectin. Even when applying the alternating model, there would only be a minor increase in possible HG allocation sites. In the following paragraphs it will be discussed how this HG surplus may be allocated in the pectin molecule.

3.3 RGII as an alternative accommodation for the HG surplus

Pectin contains, next to HG, XGA, and RGI, another important structural element: rhamnogalacturonan II (RGII). It is known that RGII is allocated within HG elements of the pectin molecule (Ishii and Matsunaga, 2001). Furthermore, two RGII elements can be linked together via a boron ester linkage (O'Neill *et al.*, 2004). To explain the surplus of HG compared to RGI, a pectin structure as schematically represented in Figure 3 could be suggested. In this pectin structure, different HG elements are linked together via RGII boron ester linkages resulting in large pectin molecules.

However, for the existence of these HG-RGII-boron ester linkages it is required that a HG element contains two RGII regions (Figure 3); in other words the RGII to HG ratio should equal 2:1. In addition, a number of HG segments should be present, without a linkage to RGI. To quantify the level of RGII segments in a pectin mixture, the 2-O-methyl fucose (2-O-Fuc) content is measured, as the RGII monomer contains only one residue of this rare sugar (Darvill *et al.*, 1978; Hilz *et al.*, 2006). In literature, ratios of 1:142 for 2-O-Fuc to GalA have been published for black current pectin (Hilz *et al.*, 2006). As the length of HG elements of various pectins (apple, citrus or beet) described in literature varied between 81-117 GalA residues (M_w 17.5 kDa) (Thibault *et al.*, 1993; Yapo *et al.*, 2007), the 2-O-Fuc to GalA ratio in black current does not equal the 2:117 ratio, which would equal to two RGII units on each HG chain. Although the 2-O-Fuc:GalA ratio of only one pectin source is compared to the HG chain length of other pectin sources, this large discrepancy makes it unlikely that the excess of HG can fit in the proposed HG-RGII-boron crosslinking model. In addition, the release of pectins with a high degree of methyl esterification, thus rather calcium insensitive pectin, with chelating agent may partly be attributed to hydrolysis of the boron diester of RGII that is linked to homogalacturonan with a high degree of methylation (Hilz *et al.*, 2005). Although RGII structures are present in the chelating agent extract of black currents pectin, this extraction does not result in a large decrease in M_w (Hilz *et al.*, 2005), which indicates that the RGII linkage is not involved in connecting relatively small backbone structures. Furthermore, the relatively high molecular weight of dilute alkali extracted pectin (Hilz *et al.*, 2005) indicates that uronyl esters are not involved in the interconnection of the surplus HG/XGA to the RGI backbone.

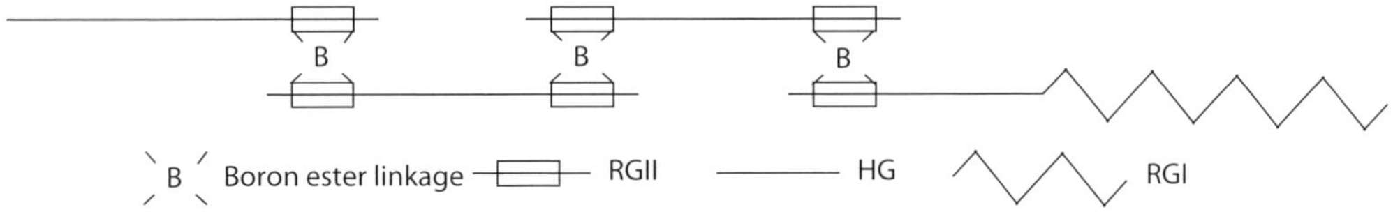

Figure 3. A pectin molecule backbone model in which homogalacturonan (HG) elements are linked to other HG segments or to HG-RGI (rhamnogalacturonan I) segments via rhamnogalacturonan II (RGII) boron ester linkages.

3.4 Both sequential and branched galacturonan substitutions to RGI?

In our study, four oligosaccharides were identified of hybrid nature, which contained both a HG and a RGI segment, where the HG segment was located at the non-reducing end. HG was only located as a backbone element at the non-reducing end of Rha, as evidenced by the absence of oligosaccharides, carrying the HG segment at the reducing end. But even when HG was located on both sides of RGI, there was a surplus of HG chains, which would advocate for more or longer HG chains per RGI unit (Coenen *et al.*, 2007; Yapo *et al.*, 2007)). Another possibility to explain the HG surplus is the presence of HG as side chains attached to RGI of the HG-RGI pectin backbone. Using Nuclear magnetic resonance (NMR) spectrometry approach Mort and Vasu (2007) reported findings supporting this hypothesis. Starting from enzyme digested saponified apple pectin obtained after acid extraction, the presence of an oligosaccharide is reported having the same structural features as the oligomer as identified by Coenen *et al.* (2007), which confirms the hypothosis that galacturonan is indeed attached in line to the non-reducing end of RGI, similar to the results reported by Coenen *et al.* (2007). Furthermore, using the ratio of the linkages as indicated by 1 dimensional proton (1H)-NMR and heteronuclear correlation (HMQC) NMR approaches, the presence of a relative high amount of terminal GalA and some terminal GalA with xylose (Xyl) attached through O-3 is reported (Mort and Vasu, 2007). Since the number of galactose and arabinose moieties is not sufficient to provide for all branched rhamnoses, these terminal (substituted) GalA sugar moieties are assumed to be connected to Rha within the RGI backbone. The number of xylosidated terminal GalA sugars indicates that in apple pectin XGA is positioned close to the RGI backbone, giving further support of the hypothesis that XGA is the galacturonan connected to RGI, where HG could than be linked to XGA.

These findings indicate that pectin consists of a RGI backbone with XGA segments linked to the RGI non-reducing end and to a number of the Rha moieties within RGI as side chains. The allocation of HG elements as side chains of the pectin backbone has also been proposed previously by Round *et al.* (1997; 2001) who used atomic force microscopy to investigate the nature of long branches attached to pectin.

Combining the observed excess of galacturonan segments and the identification of a linear connection between HG/XGA and RGI (Coenen *et al.*, 2007), with the results as presented by Mort and Vasu (2007), leads to a pectin backbone model structure as presented in Figure 4.

3.5 Possible explanation for the presence of linear HG/XGA to RGI linkage

It is not yet known whether pectin is synthesized as one polymer or whether it is assembled from the individual structural elements that become interconnected during or following insertion in the plant cell wall (Mohnen, 2002). The working hypothesis is that the structural elements are synthesized as independent polymers, which are

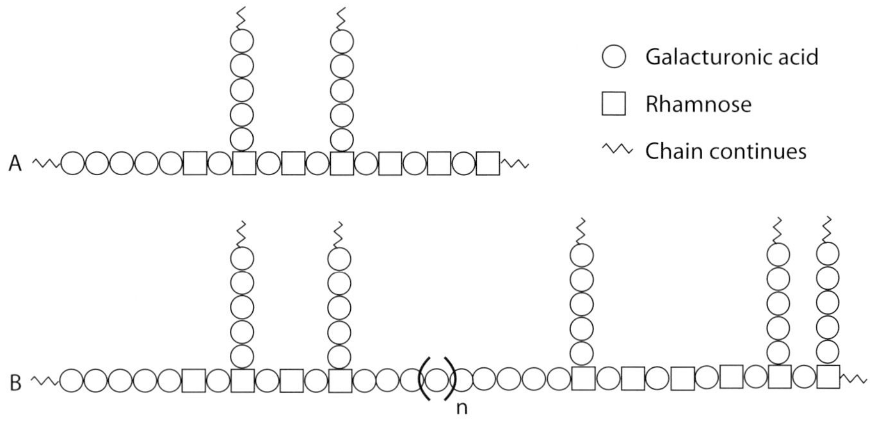

Figure 4. Putative pectin backbone models based on the observed excess of galacturonan segments and the identification of a linear connection between HG/XGA and RGI segments (Coenen et al., *2007), combined with the data provided by Mort and Vasu (2007). Xylose residues are not included, but it is expected that these are substituted to the GalA sugar moieties close to the RGI backbone to give XGA segments. A) RGI with HG side chains, connected to Rha residues. Furthermore HG is in line attached to the non-reducing end of RGI. B) Sequential backbone of alternating RGI/HG segments with additional HG side chains substituted on Rha residues. The distribution of the HG chains over the RGI segment is not taken into account.*

sublocalized in the Golgi apparatus, by different glycosyl transferases (Mohnen, 2002; Staehelin and Moore, 1995).

The identification of xyloglucan transferases in barley, which catalyzes the formation of linkages between xyloglucan and several heterologous receptor substrates (Hrmova *et al.*, 2007a), triggered the possibility/hypothesis that also transferases may exist able to link HG/XGA to RGI. Such a transferase fusion activity would most likely involve a polygalacturonase hydrolyzing enzyme, which would join HG to RGI, of which the terminal sugar moiety acceptor would be a GalA, since this would more likely fit in the active cleft of the transferase/hydrolase (Figure 5). The number of 52 identified open reading frames in Arabidopsis thaliana encoding for potential polygalacturonase genes (Benen and Visser, 2003), renders a polygalacturonase variant with a high transferase activity conceivable. In addition, based on the hypothesis that galacturonans exist as a side chain of RGI, transferase activities are expected, which realize these linkages. Consequently the linkage of galacturonan to the terminal end of RGI might be a side reaction of such putative transferase activity.

Putative transferase activities would explain the presence of galacturonan segments attached in a linear manner towards RGI. For the putative branches transferase or synthesizing enzymes remain to be identified. On the other hand, the homogeneous DP of HG from various sources as described by Yapo *et al.* (2007) does not correspond

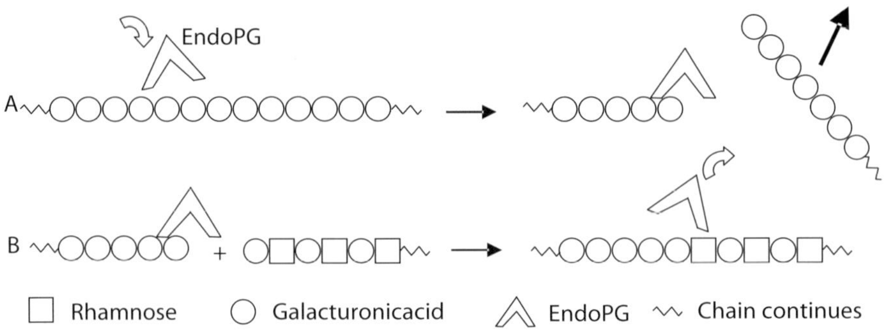

Figure 5. Schematic representation of the effect of a putative transferase activity of EndoPG, analogous to the activity of xyloglucan xyloglucosyl transferase (Hrmova et al., *2007a; Hrmova* et al., *2007b). A) Enzyme hydrolyses HG and retains a part of the polymer; B) Enzyme transfers retained oligosaccharide onto RGI acting as an acceptor substrate.*

with such putative transferase activity, which is expected to result in a more heterogeneous HG chain length.

The difference between acceptor concentration within the cell wall and in most laboratory conditions would complicate the identification of such synthesizing/transferase enzyme activity. It would be quite a challenge to test different endoPG's at high polysaccharide density for a transferase activity. RGI oligosaccharides could be labeled with a fluorescent probe and incubated with endoPG's in the presence of different HG segments/oligosaccharides, to detect putative hybrid structures emerging with an higher M_w, according to the method of Vincken *et al.* (1998).

3.6 Does one model cover all pectin structures?

The heterogeneity of pectin due to species variations and growth and climate influences is a known factor hampering pectin analysis. The unravelling of the pectin' structure is further complicated by the fact that the polymer is too large to be analyzed as a whole. To overcome this analytical challenge, pectin is enzymatically degraded in order to fit the degradation products within the M_w range of state-of-the-art techniques, like MS and NMR. However, there remains always a part of the pectin molecule which cannot be specifically degraded. Such enzyme resistant populations were identified in apple, carrot, onion, leek, potato fibre, and soy pectin (Schols *et al.*, 1990). The relevance of such populations for the pectin structure remains a question mark. Potentially, unknown structures, not accounted for, can attribute to false assumptions about the build-up of pectin structure, or withhold information which could possible explain other observations. Until now, a model based on the elucidation of the chemical fine structure of only part of pectin fragments produced in various ways has been effective in describing pectin behaviour. A high M_w pool obtained after RGH degradation of apple MHR was used to construct this pectin model described by Schols and

Voragen (1996). Therefore, one could argue that the pectin model does not cover a structure of true pectin. On the other hand, the sugar linkage composition, molecular weight profiles and enzymatic degradability of pectins from different sources often show similar trends. Unusual results can commonly be explained by the absence of particular structural elements in one source, or by an alternative distribution of the different substitutions or structural elements.

Since characterization of the low M_w undegradable pectin population is not feasible without having novel enzymes, it is (almost) impossible to make a model which covers the structural features of all types of (un)degradable pectins. Therefore, there is no other option than to realize that we are not looking at the entire pectin molecule. Consequently, new enzymatic and chemical methods should be developed, which enable the investigation of undegradable pectin structures. These methods are expected to generate complex mixtures of fragments.

4. Ongoing search for new analytical methods for use in pectin research

4.1 Possibilities and limitations of CE-LIF-MS^n

It has been reported by Coenen *et al.* (2008) that an effective separation of differently substituted galacturonic acid containing oligosaccharides can be obtained by CE-LIF analysis. By adapting this approach on line MS detection was attained, which enabled the localization of (non) sugar substituents, using nmol/ml sample quantities. Although this method was successful in the annotation of oligosaccharides present in complex RGI and XGA digests, as well as in an acid hydrolysate of Apple MHR, the observed resolution was unsatisfactory when analyzing complex enzyme digests or hydrolysates, which included (partly) methyl esterified GalA oligosaccharides.

In order to get an enhanced separation of different sub groups of acidic oligosaccharides, other CE-LIF-MS^n approaches were used as described (Coenen *et al.*, 2008). First different labels were used, to be able to differentiate between the contributions of the differently localized, moderate negatively charged carboxyl groups of GalA on the separation, which was mainly based on the three SO_3 groups in the APTS label. The alternative labels contained 0, 1, and 2 sulfonic acid groups, in combination with an amidated conjugated system, which enabled UV detection as well as coupling the label to the reducing end of the oligosaccharide. The CE-UV detection was successful for the 1- and especially for the 2- charged label, which is as well fluorescent. However, when performing CE-MS^n it was not possible to obtain a stable ion spray and ion current. Nevertheless, when using standards this CE-UV approach could be useful for quantitative analysis.

As an alternative it is possible to vary in pH, buffer composition, buffer concentration, and capillary coating, since oligosaccharides will react differently to these changes, which possibly enhance their separation. Using this approach oligosaccharides which were co-eluting performing CE-UV-MS^n as described by Coenen *et al.* (2008) might be separated. The method could be further improved by implementing LIF-detection within the CE-MS setup of APTS labelled oligosaccharides in order to perform CE-LIF-MS^n, since this gives sharper peak shapes and improved detection. The capillary coating influences the electro-osmotic flow (EOF), and is therefore of influence on the separation. Reducing the EOF by using coated fused silica capillary instead of regular fused silica could further improve this method. Possibly this could result in a stable current and ion spray, enabling CE-MS for alternative labels. Furthermore, CE-LIF-MS^n, enables quantitative (iontrap-MS) analysis of various components independent of dialysis, digestion, or (bio)synthesis procedures.

4.2 Selective chemical β-eliminative degradation

Although the β-eliminative method of Deng *et al.* (2006) was successful in liberating different neutral sugar side chains, which were previously not accessible, the method has also some drawbacks. Using this approach a small amount of galacto- and arabino-oligosaccharide series were visible in some Maldi-tof mass spectra. These series were as well present in small amounts in β-eliminative degraded potato galactan and sugar beet arabinan, used as controls. The results from the sugar composition revealed that these neutral sugars were present in low amounts in the low M_w populations, but abundantly in the Maldi-tof spectra of these samples. There are two possible explanations for this phenomenon; firstly, these oligosaccharides could originate from autolysis, caused by the high temperature during the β-eliminative treatment; secondly, there are linkages present within the neutral sugar side chains, which are also labile for β-eliminative treatment.

After β-eliminative degradation of enzyme resistant polysaccharide pools, still a substantial part (~40%) is not degraded to low molecular weight fragments, being accessible for analysis by HPAEC, CE-MS, ESI-MS^n or Maldi-tof MS. The sugar composition of this resistant parts reveals that this population contains a relative large proportion of neutral sugars. Sterical hindrance, caused by these (heavily branched) neutral sugar side chains could be a possible explanation for the resistance of these populations towards β-elimination. Alternatively, the low GalA content of these enzyme resistant polymers in relation to the neutral sugar side chain length could as well explain this persistent, high M_w population. An interesting follow-up experiment could be periodate oxidation, which specifically retains (1-3)-linked hexopyranosyl residues and produces specific degradation products for different linkage types (Perlin and Derek, 2006). Alternatively, the arabinose could be specifically removed via hydrolysis by 50-100 mM oxalic acid at 80-100 °C for 1 hour (Chambat *et al.*, 1984; Chanda *et al.*, 1950).

5. Future perspectives

Pectin is an important ingredient in the food industry as well as a key constituent of the plant cell wall. (Daas *et al.*, 2001; De Vries, 2004; May, 2000; Schols and Voragen, 1996; Voragen *et al.*, 1995). Pectin is also of interest for the pharmaceutical industry due to its health promoting effects and because of its ability to modify surfaces of medical materials. (Becker *et al.*, 2006; Guggenbichler-J-P *et al.*, 1997; Kastner *et al.*, 2002; Lim *et al.*, 1997; Morra *et al.*, 2004; Torto, 1998; Yamada *et al.*, 2003).

One thing is certain; in all of these applications the structure is responsible for the observed effect. The identification of hybrid oligosaccharides, which locate galacturonan segments on the non-reducing end of RGI, combined with the recognition of the surplus of HG, and its consequent allocation as side chain attached to Rha residues as suggested by the results of Mort and Vasu (2007) brings us one step closer for describing the correct pectin backbone structure.

The development of CE-LIF-MS^n for charged oligosaccharides introduces a new tool for the structure elucidation. The combination of this method, with the selective degradation of RGI enables the characterization of structures, which were previously not accessible for analysis. Consequently, the methods described in this thesis improve the possibilities for the structure elucidation of a broad range of polysaccharides, and therefore a better understanding of its different applications.

In order to gain more structural knowledge on pectin and other polysaccharides, it would be useful to 'combine the combined approaches' described in this contribution, with other chemical and novel enzymatic degradation methods and by implementing these methods more quantitatively.

References

Albersheim, P., A.G. Darvill, M.A. O'Neill, H.A. Schols and A.G.J. Voragen, 1996. An Hypothesis: The same six polysaccharides are components of the primary cell wall of all higher plants. In: J. Visser and A.G.J. Voragen (eds.) Pectins and Pectinases pp. 47-53. Amsterdam: Elsevier.

Becker, B.B., U.U. Kuhn and B.B. Hardewig Budny, 2006. Double-blind, randomized evaluation of clinical efficacy and tolerability of an apple pectin-chamomile extract in children with unspecific diarrhea. Arzneimittel Forschung, 56 (6): 387.

Benen, J.A. and J. Visser, 2003. Polygalacturonases. In: J.R. Whitaker, A.G.J. Voragen and D.W.S. Wong (eds.) Handbook of Food Enzymology pp. 857-866. New York: Marcel Dekker, Inc.

Braconnot, H., 1825. Recherches sur un nouvel Acide universellement répadu dans tous les végétaux. Annales de chimie et de physique, 28: 173-178.

Chambat, G., F. Barnoud and J.P. Joseleau, 1984. Structure of the Primary-Cell Walls of Suspension-Cultured Rosa-Glauca Cells. 1. Polysaccharides Associated with Cellulose. Plant Physiology, 74 (3): 687-693.

Chanda, S.K., E.L. Hirst, J.K.N. Jones and E.G.V. Percival, 1950. The Constitution of Xylan from Esparto Grass (*Stipa tenacissima*, L). Journal of the Chemical Society, 263: 1289-1297.

Coenen, G.J., 2007. Structural characterization of native pectins. Wageningen Universiteit, ISBN 978-90-8504-779-7, 152pp.

Coenen, G.-J., E.J. Bakx, R.P. Verhoef, H.A. Schols and A.G.J. Voragen, 2007. Identification of the connecting linkage between homo- or xylogalacturonan and rhamnogalacturonan type I., Carbohydrate Polymers 70: 224-235.

Coenen, G.J., M. Kabel, H.A. Schols and A.G.J. Voragen, 2008. CE-MS^n of complex pectin-derived oligomers., Electrophoresis 29: 2101-2111.

Daas, P.J.H., B. Boxma, A.M.C.P. Hopman, A.G.J. Voragen and H.A. Schols, 2001. Nonesterified Galacturonic Acid Sequence Homology of Pectins. Biopolymers, 58: 1-8.

Darvill, A.G., M. McNeil and P. Albersheim, 1978. Structure of plant cell walls part VIII: A new pectic polysaccharide. Plant Physiology, 62 (3): 418.

De Vries, J., 2004. Hydrocolloid gelling agents and their applications. In: P.A. Williams and G.O. Philips (eds.) Gums and Stabilisers for the food industry pp. 23-31. Cambridge: The Royal Society of Chemistry.

De Vries, J.A., F.M. Rombouts, A.G.J. Voragen and W. Pilnik, 1982. Enzymic degradation of apple pectins. Carbohydrate Polymers, 2: 25-33.

Deng, C., M.A. O'Neill and W.S. York, 2006. Selective chemical depolymerization of rhamnogalacturonans. Carbohydrate Research, 341 (4): 474.

Guggenbichler, J.-P., A. De Bettignies-Dutz, P. Meissner, S. Schellmoser and J. Jurenitsch, 1997. Acidic oligosaccharides from natural sources block adherence of Escherichia coli on uroepithelial cells. Pharmaceutical and pharmacological letters, 7 (1): 35.

Helmy, K.Y., J.K.J. Katschke, N.N. Gorgani, N.M. Kljavin, J.M. Elliott, L. Diehl, S.J. Scales, N. Ghilardi and M. van Lookeren Campagne, 2006. CRIg: A Macrophage Complement Receptor Required for Phagocytosis of Circulating Pathogens. Cell, 124 (5): 915-927.

Hilz, H., E.J. Bakx, H.A. Schols and A.G.J. Voragen, 2005. Cell wall polysaccharides in black currants and bilberries-characterisation in berries, juice, and press cake. Carbohydrate Polymers, 59 (4): 477.

Hilz, H., P. Williams, T. Doco, H.A. Schols and A.G.J. Voragen, 2006. The pectic polysaccharide rhamnogalacturonan II is present as a dimer in pectic populations of bilberries and black currants in muro and in juice. Carbohydrate Polymers, 65 (4): 521-528.

Hrmova, M., V. Farkas, J. Lahnstein and G. Fincher, 2007a. Xyloglucan transglycolase/hydrolase in cell wall re-modeling. Physiologia Plantarium, 130 (4): Abstract 21.

Hrmova, M., V. Farkas, J. Lahnstein and G.B. Fincher, 2007b. A Barley Xyloglucan Xyloglucosyl Transferase Covalently Links Xyloglucan, Cellulosic Substrates, and (1,3;1,4)-beta-D-Glucans. J. Biol. Chem., 282 (17): 12951-12962.

Huisman, M.M.H., C.T.M. Fransen, J.P. Kamerling, J.F.G. Vliegenthart, H.A. Schols and A.G.J. Voragen, 2001. The CDTA-soluble pectic substances from soybean meal are composed of rhamnogalacturonan and xylogalacturonan but not homogalacturonan. Biopolymers, 58 (3): 279-294.

Ishii, T. and T. Matsunaga, 2001. Pectic polysaccharide rhamnogalacturonan II is covalently linked to homogalacturonan. Phytochemistry, 57: 969-974.

Kabel, M.A., H.A. Schols and A.G.J. Voragen, 2001. Mass determination of oligosaccharides by matrix-assisted laser desorption/ionization time-of-flight mass spectrometry following HPLC, assisted by on-line desalting and automated sample handling. Carbohydrate Polymers, 44 (2): 161.

Kastner, U., S. Glasl, B. Follrich, J.P. Guggenbichler and J. Jurenitsch, 2002. Acidic oligosaccharides as active principle of aqueous carrot extracts in the prophylaxis and therapy of gastrointestinal infections. Wiener Medizinische Wochenschrift, 152 (15-16): 379-381.

Laurent, M.A. and P. Boulenguer, 2003. Stabilization mechanism of acid dairy drinks (ADD) induced by pectin. Food Hydrocolloids, 17 (4): 445.

Lim, B.O., K. Yamada, M. Nonaka, Y. Kuramoto, P. Hung and M. Sugano, 1997. Dietary Fibers Modulate Indices of Intestinal Immune Function in Rats. Journal of Nutrition, 127 (5): 663-667.

May, C.C.D., 2000. Pectins. In: G.O. Phillips and P.A. Williams (eds.) Handbook of Hydrocolloids pp. 169-189. Cambridge: Woodhead Publishing Limited.

Mohnen, D., 2002. Biosynthesis of pectins. In: G.B. Seymour and J.P. Knox (eds.) Pectins and their manipulation pp. 52-98. Oxford: Blackwell Publishing, Ltd.

Morra, M., C. Cassinelli, G. Cascardo, M.D. Nagel, C. Della Volpe, S. Siboni, D. Maniglio, M. Brugnara, G. Ceccone, H.A. Schols and P. Ulvskov, 2004. Effects on interfacial properties and cell adhesion of surface modification by pectic hairy regions. Biomacromolecules, 5 (6): 2094-2104.

Mort, A.J. and P. Vasu, 2007. Is there a link between rhamnogalacturonan and xylogalacturonan and if so, what is it like? Physiologia Plantarum, 130 (4): Abstract 23.

Mutter, M., I.J. Colquhoun, H.A. Schols, G. Beldman and A.G.J. Voragen, 1996. Rhamnogalacturonase B from Aspergillus aculeatus is a rhamnogalacturonan α-L-rhamnopyranosyl-(1,4)-α-D-galactopyranosyluronide lyase. Plant Physiology, 110 (1): 73-77.

O'Neill, M.A., T. Ishii, P. Albersheim and A.G. Darvill, 2004. Rhamnogalacturonan II: Structure and Function of a Borate Cross-Linked Cell Wall Pectic Polysaccharide. Annual Review of Plant Biology, 55 (1): 109-139.

Prade, R., D. Zhan, P. Ayoubi and A. Mort, 1999. Pectins, pectinases and plant-microbe interactions. Biotechnology and Genetic Engineering Reviews, 16: 361-391.

Ros, J.M., H.A. Schols and A.G.J. Voragen, 1998. Lemon albedo cell walls contain distinct populations of pectic hairy regions. Carbohydrate Polymers, 37: 159-166.

Round, A.N., A.J. MacDougall, S.G. Ring and V.J. Morris, 1997. Unexpected branching in pectin observed by atomic force microscopy. Carbohydrate Research, 303 (3): 251.

Round, A.N., N.M. Rigby, A.J. MacDougall, S.G. Ring and V.J. Morris, 2001. Investigating the nature of branching in pectin by atomic force microscopy and carbohydrate analysis. Carbohydrate Research, 331 (3): 337.

Schols, H.A. and A.G.J. Voragen, 1996. Complex pectins: Structure elucidation using enzymes. In J. Visser and A.G.J. Voragen. Pectins and Pectinases pp. 3-19. Amsterdam: Elsevier Science B.V.

Schols, H.A., C.C.J.M. Gereads, M.F. Searle-van Leeuwen, F.J.M. Kormelink and A.G.J. Voragen, 1990. Rhamnogalacturonase: a novel enzyme that degrades the hairy regions of pectins. Carbohydrate Research, 206: 105-115.

Schols, H.A., E.J. Bakx, D. Schipper and A.G.J. Voragen, 1995. A xylogalacturonan subunit present in the modified hairy regions of apple pectin. Carbohydrate Research, 279:, 265.

Staehelin, L.A. and I. Moore, 1995. The Plant Golgi Apparatus: Structure, Functional Organization and Trafficking Mechanisms. Annual Review of Plant Physiology and Plant Molecular Biology, 46 (1): 261-288.

Thibault, J.-F., C.M.G.C. Renard, M.A.V. Axelos, P. Roger and M.-J. Crepeau, 1993. Studies of the length of homogalacturonic regions in pectins by acid hydrolysis. Carbohydrate Research, 238: 271.

Torto, N.N., 1998. Microdialysis introduction high-performance anion-exchange chromatography/ ionspray mass spectrometry for monitoring of on-line desalted carbohydrate hydrolysates. Journal of Mass Spectrometry, 33 (4): 334.

Vincken, J.P., O.A. Zabotina, G. Beldman and A.G.J. Voragen, 1998. Xyloglucan endotransglycosylase activity in apples is ripening-related: Implications for fruit juice processing. Journal of the Science of Food and Agriculture, 78 (1): 46.

Vincken, J.-P., H.A. Schols, R.J.F.J. Oomen, G. Beldman, R.G.F. Visser and A.G.J. Voragen, 2003a. Pectin- the hairy thing: evidence that homogalacturonan is a side chain of rhamnogalacturonan I. In: A.G.J. Voragen, H.A. Schols and R.G.F. Visser (eds.) Advances in pectin and pectinase research. pp. 47-60. Dordrecht: Kluwer Academic Publishers.

Vincken, J.-P., H.A. Schols, R.J.F.J. Oomen, M.C. McCann, P. Ulvskov, A.G.J. Voragen and R.G.F. Visser, 2003b. If Homogalacturonan Were a Side Chain of Rhamnogalacturonan I. Implications for Cell Wall Architecture. Plant Physiology, 132 (4): 1781-1789.

Voragen, A.G.J., W. Pilnik, J.F. Thibault, M.A.V. Axelos and C.M.G.C. Renard, 1995. Pectins. In: A.M. Stephen (ed.) Food Polysaccharides and Their Applications pp. 287-339. New York: Marcel Dekker.

Voragen, A.G.J., G. Beldman and H. Schols, 2001. Chemistry and Enzymology of Pectins. In: B.V. McCleary and L. Prosky (eds.) Advanced Dietary Fibre Technology pp. 379-397. Oxford: Blackwell Science.

Westereng, B., 2007. Pectin from cabbage (*Brassica oleracea*): Studies on structure and bioactivity. Faculty of Mathematics and Natural Sciences (p. 185). Oslo: University of Oslo.

Westereng, B., G.J. Coenen, T.E. Michaelsen, A.G.J. Voragen, A.B. Samuelsen, H.A. Schols and S.H. Knutsen, 2009. Release and characterization of single side chains of white cabbage pectin and their complement-fixing activity. Mol. Nutr. Food Res. 53, in press. DOI 10.1002/ mnfr.200800199.

Yamada, H. and H. Kiyohara, 2007. Immunomodulating Activity of Plant Polysaccharide Structures. In: J.P. Kamerling, G.J. Boons, Y.C. Lee, A. Suzuki, N. Taniguchi and A.G.J. Voragen (eds.) Comprehensive Glycoscience: From Chemistry to Systems Biology. Volume 4 Cell Glycobiology and Development Health and Disease in Glycomedicine. pp. 663-693. Oxford: Elsevier Ltd.

Yamada, H., H. Kiyohara and T. Matsumoto, 2003. Recent studies on possible functions of bioactive pectins and pectic polysaccharides from medicinal herbs on health care. In: A.G.J. Voragen, H.A. Schols and R.G.F. Visser (eds.) Advances in Pectin and Pectinase Research, 481-490. pp. 481-490. Dordrecht: Kluwer Academic Publishers.

Yapo, B.M., P. Lerouge, J.-F. Thibault and M.-C. Ralet, 2007. Pectins from citrus peel cell walls contain homogalacturonans homogenous with respect to molar mass, rhamnogalacturonan I and rhamnogalacturonan II. Carbohydrate Polymers, 69 (3): 426-435.

Zandleven, J.S., 2006. Identification and characterization of some *Aspergillus* pectinolytic glycoside hydrolases. Agrotechnology and Food Sciences (p. 160). Wageningen: Wageningen University.

Zandleven, J., S.O. Sorensen, J. Harholt, G. Beldman, H.A. Schols, H.V. Scheller and A.G.J. Voragen, 2007. Xylogalacturonan exists in cell walls from various tissues of Arabidopsis thaliana. Phytochemistry, 68 (8): 1219-1226.

Zhan, D., P. Janssen and A.J. Mort, 1998. Scarcity or complete lack of single rhamnose residues interspersed within the homogalacturonan regions of citrus pectin. Carbohydrate Research, 308: 373-380.

Hydrodynamic properties of isolated pectin domains: a way to figure out pectin macromolecular structure?

M.-C. Ralet and J.-F. Thibault
INRA, UR1268 Biopolymères Interactions Assemblages, B.P. 71627, 44316 Nantes cedex 03, France

Abstract

Pectins are extremely complex polysaccharides that can be viewed as multiblock co-biopolymers. Although the fine structure of the different pectic blocks is now quite well known, the way in which they are combined into a macromolecular structure is still a matter of debate. Large differences in pectins quality and quantity exist between different cell types, at different stages of cellular development and even within the thickness of a given wall. A way to understand how the different structural elements are connected to one another to form variable macromolecular structures is to isolate and identify the different connection points. Another, and complementary, way is to specifically isolate the different pectic domains in order to study their hydrodynamic properties. Six monocotyledonous and dicotyledonous species were selected and pectins were extracted from cell wall materials using a four-step sequential extraction scheme. The isolation and characterization of pectin structural elements showed that structural discrepancies of pectins differing in HG/RGI balance could be due to a variable amount of HG domains decorating an RGI backbone. Our results also support the hypothesis of a hyper-variability of RGI backbone length. However, the structural complexity and heterogeneity of pectins make it very difficult to appraise the relationship between pectin fine structure and hydrodynamic properties, especially when the compared pectins are derived from different plant species and obtained with different extractants. The *Arabidopsis* pectin mutant *quasimodo2* was thereby studied. By specifically isolating HG blocks, *qua2* appeared to be specifically deficient in HG, with no change in the amount of RGI blocks. Moreover, the remaining HGs had maintained the same size as those in the wild type. The possible macromolecular build up of pectins is discussed in the light of the presented results.

Keywords: homogalacturonan, rhamnogalacturonan I, *quasimodo2*

1. Introduction

All plant cells are encapsulated in a cell wall, whose most prominent components are polysaccharides: cellulose, hemicelluloses and pectins. These polymers collectively determine cell wall shape and mechanical properties. Pectins, because of their multiple interaction properties, are likely to be key components for the architecture of plant cell walls. The term 'pectin' is a somewhat misleading since it implies the existence of a single well-defined macromolecule (Willats *et al.*, 2006). In fact, all pectins are nowadays strongly believed to contain given structural domains (Schols and Voragen, 1996). The most commonly found structural domains are: (1) homogalacturonan, (2) xylogalacturonan, (3) rhamnogalacturonan I backbone (4) rhamnogalacturonan II, (5) arabinan, (6) arabinogalactan I, and (vii) arabinogalactan II (Schols and Voragen, 2002; Coenen, 2007).

Homogalacturonan (HG) is usually the most abundant pectic structural domain. It consists of a linear backbone of (1→4)-linked α-D-GalA residues. Zhan *et al.* (1998) gave convincing arguments against the possible presence of single Rha residues within HG regions. GalA residues are commonly partly methyl-esterified at C-6 (Voragen *et al.*, 1995) and, in some plant species, partly acetyl-esterified at *O*-2 or *O*-3 (Ralet *et al.*, 2005, 2008a). Both the degree of methyl-esterification (DM) and degree of acetylation (DA) have a profound impact on associative properties. The minimum estimated length of this domain was reported to be ~ 100 GalA residues (Thibault *et al.*, 1993; Hellìn *et al.*, 2005). Several non-linear domains are derived from the unsubstituted HG. Xylogalacturonan (XGA) consists of a HG backbone with mostly single-unit substituents of xylose linked at *O*-3 of GalA residues. Rhamnogalacturonan II (RGII) is a highly conserved structure in the plant kingdom. It is a low molar mass (5-10 kDa) highly complex macromolecule with a short HG-like backbone substituted by four different side-chains encompassing several unusual sugar residues (Schols and Voragen, 2002). This structural element, although present in very limited amounts, plays a key role as it is involved in the crosslinking of two pectin molecules within the cell wall through a borate diester (Ishii *et al.*, 1999; Ishii and Matsunaga, 2001).

Another major pectic domain is rhamnogalacturonan I (RGI). The RGI backbone consists of [→2)-α-L-Rha*p*-(1→4)-α- D-Gal*p*A-(1→] repeats (McNeill *et al.*, 1980, 1984; Albersheim *et al.*, 1996). RGI domains isolated from several plant species, appeared highly acetylated at *O*-2 and/or *O*-3 of GalA units (Komavilas and Mort, 1989; Schols and Voragen, 1994; Ralet *et al.*, 2005). No strong evidence has been published that GalA units in RGI domains are methyl-esterified. The rhamnosyl residues of RGI backbone are substituted, mainly at *O*-4, with several types of arabinose and galactose-containing neutral sugars side-chains (Voragen *et al.*, 1995).

Although the structure of the different pectic domains is now quite well known, the way they are connected one with another to form a macromolecular structure is still a matter of debate. Two pectic models are considered nowadays: the '*smooth and hairy regions*' model (Schols and Voragen, 1996) and the '*RGI backbone*' model (Vincken *et al.*, 2003) (Figure 1). In the first one, 'smooth regions' (HG) alternate with 'hairy regions' (XGA and RGI encompassing neutral sugars side chains) to form the pectin macromolecule backbone. In the second one, HG, XGA, arabinan and galactan would occur as side-chains of RGI backbone, forming a kind of '*molecular brush*' (Vincken *et al.*, 2003). In both models, RGII would be an integral part of some HG domains.

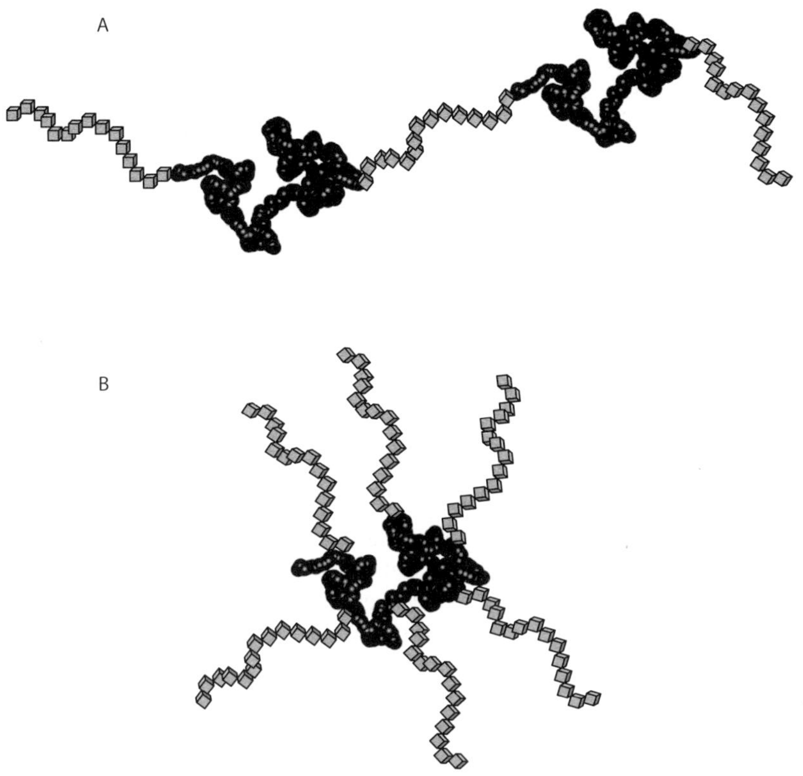

Figure 1. Pectin schematic representation. A) In the 'smooth and hairy regions' *model (Schols and Voragen, 1996), the pectic backbone consists of alternating homogalacturonan and/or xylogalacturonan (light grey) and rhamnogalacturonan-I (dark grey). Neutral sugars side chains (that are not represented on the figure) constitute the 'hairy' part of the macromolecule. B) In the* 'RGI backbone' *model (Vincken* et al.*, 2003), the RGI backbone (dark grey) is decorated, not only with neutral sugars side chains (that are not represented on the figure), but also with homogalacturonan and/or xylogalacturonan domains (light grey).*

A first question is 'how the different structural elements are connected to one another to form variable macromolecular structures?' The only way to unambiguously answer that question is to isolate and identify the different connection points between pectic domains (Coenen *et al.*, 2007; Schols *et al.*, 2009). Moreover, the amount, fine structure, and maybe length of each domain vary widely with respect to plant origin (Schols and Voragen, 1996), between different cell types, at different stages of cellular development, and even within the thickness of a given wall (Jauneau *et al.*, 1998; Willats *et al.*, 2001). So that a second question arises: 'are all domains heterogeneous with respect to length or are they invariant, the assembly mode only generating heterogeneity?' One way to answer this question is to specifically isolate the different pectic domains in order to study their hydrodynamic properties.

2. Pectins with varying GalA/Rha ratios can be isolated from cell wall materials of different plant sources

Cell wall materials (CWM), as alcohol insoluble matter, were prepared from pineapple (*Ananas comosus, Bromeliaceae*) flesh, leak (*Allium porrum, Alliaceae*), cucumber (*Cucumis sativus, Cucurbitaceae*), sugar beet root (*Beta vulgaris, Amaranthaceae*), fennel (*Foeniculum vulgare, Apiaceae*) bulb, and lemon (*Citrus medica, Rutaceae*) albedo. Pectins were sequentially extracted from the different CWM as follows: (1) Water (pH 4.5, 25 °C, 3 × 30 min), (2) 1% w/v K-oxalate (pH 4.5, 25 °C, 3 × 30 min), (3) HCl (0.05M, 85 °C, 3 × 30 min), (4) NaOH (0.05M, 4 °C, 3 × 30 min). This sequential extraction yielded four extracts that were brought to pH5, concentrated, extensively dialysed against distilled water and freeze-dried to give water-extracted pectins (WEP), K-oxalate-extracted pectins (OEP), acid-extracted pectins (HEP) and alkali-extracted pectins (OHEP) (Yapo *et al.*, 2007).

GalA extractability (g GalA extracted/100g of GalA in the initial CWM) varied from 66 (sugar beet) to 95 (leek and lemon) (Table 1). Whatever the plant source, water was a poor extractant while HCl was the most efficient one. GalA extractability by K-oxalate and by NaOH varied widely with respect to plant source, 3-31% and 7-26% for K-oxalate and NaOH, respectively (Table 1).

Table 1. GalA extractability (g of GalA extracted for 100g of GalA present in the initial CWM) and GalA/Rha molar ratios of the sequentially extracted pectins (water-extracted pectin, WEP; K-oxalate-extracted pectin, OEP, acid-extracted pectin, HEP and alkali-extracted pectin, OHEP) from leek, pineapple, sugar beet, cucumber, lemon and fennel CWM. (Values are from Yapo, 2007).

	WEP	OEP	HEP	OHEP
GalA extractability				
Leek	9	16	41	17
Pineapple	16	11	42	26
Sugar beet	3	3	35	25
Cucumber	8	15	35	20
Lemon	11	31	45	7
Fennel	5	19	28	17
GalA/Rha				
Leek	35	44	13	10
Pineapple	40	55	29	24
Sugar beet	40	48	11	10
Cucumber	21	27	11	9
Lemon	57	65	29	12
Fennel	29	41	9	10

Obviously, the 'type' of pectins extracted varied with respect to the extractant used with high GalA/Rha molar ratios for pectins extracted by smooth agents (water and K-oxalate) and lower ones for pectins extracted by stronger agents (HCl and NaOH) (Table 1). Whatever the plant origin, HEP and OHEP are therefore probably significantly richer in RGI domains than WEP and OEP.

3. Isolated pectins from different plant sources encompass various amounts of HG domains of very similar degrees of polymerisation

After saponification of the different pectins, HG domains were isolated exploiting the differences in the susceptibility to acid hydrolysis of the glycosidic linkages. Indeed, linkages between two unmethylated GalA residues are much more stable than aldobiuronic (GalA-Rha) or pseudo-aldobiuronic (Rha-GalA) linkages. Under the conditions used, resistant HG domains were recovered as acid-insoluble material while hydrolyzed RGI domains became acid-soluble (Thibault *et al.*, 1993; Yapo *et al.*, 2007). Acid-insoluble fractions consisting of almost pure GalA (> 98% w/w) were recovered. The partitioning of the total amount of GalA between acid-insoluble and acid-soluble fractions was used as a measure of the ratio between HG and RG domains of the pectins (Table 2).

In agreement with the previously observed pectins GalA/Rha molar ratios, pectins extracted by smooth agents (water and K-oxalate) appear to be rich in HG domains while pectins extracted by stronger agents (HCl and NaOH) are significantly richer in RGI domains.

HG domains were analysed for their hydrodynamic properties by high performance size-exclusion chromatography (HP-SEC) combined with multiple-angle laser light-scattering detection. HP-SEC was performed at room temperature on a system constituted of one Shodex OH SB-G pre-column followed by two columns in series

Table 2. GalA recovery in HG domains (g GalA recovered in HG domains/100 g of GalA present in the initial pectins) of the sequentially extracted pectins (water-extracted pectin, WEP; K-oxalate-extracted pectin, OEP, acid-extracted pectin, HEP and alkali-extracted pectin, OHEP) from leek, pineapple, sugar beet, cucumber, lemon and fennel CWM. (Values are from Yapo, 2007).

GalA recovery in HG domains	WEP	OEP	HEP	OHEP
Leek	90	93	80	84
Pineapple	93	95	88	85
Sugar beet	88	92	86	85
Cucumber	91	93	78	84
Lemon	98	98	96	85
Fennel	88	95	81	81

(Shodex OH-Pack SB-804 HQ and OH-Pack SB-805 HQ, Shodex, Showa Denko KK, Miniato, Japan) eluted with 0.05M $NaNO_3$ buffer containing 0.02% NaN_3 as preservative at a constant flow rate of 42 ml/h. A refractometer (RI) and a multiple-angle laser light-scattering device (MALLS) (Mini Dawn, Wyatt, Santa Barbara, CA) operating at three angles (41°, 90° and 138°) were used as detectors. Samples were solubilized in 0.05 M $NaNO_3$ buffer containing 0.02% NaN_3 (~ 10 mg/ml), left overnight with tail-over-head continuous mixing, centrifuged and filtered (0.45 µm) before injection of 50 µl. Data for molar mass determinations were analyzed using Astra 1.4 software (Wyatt, Santa Barbara, CA) taking dn/dc 0.146.

HG domains isolated from the different pectin samples gave one single narrow log-norm type RI peak that appeared well resolved between the exclusion limit and the total volume of the column. Limited amounts of salts were eluted at the total volume of the column (Figure 2).

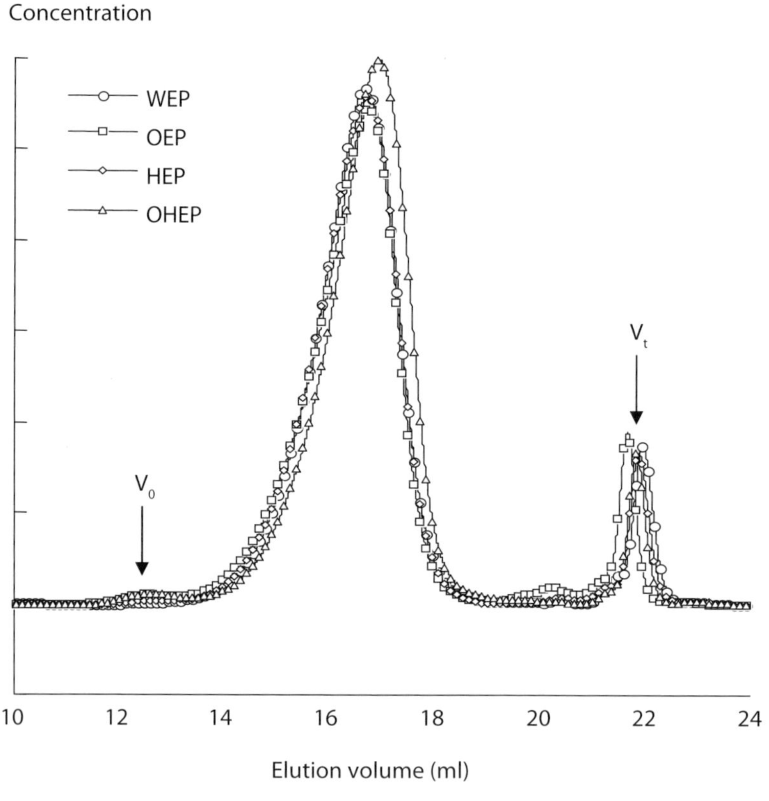

Figure 2. HPSEC chromatograms of HG domains isolated from the sequentially extracted lemon pectins (water-extracted pectin, WEP; K-oxalate-extracted pectin, OEP, acid-extracted pectin, HEP and alkali-extracted pectin, OHEP).

For a given plant species, HG domains exhibited similar average molar masses and low polydispersity index ($<M_w>/<M_n>$), whatever the extractant used (Table 3; Figure 2).

Moreover, average degrees of polymerisation were quite similar whatever the plant species. These average degrees of polymerisation are in agreement with previously published results on *Rosa glauca* cultured-cells and cotton cotyledons HG domains (Chambat and Joseleau, 1980; Zhang *et al.*, 2007) (dp ~ 70-100).

Although pectins extracted by smooth agents (water and K-oxalate) and pectins extracted by stronger agents (HCl and NaOH) exhibited different HG/RGI ratios, their constitutive HG domains were of very similar degrees of polymerisation and of low polydispersity. The hypothesis of a HG length periodicity (Thibault *et al.*, 1993) is here confirmed.

4. Isolated RGI backbone domains exhibit variable degrees of polymerisation depending on the plant source

Lemon and sugar beet acid-extracted pectins were saponified prior to hydrolysis by endo-polygalacturonase II (Yapo, 2007; Yapo *et al.*, 2007). RGI domains were separated from oligogalacturonates arising from HG domains by anion-exchange chromatography on DEAE-Sepharose CL-6B (32 × 2.6 cm, flow rate 90 ml/h, NaCl gradient 0-0.6 M). GalA monomers, dimers and trimers arising from HGs were eluted between 0 and 0.1 M NaCl while RGI domains, together with some oligogalacturonates of higher degrees of polymerisation, were eluted for higher NaCl concentrations (0.1-0.6 M). The crude RGI recovered after anion-exchange chromatography was dialyzed, concentrated, and further purified by size-exclusion chromatography on Sephacryl S-200 (90 × 1.6 cm, flow rate 60 ml/h, 0.05 M Na-succinate buffer pH 4.5). Three peaks were recovered, the first one corresponding to RGI domains, the second one to RGII domains and the third one to residual oligogalacturonates (Yapo *et al.*, 2007). Whole

Table 3. HG domains number-average degrees of polymerisation (i.e. number of GalA resiues) of the sequentially extracted pectins (water-extracted pectin, WEP; K-oxalate-extracted pectin, OEP, acid-extracted pectin, HEP and alkali-extracted pectin, OHEP) from leek, pineapple, sugar beet, cucumber, lemon and fennel. Polydispersity index (M_w/M_n) were ≤ 1.25 for all samples. (Values are from Yapo, 2007).

HG degree of polymerisation	WEP	OEP	HEP	OHEP
Leek	67	70	82	64
Pineapple	61	63	72	73
Sugar beet	100	67	89	70
Cucumber	95	90	87	90
Lemon	94	95	86	80
Fennel	76	78	72	69

RGI domains (encompassing neutral sugars side-chains) were recovered and an aliquot was dissolved in 0.05 M succinate buffer pH 4.5 and degraded by an enzymatic mixture containing endo-arabinanase, endo-galactanase, α-arabinofuranosidase and β-D-galactosidase. The hydrolysates were dialysed (1kDa MWCO dialysis tubing) and freeze-dried to give RGI backbone domains. Lemon and sugar beet whole RGI and RGI backbone domains were analysed for their sugar composition and for their hydrodynamic properties (Table 4).

In agreement with the GalA/Rha ratios of initial pectins (11 for sugar beet and 29 for lemon; Table 1) and the GalA recovery in HG domains (86% for sugar beet and 96% for lemon; Table 2), GalA recovery in RGI domains was three times higher for sugar beet than for lemon (Table 4). As previously reported by Zhan *et al.* (1998) on commercial citrus pectins, most of the Rha residues were recovered in RGI domains. GalA/Rha molar ratios were close to 1, in agreement with a strict alternance of Rha and GalA residues in RGI domains. These findings perfectly agree with Zhan *et al.* (1998) findings that 85% of the Rha in citrus pectins was present in a strictly repeating $[GalA\text{-}Rha]_n$ rhamnogalacturonan. Both lemon and sugar beet whole RGI domains were rich in Ara and Gal (50-60 mol%). Whole RGI domains exhibited a high weight-average molar mass, in agreement with previous estimations using pullulan standards

Table 4. Characteristics of RGI domains recovered from lemon and sugar beet acid-extracted pectins. Whole RGI corresponds to RGI domains encompassing neutral sugars side-chains. RGI backbone refers to whole RGI domains hydrolysed by neutral sugars side-chains-degrading enzymes. Values are from Yapo et al. *(2007) and Ralet* et al. *(2008b).*

	Lemon		Sugar beet	
	Whole RGI	RGI backbone	Whole RGI	RGI backbone
GalA recovery[1]	3	3	9	9
Rha recovery[1]	93	95	>95	>95
Sugar composition (mol%)				
GalA	20	46	26	42
Rha	19	43	22	35
Ara	36	2	27	6
Gal	24	8	23	17
GalA/Rha	1.1	1.1	1.2	1.2
$<M_n>$[2]	52,500	12,500	83,000	55,000
Average number of Rha-GalA repeats[3]	67	34	121	120

[1]g GalA (or Rha) recovered in RGI domains/100 g of GalA (or Rha) present in the initial deesterified pectins.

[2]$<M_n>$, weight-average molar mass.

[3]calculated as follows: ($<M_n>$*Rha)/ [(176*GalA)+(146*Rha)+(132*Ara)+(162*Gal)].

(McNeil *et al.*, 1980; Zhan *et al.*, 1998). An average number of Rha-GalA repeats of 67 and 121 was calculated for lemon and sugar beet RGI whole domains, respectively (Table 4).

Neutral sugars side-chains degrading enzymes efficiently removed Ara and Gal residues from lemon whole RGI domains leaving an RGI backbone consisting of 89 mol% GalA + Rha. Sugar beet whole RGI was less prone to side-chains degradation since the RGI backbone recovered still exhibited high amounts of Gal (Table 4). Gal residues are present as rather long 'degradable' side chains in lemon (Ralet and Thibault, 1994) and rather short 'non degradable' ones in sugar beet (Guillon and Thibault, 1989; Oosterveld *et al.*, 1996). Furthermore, the proportion of branched Rha residues is higher in sugar beet (~ 60%) than in lemon (~ 30%) (Guillon and Thibault, 1989; Ralet and Thibault, 1994). After enzymatic degradation by side chains degrading enzymes, more residual monomeric Gal residues attached to Rha residues are likely to be present in sugar beet than in lemon. RGI backbone domains exhibited an average number of Rha-GalA repeats of 34 for lemon and 120 for sugar beet. It is noteworthy that, in the case of lemon, the calculated average number of Rha-GalA repeats differs before and after enzymatic hydrolysis of neutral sugars side chains (67 before enzymatic hydrolysis and 34 after). This was not the case for sugar beet (121 calculated Rha-GalA repeats before enzymatic hydrolysis and 120 after). The presence of residual rhamnogalacturonase activity was indeed detected in the β-galactosidase preparation used. Residual rhamnogalacturonase most likely acted on the slightly branched lemon RGI domains and not on the highly branched sugar beet ones. An average number of Rha-GalA repeats of ~ 60 for lemon and of ~ 120 for sugar beet is most probable.

5. The *Arabidopsis quasimodo2* mutant is HG-deficient but the remaining HG have maintained the same size as those in the wild type

To further investigate pectic domains variability, the *Arabidopsis* mutant *quasimodo2* (*qua2*) was studied. This mutant clustered with the *quasimodo1/gaut8* pectic mutant investigated by Bouton *et al.* (2002). *qua2* mutants exhibited short dark-grown hypocotyls, reduced cell adhesion, a dwarfed mature plant, and were not allelic to *quasimodo1/gaut8* (Mouille *et al.*, 2007). Qua2 (At1g78240) was shown to be a Golgi-localized protein with a putative methyltranferase domain (Mouille *et al.*, 2007). Pectic polysaccharides were sequentially extracted from WT and *qua2* cell wall materials with hot dilute acid (H^+-extract) and cold dilute alkali (OH^--extract). Those extracts were purified by anion-exchange chromatography and purified pectins were characterised. Sugars were primarily GalA, Gal and Rha. *qua2* pectins were significantly poorer in GalA than WT ones (Ralet *et al.*, 2008b). HG domains were then isolated by acidic means from saponified purified pectins (Thibault *et al.*, 1993). Acid-insoluble fractions consisting of almost pure GalA (> 98% w/w) were recovered. The partitioning of the total amount of GalA between acid-insoluble and acid-soluble fractions was used as a measure of the ratio between HG and RGI domains of the pectins. HG domains isolated as acid-insoluble material represented 60-70% of the GalA initially present in the pectins for WT and only 50% for *qua2*. When referring to whole cell wall material, calculation showed that the amount of RGI-derived GalA did not differ significantly between WT and *qua2*. Instead, HG-derived GalA showed a ~ 50% reduction in *qua2* compared to WT (Mouille *et al.*, 2007; Ralet *et al.*, 2008b). Interestingly, the remaining HG in *qua2* showed an unaltered degree of methyl-esterification (Mouille *et al.*, 2007).

HG domains were analysed for their hydrodynamic properties by high performance size-exclusion chromatography (HP-SEC) combined with multiple-angle laser light-scattering detection. Molar mass distributions for Arabidopsis HG domains did not reveal significant differences between WT and qua2 and values for $<M_n>$ and $<M_w>$ were similar (Ralet *et al.*, 2008b). It can be concluded that for *qua2* pectins, which exhibited a 50% reduced number of HG domains compared to WT pectins, the remaining HG domains were not altered in their length (number-average degrees of polymerisation 77).

Purified acid-extracted pectins were analysed by high performance size-exclusion chromatography (HP-SEC) combined with multiple-angle laser light-scattering and viscometric detection. Indeed, information about chain shape and extension may be obtained by analysing the dependency of intrinsic viscosity upon molar mass along the elution profile (Bohdanecky, 1983). Although average molar masses and molar mass distributions were similar for WT and *qua2* pectins, their average intrinsic viscosity values differed significantly, revealing some possible conformational differences between those two pectins (Ralet *et al.*, 2008b). The quantitative estimate of the conformational flexibility using Bohdanecky (1983) representation revealed that *qua2* pectins were significantly more flexible than WT ones (Ralet *et al.*, 2008b).

6. Conclusions

Pectins differing in their HG/RGI ratios were sequentially extracted from (1) cell wall materials from various plant sources (pineapple flesh, leak, cucumber, sugar beet root, fennel bulb, and lemon albedo), (2) *Arabidopsis* WT and *qua2* cell wall materials. HG domains were specifically isolated using chemical means and their hydrodynamic properties were investigated. All pectins studied, although some of them differed a lot with respect to their HG/RGI balance, encompassed HG domains of degrees of polymerisation between 60-100. No relationship between HG/RGI GalA partitioning and HG length was found. For lemon and sugar beet acid-extracted pectins, the isolation and characterization of pectin structural elements showed that (1) HG domains exhibited very similar degrees of polymerisation (~ 90 for both lemon and sugar beet), (2) RGI backbone domains exhibited variable degrees of polymerisation (~ 120 for lemon, i.e. 60 Rha-GalA repeats and ~ 240 for sugar beet, i.e. 120 Rha-GalA repeats).

Taking into account the HG/RGI GalA partitioning and the length of the isolated pectic domains, it can be calculated that there are in average 17 HG domains for 1 RGI domain in lemon acid-extracted pectins and 8 HG domains for 1 RGI domain in sugar beet acid-extracted pectins. The theoretical GalA/Rha ratios of those macromolecules would be 25 for lemon and 7 for sugar beet, in reasonably good agreement with experimental values (29 and 11 for lemon and sugar beet, respectively). It is clear that there is a huge excess of HG domains to build up a pectin macromolecule following the '*smooth and hairy regions*' alternating model (Schols and Voragen, 1996). Connecting elements between XGA or HG and RGI that were recently isolated (Coenen *et al.*, 2007), agree, however, with this alternating model. As fully discussed in Coenen (2007), the surplus of HG domains is likely to be explained by their allocation as side chains attached to RGI of an HG-RGI backbone, as in the '*RGI backbone*' model (Vincken *et al.*, 2003). A schematic representation of such a structure is given in Figure 3.

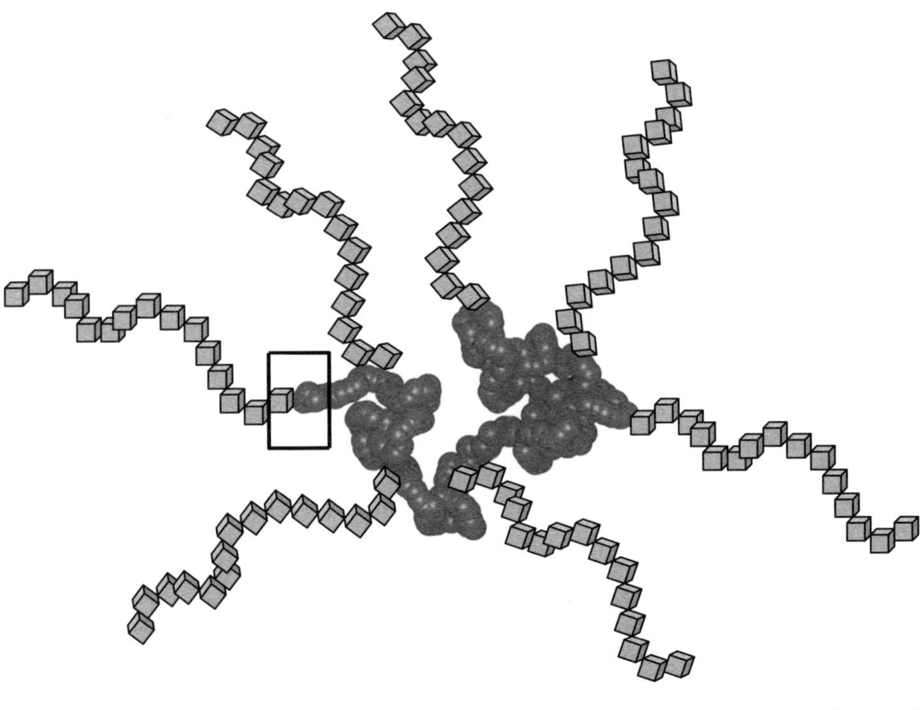

Figure 3. Pectin schematic representation. One homogalacturonan or xylogalacturonan domain (light grey) is attached in line to the non-reducing end of one RGI domain (dark grey). 'Hybrid oligosaccharides' *isolated by Coenen* et al. *(2007) are in the square. The RGI domain is further decorated with extra homogalacturonan and/or xylogalacturonan domains. Neutral sugars side chains are not represented in the figure.*

It can be hypothesized that structural discrepancies of pectins differing in their HG/RGI balance could be due to a variable amount of HG domains decorating the HG-RGI backbone. Our results also support the hypothesis of a hyper-variability of RGI backbone length. The understanding of the length and conformation of the different pectic domains, together with the identification of '*hybrid oligosaccharides*' (Coenen, 2007), will hopefully soon lead to the elucidation of the macromolecular build up of pectins. It has however to be kept in mind that the structure of 'mature' pectins does not necessarily reflect the structures as they are synthesised (Mohnen, 2008).

References

Albersheim, P., A.G. Darvill, M.A. O'Neill, H.A. Schols and A.G.J. Voragen, 1996. An hypothesis: the same six polysaccharides are components of the primary cell wall of all higher plants. In: J. Visser and A.G.J.Voragen (eds.) Pectins and Pectinases, Elsevier Science B.V., Amsterdam, NL, pp 47-53.

Bohdanecky, M., 1983. New method for estimating the parameters of the wormlike chain model from the intrinsic viscosity of stiff-chain polymers. Macromolecules 16: 1483-1492.

Bouton, S., E. Leboeuf, G. Mouille, M.-T. Leydecker, J. Talbotec, F. Granier, M. Lahaye, H. Höfte and H.N. Truong, 2002. *QUASIMODO1* encodes a putative membrane-bound glycosyltransferase required for normal pectin synthesis and cell adhesion in *Arabidopsis*. Plant Cell 14: 2577-2590.

Chambat, G. and J.-P. Joseleau, 1980. Isolation and characterization of a homogalacturonan in the primary cell walls of Rosa cell cultures *in vitro*. Carbohydrate Research 85: C10-C12.

Coenen, G.J., 2007. Structural characterization of native pectins. Ph.D. thesis, Wageningen University, NL. ISBN 978-90-8504-779-7.

Coenen, G.J., E.J. Bakx, R.P. Verhoef, H.A. Schols and A.G.J. Voragen, 2007. Identification of the connecting linkage between homo- or xylogalacturonan and rhamnogalacturonan type I. Carbohydrate Polymers 70: 224-235.

Guillon, F. and J.-F. Thibault, 1989. Methylation analysis and mild acid hydrolysis of the 'hairy' fragments of sugar-beet pectins. Carbohydrate Research 190: 85-96.

Hellìn, P., M.-C. Ralet, E. Bonnin and J.-F. Thibault, 2005. Homogalacturonans from lime pectins exhibit homogeneous charge density and molar mass distributions. Carbohydrate Polymers 60: 307-317.

Ishii, T. and T. Matsunaga, 2001. Pectic polysaccharide rhamnogalacturonan II is covalently linked to homogalacturonan. Phytochemistry 57: 969-974.

Ishii, T., T. Matsunaga, P. Pellerin, M.A. O'Neill, A.G. Darvill and P. Albersheim, 1999. The plant cell wall polysaccharide rhamnogalacturonan II self-assembles into a covalently cross-linked dimer. Journal of Biological Chemistry 274: 13098-13104.

Jauneau, A., S. Roy, D. Reis and B. Vian, 1998. Probes and microscopical methods for the localization of pectins in plant cells. International Journal of Plant Sciences 159: 1-13.

Komavilas, P. and A.J. Mort, 1989. The acetylation at O-3 of galacturonic acid in the rhamnose-rich portion of pectins. Carbohydrate Research 189: 261-272.

McNeil, M., A.G. Darvill and P. Albersheim, 1980. Structure of plant cell walls X. Rhamnogalacturonan I a structurally complex pectic polysaccharide in the walls of suspension cultured sycamore cells. Plant Physiology 66: 1128-1134.

McNeil, M., A.G. Darvill, S.C. Fry and P. Albersheim, 1984. Structure and functions of primary cell walls of plants. Annual Review of Biochemistry 53: 625-663.

Mohnen, D., 2008. Pectin structure and biosynthesis. Current Opinion in Plant Biology 11: 266-277.

Mouille, G., M.-C. Ralet, C. Cavelier, C. Eland, D. Effroy, L. McCartney, H.N. Truong, V. Gaudon, J.-F. Thibault, A. Marchant and H. Höfte, 2007. A Golgi-localized putative methyltransferase is required for normal homogalacturonan synthesis in *Arabidopsis thaliana*. Plant Journal 50: 605-614.

Oosterveld, A., G. Beldman, H.A. Schols and A.G.J. Voragen, 1996. Arabinose and ferulic acid rich pectic polysaccharides extracted from sugar beet pulp. Carbohydrate Research 288: 143-153.

Ralet, M.-C. and J.-F. Thibault, 1994. Extraction and characterisation of very highly methylated pectins from lemon cell walls. Carbohydrate Research 260: 283-296.

Ralet, M.-C., J.C. Cabrera, E. Bonnin, B. Quéméner, P. Hellìn and J.-F. Thibault, 2005. Mapping sugar beet pectin acetylation pattern. Phytochemistry 66: 1832-1843.

Ralet, M.-C., M.-J. Crépeau and E. Bonnin, 2008a. Evidence for a blockwise distribution of acetyl groups onto homogalacturonans from a commercial sugar beet (*Beta vulgaris*) pectin. Phytochemistry 69: 1903-1909.

Ralet, M.-C., M.-J. Crépeau, J. Lefèbre, G. Mouille, H. Höfte and J.-F. Thibault, 2008b. The reduced number of homogalacturonan domains in pectins of an Arabidopsis mutant enhances the flexibility of the polymer. Biomacromolecules 9: 1454-1460.

Schols, H.A. and A.G.J. Voragen, 1994. Occurence of pectic hairy regions in various plant cell wall materials and their degradability by rhamnogalacturonase. Carbohydrate Research 256: 83-95.

Schols, H.A. and A.G.J. Voragen, 1996. Complex pectins: Structure elucidation using enzymes. In: J. Visser and A.G.J. Voragen (eds.) Pectins and Pectinases. Elsevier Science B.V., Amsterdam, NL, pp 3-19.

Schols, H.A. and A.G.J. Voragen, 2002. The chemical structure of pectins. In: G.B. Seymour and J.P. Knox (eds.) Pectins and their Manipulation. Blackwell Publishing, Oxford, UK, pp 1-29.

Schols, H.A., G.-J. Coenen and A.G.J. Voragen, 2009. Revealing pectin's structure. In: H.A. Schols, R.G.F. Visser and A.G.J. Voragen (eds.) Pectins and pectinases. Wageningen Academic Publishers, Wageningen, the Netherlands, pp 17-32.

Thibault, J.-F., C.M.G.C. Renard, M.A.V. Axelos, P. Roger and M.-J. Crépeau, 1993. Studies of the length of homogalacturonic regions in pectins by acid hydrolysis. Carbohydrate Research 238: 271-286.

Vincken, J.-P., H.A. Schols, R.J.F.J. Oomen, M.C. McCann, P. Ulvskov, A.G.J. Voragen and R.G.F. Visser, 2003. If homogalacturonan were a side chain of rhamnogalacturonan I. Implications for cell wall architecture. Plant Physiology 132: 1781-1789.

Voragen, A.G.J., W. Pilnik, J.-F. Thibault, M.A.V. Axelos and C.M.G.C. Renard, 1995. Pectins. In: A.M. Stephen (ed.) Food Polysaccharides and Their Applications. Marcel Dekker, New-York, USA, pp 287-339.

Willats, W.G.T., L. McCartney, W. Mackie and J.P. Knox, 2001. Pectin: cell biology and prospects for functional analysis. Plant Molecular Biology 47: 9-27.

Willats, W.G.T., J.P. Knox and J.D. Mikkelsen, 2006. Pectin: new insights into and old polymer are starting to gel. Trends in Food Science and Technology 17: 97-104.

Yapo, B.M., 2007. Etude de la variabilité structurale des pectines. Ph.D. thesis, University of Nantes, F.

Yapo, B.M., P. Lerouge, J.-F. Thibault and M.-C. Ralet, 2007. Pectins from citrus peel cell walls contain homogalacturonans homogeneous with respect to molar mass, rhamnogalacturonan I and rhamnogalacturonan II. Carbohydrate Polymers 69: 426-435.

Zhan, D., P. Janssen and A.J. Mort, 1998. Scarcity or complete lack of single rhamnose residues interspersed within the homogalacturonan regions of citus pectin. Carbohydrate Research 308: 373-380.

Zhang, Z., M.L. Pierce and AJ Mort, 2007. Changes in homogalacturonans and enzymes degrading them during cotton cotyledon expansion. Phytochemistry 68: 1094-1103.

Structural elucidation of cell wall polysaccharides from okra (*Abelmoschus esculentus* (L.) Moench)

N. Sengkhamparn[1,2], R. Verhoef[1], E.J. Bakx[2], H.A. Schols[1], T. Sajjaanantakul[2] and A.G.J. Voragen[1]
[1]Laboratory of Food Chemistry, Department of Agrotechnology and Food Sciences, Wageningen University, P.O. Box 8129, 6700 EV Wageningen, the Netherlands
[2]Department of Food Science and Technology, Faculty of Agro-Industry, Kasetsart University, Ladyao, Jatujak, Bangkok 10900, Thailand

Abstract

Polysaccharides making up the walls of okra tissue where isolated by a sequential extraction procedure. This yielded different polysaccharide fractions. The Hot Buffer Soluble Solids (HBSS) fraction contained a regular structure which was predominately composed of highly branched rhamnogalacturonan (RG) I elements with short galactan side chains (1-2 galactose units). The degree of acetylation of this rhamnogalacturonan was relatively high. NMR studies of partially hydrolysed HBSS using rhamnogalacturonan hydrolase revealed that the rhamnose residues were *O*-acetylated and/or substituted with a terminal alpha-linked galactose moiety. The CHelating agent Soluble Solids (CHSS) fraction contained mainly relatively high methyl esterified homogalacturonan segments carrying minor quantities of RGI with galactan side chains (2-3 galactose units) which were slightly longer than found for HBSS RGI. The CHSS fraction contained arabinan and arabinogalactan (AG) type II structures as neutral side chains of the rhamnogalacturonan structural element. The Diluted Alkali Soluble Solids (DASS) fraction was characterized by high amounts of rhamnose, galactose, galacturonic acid, next to significant levels of arabinose, indicating that this RGI carried arabinans, AG type I and AG type II structures as neutral side chains. The hemicellulose fraction from the okra AIS represented mainly xyloglucan and xylan as identified in the Concentrated Alkali Soluble Solids (CASS) fraction.

Keywords: okra, cell wall polysaccharide, acetylated rhamnogalacturonan I, sequential extraction, alpha-galactose

1. Introduction

The okra plant, *Abelmoschus esculentus* (L.) Moench, family Malvaceae, is a native plant from Africa and is now cultivated in many other areas such as Asia, Middle East and southern states of the USA. The okra pods are often used as a vegetable and in folk medicine as a diuretic agent and for treatment of dental disease (Ndjouenkeu *et al.*, 1996). Water extracted polysaccharides from okra can act as fat substitute in many product (Costantino and Romanchick-Cerpoviez, 2004; Romanchik-Cerpovicz *et al.*, 2002, 2006).

The okra polysaccharide is an acidic polysaccharide consisting of galactose, rhamnose and galacturonic acid (Whistler and Conrad, 1954). Okra polysaccharide has been reported to contain a repeating unit of alternating rhamnose and galacturonic acid moieties with a disaccharide side chain of (1,4) – linked galactosyl moieties attached to O-4 of rhamnosyl moieties and to contain rather high levels of acetyl groups (5.5% w/w) (Tomada *et al.*, 1980). However, the precise position of the acetyl groups within the pectin has not been reported.

The acidic polysaccharide showed typical chemico-physical properties deviating from similar extracts of other plants. It showed highly viscous, shear thinning behaviour and weak gel behaviour at low concentration (0.1% w/w) (Sengkhamparn, unpublished data).

Apart from the chemical characteristics of okra pectin, the information about other polysaccharides like hemicelluloses is still missing. In this study, we present the characterisation of all main cell wall polysaccharide in okra by using sequential extraction with different aqueous extractants with emphasis on the novel pectin structures. With this information we hope to better understand the characteristics of the extracted polysaccharides.

2. Material and methods

2.1 Material

Soft and mature okra (*Abelmoschus esculentus* (L.) Moench) pods, 5-10 cm in length were collected at local market in Thailand. The okra Alcohol Insoluble Solids, (okra AIS) were prepared and sequentially extracted with 0.05 M sodium acetate buffer (Hot Buffer Soluble Solids, HBSS), 0.05 M EDTA and 0.05 M sodium acetate in 0.05 M sodium oxalate (CHelating agent Soluble Solids, CHSS), 0.05 M sodium hydroxide and 20mM $NaBH_4$ (Diluted Alkali Soluble Solids, DASS), and 6 M sodium hydroxide and 20mM $NaBH_4$ (Concentrated Alkali Soluble Solids, CASS) according to Sengkhamparn *et al.* (2009a).

2.2 Chemical characterization

Of all extracted fractions the sugar compositions was determined by gas chromatography (Englyst and Cummings, 1984) and by HPAEC. Glycosidic linkage composition were analyzed as described by Hakomori and modified by Verhoef *et al.* (2002).The degree of acetylation and methyl esterification of samples were determined by using HPLC (Voragen *et al.*, 1986).

2.3 Enzyme degradation

The HBSS were saponified by adding NaOH in order to remove the methyl esters and acetyl groups before incubating it with enzymes. The HBSS and saponified HBSS were incubated with rhamnogalacturonan hydrolase (RG Hydrolase) from *Aspergillus aculeatus* (Sengkhamparn *et al.*, 2009b). The change in molecular weight (Mw) distribution and released oligomeric products were monitored using HPSEC (Sengkhamparn *et al.*, 2009b).

For NMR analysis, the RG Hydrolase HBSS digest with a Mw of 20 kDa obtained from preparative SEC, Sephacryl S500 column followed by Sephacryl S300 column, was exchanged twice with 99.96% D_2O and the NMR spectra were recorded according to Sengkhamparn *et al.* (2009b).

3. Results and discussions

3.1 Sugar composition of okra AIS and okra AIS extracts

Okra cell wall material and fractions derived from it were firstly characterized by their sugar composition (Table 1). It is shown that okra AIS consisted of mainly Glc, Gal and GalA suggesting the presences of different types of polysaccharides including pectin, hemicelluloses such as xylan and xyloglucan, and cellulose.

The HBSS fraction was composed of GalA, and high amounts of Rha and Gal while, the CHSS fraction contained much lower levels of Gal and Rha, next to the predominantly present GalA residues. Based on the ratio of Rha:GalA within a rhamnogalacturonan (RG) I backbone being 1:1 (Schols and Voragen, 2002), the HBSS fraction contained

Table 1. Sugar composition of okra AIS and okra polysaccharide extracts and residues in mol% (Sengkhamparn et al., *2009a).*

	Rha	Ara	Fuc	Xyl	Man	Gal	Glc	GalA	GlcA	DA (%)[1]	DM (%)[1]	Carbohydrate content[2]
AIS	3	5	0	5	3	17	44	16	7	40	59	57.5
HBSS	26	0	0	0	0	34	1	35	3	58	24	90.0
CHSS	14	3	0	0	0	17	1	63	2	18	48	86.2
DASS	13	13	0	0	0	19	4	48	2	-[3]	-[3]	76.9
CASS	1	3	2	27	3	8	52	3	1	-[3]	-[3]	86.8

[1]moles acetyl or methanol per 100 moles of galacturonic acid.
[2]gram qualities per 100 g of fraction.
[3]Not determined.

predominately RGI segments (85%), while the CHSS contained predominately homogalacturonan (HG) segments (74%). The high amounts of RGI segments in the water extract is not very common but has been reported in detail by Sengkhamparn *et al.* (2009a). The sugar linkage composition indicated the presence of highly branched RGI structures with short galactan side chains (1-2 Gal moieties) for the HBSS fraction while the RGI structure present in the CHSS fraction appears slightly less branched and carries slightly longer galactan side chains (Sengkhamparn *et al.*, 2009a).

The degree of methyl esterification (DM) of the HBSS was lower than the DM of the CHSS (Table 1) which might be explained by the fact that only the HG segments of the pectic molecules are methyl esterified (Sengkhamparn *et al.*, 2009a). The degree of acetylation (DA) of the HBSS was higher than the CHSS fraction.

The DASS fraction contained high amounts of Gal and GalA and 43% of RGI segments. Moreover, the DASS fraction was relatively rich in Ara. Linkage analysis showed the presence of branched arabinans and both arabinogalactan type I and II side chains. The CASS fraction is especially rich in Glc and Xyl, indicating the presence of xyloglucans in this hemicellulose fraction, while the small amount of GlcA residues may point to the presence of acidic xylans as well. The presence of these polymers was confirmed by specific enzyme degradation studies followed by detailed analysis of the degradation products as reported by Sengkhamparn *et al.* (2009a).

3.2 The position of acetyl group within the HBSS RGI

3.2.1 Rhamnogalacturonan hydrolase (RG Hydrolase) treatment

To gain more information about the RGI structure within the HBSS, the samples were incubated with the enzyme RG Hydrolase from *Aspergillus aculeatus*. The HPSEC patterns of the HBSS digest (Figure 1) clearly show that the high Mw population is degraded with the enzyme resulting in degradation products eluting in a broad Mw range. When the saponified HBSS fraction was treated by the enzyme, degradation was much more pronounced illustrating that RG Hydrolase (Sengkhamparn *et al.*, 2009b) requires the removal of acetyl groups for an optimal performance (Searle-van Leeuwen *et al.*, 1996).

The MALDI-TOF mass spectrum of the HBSS digest shows the presence of acetylated oligomer fragments ($Rha_1GalA_1Gal_1Ac_3$, Rha_2GalA_2Ac, $Rha_2GalA_2Gal_2Ac_{1-2}$ and $Rha_2GalA_2Gal_3Ac_3$). According to the observation by Searle-van Leeuwen *et al.* (1996) that RG hydrolase is not active on acetylated RGI, this is quite surprising and may suggest that the acetyl groups probably are substituted to other sugar residues in RGI backbone rather than to the galacturonic acid residues (Sengkhamparn *et al.*, 2009b).

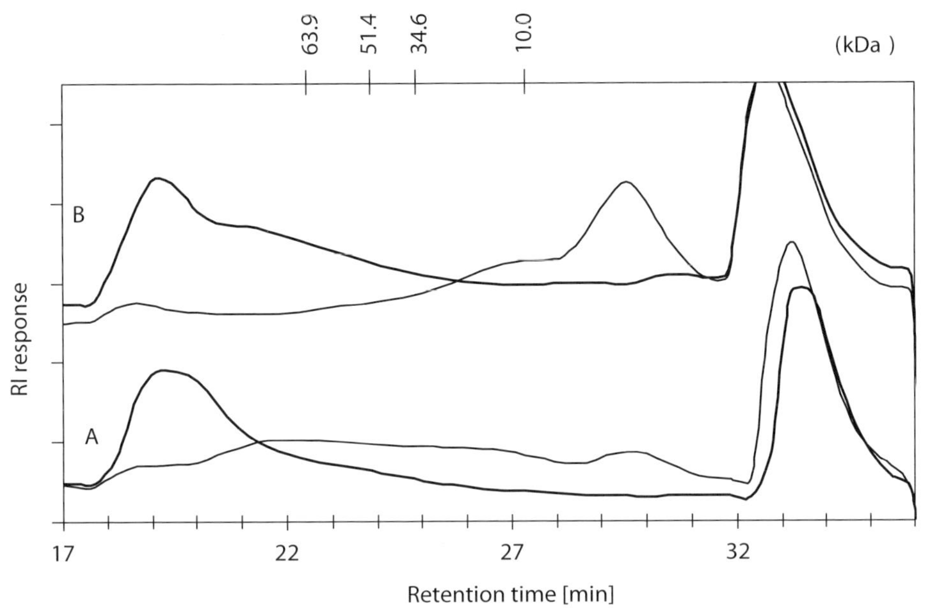

Figure 1. HPSEC-elution patterns of HBSS (A) and saponified HBSS (B) before (thick line) and after (thin line) incubation with RG Hydrolase (the molecular weight indication are based on pectin standard).

3.2.2 Electron spray ionization Ion-trap Mass Spectrometry

To get an insight in the structure of RG oligomers and the precise location of the acetyl group, electron spray ionization ion-trap mass spectrometry (ESI-Iontrap-MS) was performed. Mass spectrometric fragmentation in the MS^3 spectrum of m/z = 533 ($Rha_1GalA_1Rha_1Ac_1$) yielding the formation of Rha_1Ac_1 (m/z= 327 and 345 (-H_2O)) indicated that the acetyl groups were at least partly linked to the rhamnose residues within the okra RGI. It can be concluded that rhamnose residues in HBSS RGI were *O*-acetylated which have no hindrance effect to the RG Hydrolase activity (Sengkhamparn *et al.*, 2009b).

3.2.3 Nuclear Magnetic Resonance

To confirm the position of the acetyl groups on rhamnose within the rhamnogalacturonan I polymer, the HBSS RG Hydrolase digest with a Mw about 20 kDa was analyzed by NMR. Within the anomeric region of the ^1H NMR spectrum recorded, 5 major sugar moieties could be observed (Table 2) and a signal at 2.11 ppm generally belonging to an *O*-acetyl substituent was observed. The *O*-acetyl substituent was found to be attached to O-3 of the rhamnose residues as shown by the significant

Table 2. Proton chemical shifts (ppm) of HBSS treated RG Hydrolase.

	H-1	H-2	H-3	H-4	H-5	H-6	Ref
(A) 1,2,4-α-L-rha[1]	5.24	4.23	5.25	3.90	3.97	1.33	
(B) 1,4,-α-D-GalA (RGI)	4.98	3.95	4.11	4.44	4.56		(Colquhoun *et al.*, 1990; Deng *et al.*, 2006; Mutter *et al.*, 1994; Renard *et al.*, 1999)
(C) t-α-D-Gal	4.95	3.81	3.92	4.04	4.40	3.70	(Gronberg *et al.*, 1994; Strecker *et al.*, 1995)
(D) 1,4-β-D-Gal	3.60	3.48	3.74	4.03	-[2]	-[2]	(Colquhoun *et al.*, 1990; Habibi *et al.*, 2004; Huisman *et al.*, 2001)
(E) t-β-D-Gal	4.49	3.43	3.65	3.89	-[2]	-[2]	(Colquhoun *et al.*, 1990; Habibi *et al.*, 2004; Huisman *et al.*, 2001)
CH_3 (O-acetyl)	2.11						
COO	174.91						

[1]Arabic letters correspond to annotation.

[2]not determined; values are expressed against internal acetone at 2.23 and 31.55 ppm.

down field shift of H-2 (4.23 ppm) and H-3 (5.25 ppm) and the absence of the spin systems found before for acetylated GalA units (Lerouge *et al.*, 1993, Ishii, 1997).

The spin systems of the galactose moiety gave some problems for identification since they were quite different from the usual signals for a β-1,4-linked gal residue. However, using the SweetDB database of the German Cancer Research Center Heidelberg (http://www.glycosciences.de/sweetdb/), enabled us to identify the galactose residue to be an terminal α-galactosyl residue linked to O-4 of a β-1,4-linked gal and O-4 of a rhamnose (Gronberg *et al.*, 1994; Strecker *et al.*, 1995). The presence of a α-galactosyl residues at the terminus of the RGI side chains has not been observed before in any other pectin (Sengkhamparn *et al.*, 2009b).

4. Conclusions

The sequential extraction of okra AIS yielded different polysaccharide fractions of which the pectic fractions showed some structural features which have not been described before. The HBSS consisted of a regular structure of highly branched RGI with very short galactan side chains. The rhamnosyl residues in the HBSS RGI were O-acetylated at position O-3 and were also (partly) substituted with terminal alpha-linked galactose residues. The CHSS contains mainly HG and less RGI with slightly longer galactan side chain. Similar to the HBSS, the RGI of CHSS also contained acetylated rhamnose residues but even more acetylated galacturonic acid residues. The DASS fraction contains complex pectins with arabinans and arabinogalactan I and II type structures as neutral side chains. Besides pectic material, the okra AIS also contains xyloglucans and xylans. Further investigations will be perfomed to see if some typical characteristics of pectic okra extracts can be explained by the unique structural features established for these acidic polysaccharides.

References

Costantino, A.J. and J.E. Romanchick-Cerpoviez, 2004. Physical and sensory measures indicate moderate fat replacement in frozen dairy dessert is feasible using okra gum as a milk-fat ingredient substitute. J. Am. Diet. Assoc. 104: 44.

Colquhoun, I.J., G.A. De Ruiter, H.A. Schols and A.G.J. Voragen, 1990. Hairy (ramified) regions of pectins 3. Identification by nmr-spectroscopy of oligosaccharides obtained by treatment of the hairy regions of apple pectin with rhamnogalacturonase. Carbohydr. Res. 206: 131-144.

Deng, C., M.A. O'Neill and W.S. York, 2006. Selective chemical depolymerization of rhamnogalacturonans. Carbohydr. Res. 341: 474.

Englyst, H.N. and J.H. Cummings, 1984. Simplified method for the measurement of total non-starch polysaccharides by gas-liquid chromatography of constituent sugars as alditol acetates. Analyst. 109: 937-942.

Gronberg, G., U. Nilsson, K. Bock and G. Magnusson, 1994. Nuclear-magnetic-resonance and conformational investigations of the pentasaccharide of the forssman antigen and overlapping disaccharide, trisaccharide, and tetrasaccharides sequences. Carbohydr. Res. 257: 35-54.

Habibi, Y., A. Heyraud, M. Mahrouz and M.R. Vignon, 2004. Structural features of pectic polysaccharides from the skin of Opuntia ficus-indica prickly pear fruits. Carbohydr. Res. 339: 1119-1127.

Huisman, M.M.H., C.T.M. Fransen, J.P. Kamerling, J.F.G. Vliegenthart, H.A. Schols and A.G.J. Voragen, 2001. The CDTA-soluble pectic substances from soybean meal are composed of rhamnogalacturonan and xylogalacturonan but not homogalacturonan. Biopolym. 58: 279-294.

Ishii, T., 1997. O-acetylated oligosaccharides from pectins of potato tuber cell walls. Plant Physiol. 113: 1265-1272

Lerouge, P., M.A. O'Neill, A.G. Darvill and P. Albersheim, 1993. Structural Characterization of Endo-Glycanase-Generated Oligoglycosyl Side-Chains of Rhamnogalacturonan-I. Carbohydr. Res. 243: 359-371.

Mutter, M., G. Beldman, H.A. Schols and A.G.J. Voragen, 1994. Rhamnogalacturonan α-L-rhamnopyranosylhydrolase. A novel enzyme specific for the terminal nonreducing rhamnosyl unit in rhamnogalacturonan regions of pectin. Plant Physiol. 106: 241-250.

Ndjouenkeu, R., F.M. Goycoolea, E.R. Morris and J.O. Akingbala, 1996. Rheology of okra (*Hibiscus esculentus* L.) and dika nut (*Irvingia gabonensis*) polysaccharides. Carbohydr. Polym. 29: 263-269.

Renard, C.M.G.C., M.J. Crepeau and J.F. Thibault, 1999. Glucuronic acid directly linked to galacturonic acid in the rhamnogalacturonan backbone of beet pectins. Eur. J. Biochem. 266: 566-574.

Romanchik-Cerpovicz, J.E., A.C. Costantino and H.G. Laura, 2006. Sensory evaluation ratings and melting characteristics show that okra gum is an acceptable milk-fat ingredient substitute in chocolate frozen dairy dessert. J. Am. Diet. Assoc. 106: 594-597.

Romanchik-Cerpovicz, J.E., R.W. Tilmon and K.A. Baldree, 2002. Moisture retention and consumer acceptability of chocolate bar cookies prepared with okra gum as a fat ingredient substitute. J. Am. Diet. Assoc. 102: 1301-1303.

Schols, H.A. and A.G.J. Voragen, 2002. The chemical structure of pectins. In: G.B. Seymour and J.P. Knox (eds.) Pectins and their Manipulation. Blackwell Publishing, Oxford, UK, pp 1-29.

Searle-van Leeuwen, M.J.F., J.P. Vincken, D. Schipper, A.G.J. Voragen and G. Beldman, 1996. Acetyl esterases of Aspergillus niger: purification and mode of action on pectins. In: J.J. Visser and A.G.J. Voragen (eds.) Pectins and Pectinases. Elsevier, Amsterdam, pp. 793-798.

Sengkhamparn, N., R. Verhoef, H.A. Schols, T. Sajjaanantakul and A.G.J.Voragen, 2009a. Characterization of Cell Wall Polysaccharide from Okra (*Abelmoschus esculentus* (L.) Moench). Carbohydr. Res. Doi.10.1016/j.carres.2008.10.012

Sengkhamparn, N., R. Verhoef, E.J. Bakx, H.A. Schols, T. Sajjaanantakul and A.G.J.Voragen, 2009b. Okra pectin contains an unusual substitution of its rhamnosyl residues with acetyl and alpha linked galactosyl groups. Carbohydr. Res. in press

Strecker, G., J.M. Wieruszeski, Y. Plancke and B. Boilly, 1995. Primary structure of 12 neutral oligosaccharide-alditols released from the jelly coats of the anuran xenopus-laevis by reductive beta-elimination. Glycobiol. 5: 137-146.

Tomada, M., K. Shimada, Y. Saito and M. Sugi, 1980. Plant mucilages. XXVI. Isolation and Structural Features of a mucilage, 'Okra mucilage', from the immature fruit of Abelmoschus esculentus. Chem. Pharm. Bull. 28: 2933-2940.

Verhoef, R., P. de Waard, H.A. Schols, M. Rättö, M. Siika-aho and A.G.J. Voragen, 2002. Structural elucidation of the EPS of slime producing BrevundimonasVesicularis sp. isolated from a paper machine. Carbohydr. Res. 337: 1821-1831.

Voragen, A.G.J., H.A. Schols and W. Pilnik, 1986. Determination of the degree of methylation and acetylation of pectins by hplc. Food Hydrocolloid. 1: 65-70.

Whistler, R.L. and H.E. Conrad, 1954. A Crystalline galactobiose from acid hydrolysis of okra mucilage. J. Am. Chem. Soc. 76: 1673-1974.

Molecular shape and functionality of HM pectin

K. Meyer-Hansen, H.L. Nielsen and C. Rolin
CP Kelco ApS, Ved Banen 16, 4623 Lille Skensved, Denmark

Abstract

Pectin samples from different botanical sources have different shapes when analysed by size exclusion chromatography in combination with light scattering, viscosity and refractive index detections. Lemon pectin has fairly high $[\eta]/M_W$ proportion (weight-averaged intrinsic viscosity per weight-averaged molecular weight) while this proportion is lower with typical orange pectin and even lower with sugar beet pectin. HM pectin may under appropriate circumstances and in the presence of calcium ions produce thickened or semi-gelled solutions by a mechanism that in some respects resembles the commercially widely used gelling of LM pectin. Besides obviously being dependent upon the intramolecular distribution of free pectin carboxyl groups, this functionality also depends upon the molecular size. The ambition of our paper is to demonstrate this structure-function relationship.

Keywords: molecular shape, degree of blockiness, intrinsic viscosity, strength, yoghurt serum

1. Introduction

Commercial pectin (European Union, 1998, United Nations FAO and WHO, 2001, Food Chemicals Codex, 2004, United States Phamacopoeia and National Formulary, 2004) is used for contributing desirable physical characteristics to the compositions in which it is used as an ingredient. Normally, what a pectin user would perceive as the *'strength'* of the pectin is inversely proportional to the dosage that is required for attaining the desired physical characteristic. Commercial definitions of pectin grading strength thus specify on one hand a reference composition that more or less emulates the relevant commercial use, and on the other hand a numerical standard result for the measurement of some physical property of the composition. This could, for example, be a certain firmness if the reference composition is a gel (IFT, 1959), or it could be a specified amount of sediment if the composition is a suspension (Glahn, 1982). The pectin grading strength is then inversely proportional to the pectin dosage that brings exactly the stipulated standard result.

Establishing a commercial grading strength method that is resource-efficient and reproducible and at the same time is a fair value-metric for a diverse group of customers can be a formidable challenge in itself. Once a method is in place, however, the next problem for the pectin manufacturer is to understand how to control this grading strength through manipulations in the pectin production process. Empirical testing based solely upon observing the ensuing strength vis-à-vis production conditions has some character of 'needle in a haystack', because the grading strength can depend in a complex way upon a multitude of circumstances, including, of course, the selection of the botanical raw material. Thus, structure-function relationships become interesting, because it is often easier to describe how the individual production parameters impact the molecular weight, for example, than it is to describe what they mean for a grading strength.

An obvious parameter to consider for mapping out pectin structure-function relationships is the Degree of Methyl esterification (DM). Commercial pectin is classified into High Methyl ester pectin (HM pectin) and Low Methyl ester pectin (LM pectin) according to whether the DM is above or below 50. The traditional – and until about 30 years ago entirely dominant – use for HM pectin is for gelling jam of about 65 brix and pH 3.1. Successful gelling only takes place if the brix and pH are in the vicinity of these prototypical conditions, but it does ideally not depend much upon the presence of Ca^{++}, although in practice some effect of ions is seen (Neidhart *et al.*, 1996). It is thought to be brought about by hydrogen bonding and hydrophobic interactions (Oakenful and Scott, 1984). LM pectin, in contrast, is typically used for jam of lower brix than what would allow for HM pectin gelling. This LM pectin use depends strongly upon the presence of Ca^{++}, and the mechanism has been hypothetically described as molecular chains being held together by bridges of Ca^{++} ions (Grant *et al.*, 1973, Braccini and Pérez, 2001). The above two ways of making jam with pectin have thus been described as separate phenomena governed by each their characteristic molecular mechanism.

On the other hand, it is also known that various plant pectin esterases occur in the raw materials for commercial pectin, and that these enzymes may create so-called blocks, i.e. segments of solely non-esterified galacturonan repeating units. In this way an HM pectin molecule may possess local domains that resemble LM pectin, and it may also happen that the inter-molecular distribution is broad so an HM pectin sample may contain a significant fraction of molecules that individually would be LM pectin molecules. Pectin samples differ in this respect depending upon their raw material – the species as well as the ripeness – and in general upon their prior exposure to plant pectin esterase (Speiser and Eddy, 1946, Kohn *et al.*, 1968). HM pectin that has been significantly impacted by plant pectin esterase is calcium-sensitive, meaning that the rheology of its solutions and gels depends upon (is sensitive towards) the concentration of Ca^{++}. Accordingly, for mapping out pectin structure-function relationships one may in addition to the (average) DM have to consider how the esterified and non-esterified carboxyl groups are situated in the molecules. A useful metric in this respect (though it does not capture all possibilities for variation) is the Degree of Blockiness (DB; Daas *et al.*, 1999).

The molecular size strongly influences pectin functionality, and some appropriate metric for it should thus be included in structure-function descriptions. The strength of HM pectin with respect to bringing about firmness of jam-emulating gels was found to be proportional to the intrinsic viscosity, [η], of the pectin in a good solvent (Christensen, 1954). With the instrumental possibilities that were commonly available until about 1995, it was too cumbersome getting a sufficient amount of true molecular weight data, and viscosity measurements were in some cases used in the place of molecular weight for mapping out structure-function relationships (Smith and Bryant, 1967). In other cases, the molecular weight that could be inferred from size exclusion chromatography while calibrating against other materials of known molecular weight, typically pollulan or dextran, was reported (Deckers *et al.*, 1986). Nowadays, size exclusion chromatography with combinations of several detection systems, like refractive index plus light scattering plus viscosity, provides a basis for routinely characterizing the molecular populations of polydisperse materials. It is thus understood that pectin molecules differ, not only with respect to their degree of polymerization, but also in shape, and that pectin samples of different origin may have different 'translations' between molecular weight and [η] (Fishman *et al.*, 2006).

The objective of the present study is to establish a metric for the strength of pectin with respect to thickening the serum phase of yoghurt beverages, and to relate this strength to molecular structural data. It has previously been found that aqueous solutions of HM pectin and calcium salts are shear-thinning and short-term shear-reversible (Rolin, 1994). Their rheology does, on the other hand, gradually change over some longer time-scale, and occasionally, lumps of soft gels in a thinner liquid develop. This syneresis-resembling phenomenon has even been suggested as a tool for separating the pectin into a calcium-sensitive and a less calcium-sensitive fraction (Glahn, 1995; Glahn and Rolin, 1996). In spite of this time- and history-dependence of the rheology, we have succeeded in developing a reproducible strength method and relating the strength to pectin M_W, [η] (in good solvent), DM, and DB.

2. Materials and methods

2.1 Pectin samples

Different pectin samples were used for different parts of the study. All pectin samples for studies of degree of blockiness were from dried peel of citrus fruit extracted commercially, or in pilot scale under conditions emulating commercial conditions, as described in general reviews (Rolin, 2002). The acidity, extraction temperature and extraction duration were chosen according to commercial practice so the ensuing pectin samples got a DM that is typical for acidified milk stabilizing grades, in most cases between 70 and 77. Based on the prior impact of the esterases that are native to the citrus fruit, commercial dried peel for pectin production may generally differ with respect to the DB that the extracted pectins will acquire; the pectins of the present study differed for this reason in spite of having rather similar DM. Some of the pectin samples were further de-esterified by exposing the liquid pectin extract either to pectin esterase from *Aspergillus*, or pectin esterase from papaya.

The reported Mark-Houwink plot was made with pectin samples extracted as above described from different botanical materials, lemon, orange, apple, sugar beet, and potato. These samples were not additionally de-esterified.

For the study of viscosity in good solvents (Nielsen, 2000) a total of 17 different pectin samples were used. These were all based on citrus and included experimental as well as commercial samples. The samples were initially prepared with different DM and degree of amidation, DA, brough about by usual commercial processes (Rolin, 2002), and some of them were later modified by exposing aqueous solutions of them either to pectin esterase or pectin lyase.

2.2 Measurement of degree of blockiness

This was done as described in Daas *et al.*, 1999.

2.3 Measurement of M_w and [η]

Molecular parameters were determined by size exclusion chromatography followed by three detectors: refractive index, right angle light scattering, and viscosity.

Equipment: Viscotek Tri-Sec instrument equipped with 3 TSK Columns GMPWXL (Supelco), Eluent: 0.3 M Li-acetate buffer pH 4.8mM, temperature 40 °C.

The molecular parameters were deduced from the detector signals by the software supplied by Viscotek, assuming dn/dc for pectin = 0.145

2.4 Measurement of pectin viscosities in acetate buffer

For these studies, (40-200mg) pectin was moistened with 96% ethanol (4μl/mg pectin) and dissolved in solvent (0.3M Li-acetate buffer pH 4.8mM or 0.083M K-acetate buffer pH 4.4) (Nielsen, 2000) to the needed concentration. The mixture was placed on a water bath (70 °C) with magnetic stirrer and mixed for 30 min., then transferred to 40 °C for 60 min. The solution was transferred to a 100 ml measuring flask and solvent was added to the marker. Samples of low viscosity (< 2mPa·s) were measured by capillary viscometers (Cannon Fenske #25, #50 or #75), while viscosities from approx. 1mPa·s to approx. 3000mPa·s were measured by Haake RS100 rheometer equipped with a DG41 (>1000mPa·s) or Z20 geometry. Increasing stress was applied from 0.1 to 446Pa, however, if shear rate exceeded $800s^{-1}$ the measurement was stopped and the zero shear viscosity was estimated by extrapolating to low shear rate $10s^{-1}$, only relevant for the high viscosity samples that exhibit shear thinning behaviour. Temperature was maintained at 40 °C.

2.5 Measurement of $c^{\dagger}$ in beverage-simulating buffer

A buffer was prepared from in g/l: sucrose, 150; lactose, 20; Ca-lactate ($C_6H_{10}O_6Ca$, $5H_2O$), 6.52; 90% lactic acid, 3.62; K_2HPO_4, 1.40; KH_2PO_4, 1.02; sodium benzoate, 1.00; NaOH, 0.22; citric acid until pH = 3.75.

For each pectin sample, a 1.2% pectin solution was prepared by moistening 1.20 g of the pectin with 5.00 ml 2-propanol in a tared blue-cap bottle, then adding 95 ml of boiling de-ionized water, stirring at 75 °C for 30 minutes, cooling to room temperature, and adjusting the net weight to 100.0 g with deionized water.

For each final test system, the above pectin solution was diluted with the appropriate amount of water to attain twice the intended final concentration. 15 ml of this was placed in a 100 ml blue-cap bottle with a stirring-magnet and kept stirring at such intensity, that an about 1 cm deep vortex resulted. While starting time-taking, 15 ml of the buffer was then added from a pipette during about 13 seconds, and stirring was continued for another about 12 seconds.

The liquid was transferred to a Brookfield LVT with adaptor, and the measurement was started at 6 rpm 60 seconds after the instant when the buffer addition was initiated. The viscosity was read after another 60 seconds.

For each pectin sample, different concentrations were tried until there was at least one measurement with viscosity between 18 and 25 mPa·s, and at least one measurement between 25 and 35 mPa·s. Then the pectin concentration, $c^{\dagger}$, corresponding to 25 mPa·s was determined by interpolation whilst assuming that the logarithm to the viscosity reading related linearly to the pectin concentration.

Notes: the final composition was designed for emulating a commercial beverage with 3% MSNF from yoghurt. (For comparison, pure bovine yoghurt has about 8.5% MSNF.) The buffer could typically keep for about a week at room temperature.

3. Results

A characteristic example of a Mark-Houwink plot for diverse pectin samples is shown in Figure 1. Generally speaking the $[\eta]/M_W$-proportion is large and fairly constant for all M_Ws of the population for lemon pectin, while for the orange pectin, the $[\eta]/M_W$ is lower on average and in particular for the highest represented M_Ws. This is even more pronounced for pectin from sugar beet and potato. Figure 2 shows the M_W vis-à-vis the weight averaged $[\eta]$ for a large number of pectin samples. The orange pectin samples are highlighted; it is evident that a lower $[\eta]/M_W$ proportion is typical of the orange pectin samples of our study as compared to the lemon pectin samples.

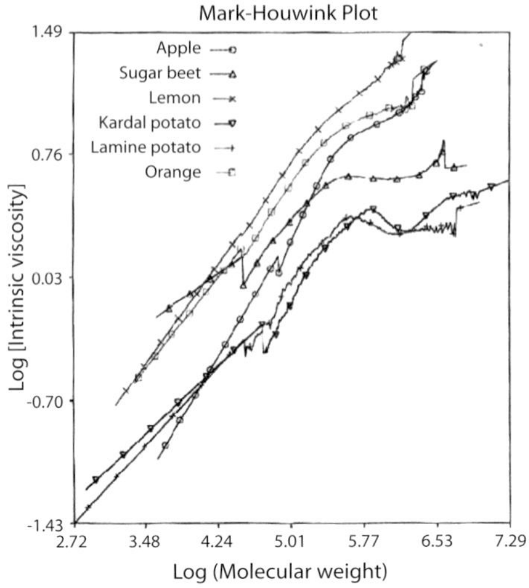

Figure 1. Mark-Houwink plot for diverse pectin samples.

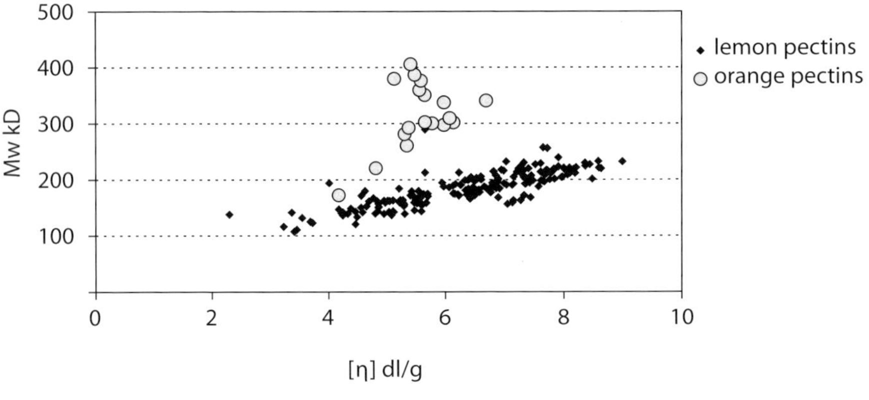

Figure 2. Intrinsic viscosity and weight-averaged molecular weight for lemon pectins and orange pectins.

The Degree of Blockiness of the same pectin samples as those of Figure 2 have been shown together with the DM as Figures 3a-d. In Figure 3a all samples are represented with the same symbol, while pectins that were treated with *Aspergillus* esterase are highlighted in Figure 3b, pectins treated with plant pectin esterase are highlighted in Figure 3c, and orange pectins are highlighted in Figure 3d.

The concentration dependence of the specific viscosity of pectin solutions in (for most samples) the same Li-acetate buffer as is used for the Viscotek chromatography is shown as Figure 4. This buffer is considered a good solvent for pectin with minimum

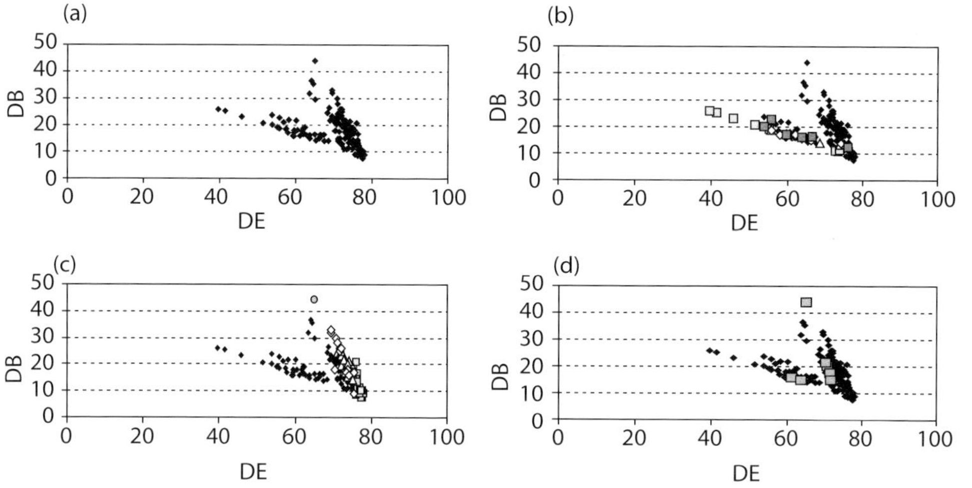

Figure 3. Degree of esterification and degree of blockiness. a) all pectins shown with same symbol; b) pectins de-esterified with Aspergillus *esterase highlighted, samples with the same symbol were de-esterified from the same originating material; c) pectin de-esterified with plant pectin esterase highlighted, samples with the same symbol were de-esterified from the same originating material; d) orange pectins highlighted.*

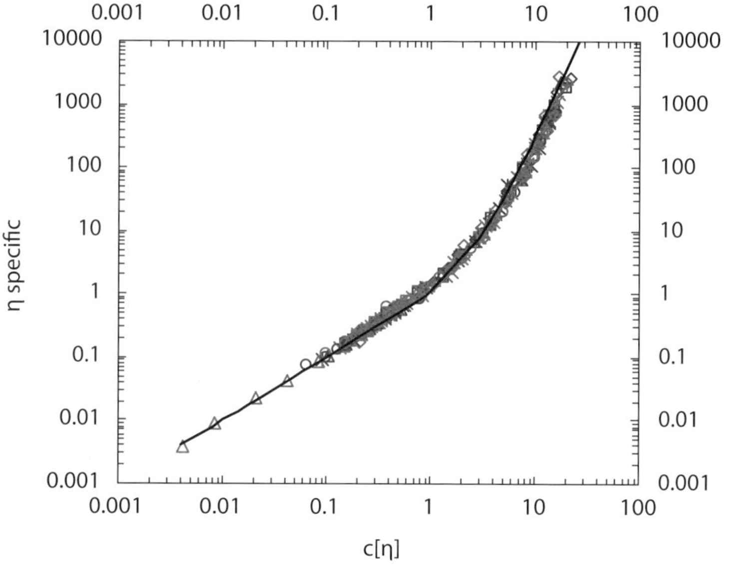

Figure 4. Entanglement overlap plot for seventeen different pectin samples. The samples have DM ranging from 28 through 76 and [η] between 0.92 and 5.4 dl/g. For further detail see Nielsen (2000).

tendency for clustering of pectin molecules (Fishman *et al.*, 2001). The pectin samples are not the same as the pectin samples of the other figures, but they represent a diverse group of pectin samples of very different compositions and properties. The data has been fitted to the equation (Ren *et al.*, 2003; see also Doublier and Wood, 1995):

$$\eta_{sp} = c \cdot [\eta] + K \cdot (c \cdot [\eta])^n \quad (1)$$

In this equation, η_{sp} means the specific viscosity, while K and n are adjustable parameters. We find as best estimates K = 0.1 and n = 3.5. Different measurement techniques were used for the various regimes of viscosity represented in the figure (see the *Materials and Methods* section), but the entire data collapses into this single graph without technique-dependent systematic deviations.

The strength of the pectin with respect to thickening yoghurt-beverage-simulating buffer has been put in context with molecular structure in Figure 5, which shows a relation between $c^\dagger$, $[\eta]$, and the degree of blockiness. These are the same samples as Figures 2 and 3 with orange pectin being highlighted. For reasons to be explained in the *discussion section*, we have chosen to fit the data to the equation:

$$[\eta] \cdot c^\dagger = Q \cdot \exp(-\alpha \cdot B) \quad (2)$$

in which B stands for the degree of blockiness and Q and α are adjustable parameters. We find the estimates Q = 5.69 and α = 0.0556 with R^2 = 0.71. It is evident from Figure 5, that the orange pectin samples on average are a bit stronger than the relation predicts.

Figure 6 corresponds to Figure 5 but for using the M_W in place of $[\eta]$. In the corresponding equation:

$$M_W \cdot c^\dagger = Q^* \cdot \exp(-\alpha^* \cdot B) \quad (3)$$

we find the estimates $Q^* = 1.0 \cdot 10^6$ $g^2/mol \cdot L$ and $\alpha^* = 0.0456$ with $R^2 = 0.55$. All orange pectins are weaker than this relation predicts.

4. Discussion

As was also mentioned in the introduction, the original objective of the present study was to establish a metric for the strength of pectin with respect to thickening the serum phase of yoghurt beverages. Accordingly, the bulk of the studied pectin samples were produced with the application for low-pH protein suspensions, in particular yoghurt-containing beverages, in mind (except those used for the η_{sp} versus $c \cdot [\eta]$ relation in good solvent, Figure 4). This means that raw materials and production conditions were chosen in a way that empirically gives good properties for this purpose. It also means that the DM with most samples falls within a fairly narrow range: two

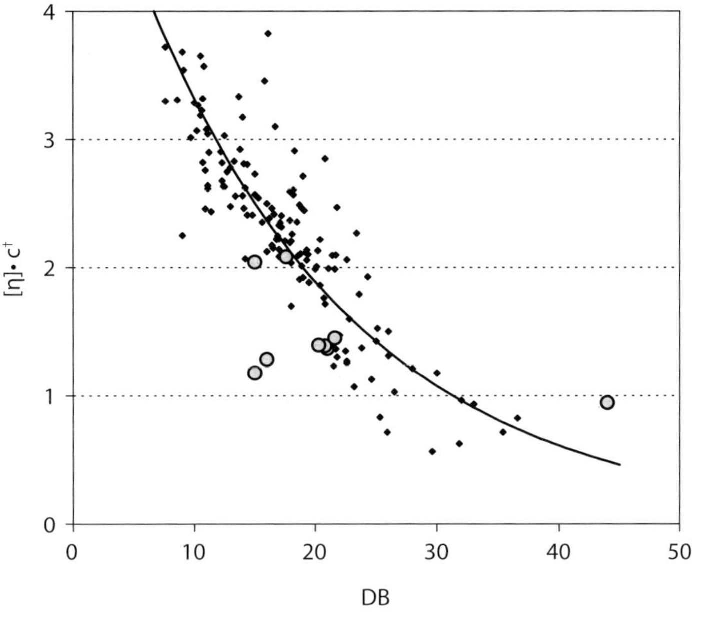

Figure 5. The product of intrinsic viscosity and the concentration c† depicted against degree of blockiness. Little dots are lemon or lime pectins, large dots are orange pectin and the solid line is the model of Equation 2.

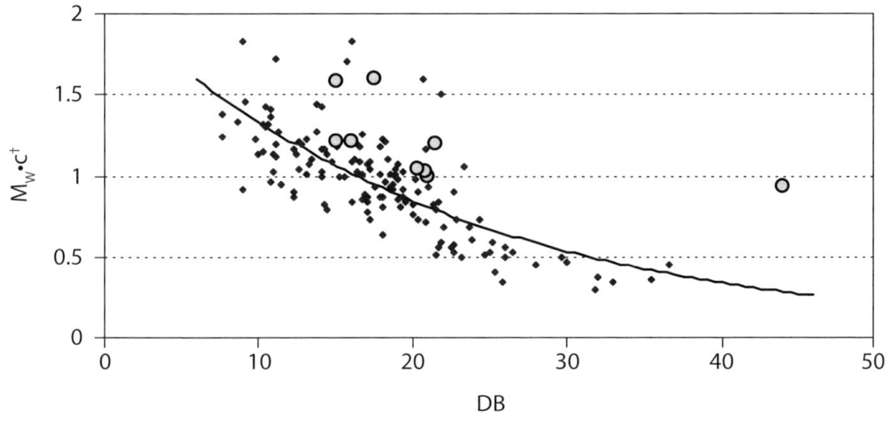

Figure 6. The product of c† and the weight-averaged molecular weight depicted against the degree of blockiness. Little dots are lemon or lime pectins, larger dots are orange pectins and the solid line is the model of Equation 3.

thirds of the studied samples have DM between 68 and 77. Ideally, when mapping out structure-function relationships there should be an about even representation of samples within the entire range of the structure-parameters, and the 'independent' parameters should not be dependent upon each other. The studied population does, unfortunately, not fully comply with these ideal qualifications. For example the DM and the DB are correlated. In principle it could be imagined that differently looking structure-function relations could be found if the samples for study were selected in other ways.

In the studied set of samples, the orange pectins clearly have lower $[\eta]/M_W$ proportion than the lemon and lime pectins. In general, when looking to two materials that have different $[\eta]/M_W$ proportion, and if assuming that the solvent has about equally good solvent-quality for the two materials, the usual interpretation would be that the sample having the highest $[\eta]/M_W$ proportion is the more elongated of the two. I.e., the sample of low $[\eta]/M_W$ would have a relatively more flexible backbone – in other words a backbone of shorter persistence length – or it would be more branched. It is known that pectin molecules may cluster even in relatively good solvents (Fishman *et al.*, 2001), so it cannot be ruled out that the apparent molecular size that is picked-up by the Viscotek detector system is influenced by clustering, and it could be imagined that this tendency for clustering might be different for orange pectins versus lemon or lime pectins. However, our data clearly reveals that the $[\eta]/M_W$ proportion does not depend upon the DB.

The findings of Figure 3, that the degree of blockiness increases only moderately when pectin is de-esterified with *Aspergillus* esterase, while it increases more strongly for the samples that have been exposed to plant pectin esterase, is not surprising and has previously been described by a multitude of authors (examples: Kohn *et al.*, 1968, Ishii *et al.*,1979). In older literature, pronounced blockiness is seen as a typical trait of orange pectin. In our set of samples with which there are many lemon or lime pectin samples that have been exposed to esterase, high DB cannot be said to be particular to the orange pectin.

The dependence of the specific viscosity of pectin in good-solvent solutions upon its concentration and $[\eta]$ in the same solvent has been described previously (Axelos *et al.*,1989), but the present study covers a larger range of viscosities and pectin compositions resulting in different intrinsic viscosities than previous studies, and it confirms that the intrinsic viscosity is a good and unambiguous metric for the strength of pectin with respect to thickening good solvents.

In the beverage-simulating serum, the viscosity also depends upon ionic interactions, hypothetically egg-box cross-linking of molecules. We have chosen to quantify this phenomenon by reporting the concentration of pectin that creates a specific rheology, rather than reporting the rheology that is created by a fixed concentration of pectin. Our chosen way is cumbersome, because it requires at least two measurements with different pectin concentrations, and an interpolation, rather than just one measurement. The rationale for accepting this added burden is

- in our preferred method, all rheological measurements are done under similar conditions that are suitable for the instrument,
- all pectin materials, no matter how different they may be, can be compared with the same method,
- the population of final results can reasonably be approximated with the normal distribution, and
- the method links to the historical grading strength concepts for pectin.

In our opinion, it is not practical to use only one fixed pectin concentration for comparing the thickening ability of diverse pectin samples, because such method would not have good sensitivity with respect to differentiating between the relatively weak samples, while, on the other hand, the strongest samples would have excessively high thickness and be torn apart by the measurement. In particular, the results would be far from being normally distributed.

For modelling $c^{\dagger}$ as a function of pectin structural/compositional parameters we have chosen to take inspiration from classical work with HM pectin/sucrose gels (Christensen, 1954) as well as from the functional expression of the good solvent behaviour, Equation 1, by using the formalism

$[\eta]\cdot c^{\dagger} = f(\text{molecular structure})$.

Next to $[\eta]$, the parameter that most strongly influences $c^{\dagger}$ is the content of calcium-reactive blocks, and we have in Figure 5 chosen to depict $[\eta]\cdot c^{\dagger}$ in function of DB. In our work, $[\eta]$ and $c^{\dagger}$ are not determined in the same solvents, but the idea behind the concept 'calcium sensitivity' is that the two solvents should be about equally good for the samples that are *not* calcium sensitive. Thus, in the limit for DB = 0, one should expect the product $[\eta]\cdot c^{\dagger}$ to attain about the same value as predicted from Equation 1. Since $c^{\dagger}$ by definition is the concentration that yields 25 mPa·s, the η_{sp} at measurement becomes 24 (the solvent has viscosity ≈ 1 mPa·s), and our expectation for DB = 0 becomes $[\eta]\cdot c^{\dagger} = 4.5$. Assuming that the graph of Figure 5 should extrapolate to about this value at its intersection with the y-axis, it becomes evident that the data cannot be fitted to a straight line, because the $[\eta]\cdot c^{\dagger}$ would then become negative at DB ≈ 40 (which would be physically unreasonable). The curve has to approach asymptotically to a horizontal line of $[\eta]\cdot c^{\dagger} \geq 0$. The exponential function is a simple and convenient function that has this feature. Thus we arrived at Equation 2. We also tried using DB·(100-DM) in the place of DB, but this did not improve the R^2 of the fit. Further, we verified that the prediction error did not correlate to DM, or to the $[\eta]/M_W$-proportion, so, in effect, we could not improve the model by including these variables.

In the Li-acetate buffer, i.e. the relatively good solvent, we estimate the threshold for onset of entanglement to be $[\eta]\cdot c \approx 2.5$ and $\eta_{sp} \approx 5$. In the beverage simulating buffer, which is a poorer solvent for the calcium sensitive pectins, we find $\eta_{sp} = 24$ at values of [η]·c less than 2.5. The value of [η]·c for attaining $\eta_{sp} = 24$ progressively decreases with DB, and this must be because chains adjoin via Ca^{++} ions. Our interpretation is that a network can form at polymer concentrations below those that cause molecular entanglement; a point which has also been noted by others (Morris, 2007). In turn, this means that the molecules arrange themselves relatively to each other in some systematic way so that they embrace the liquid more efficiently than they would have been able to if they had been randomly orientated. This increased efficiency of embracing the solvent takes place at least while the viscosity is being measured about two minutes after the mixing of ingredients. If said systematic arrangement means that the molecules align in tubular structures like those observed with microscopy (Löfgren *et al.*, 2002, Fishman *et al.*,2004), the average pore size of the network must become relatively large, viz. larger the more molecules per tubular cross-section. It is thus, perhaps, not surprising that the ensuing rheology is fragile with tendency for lump formation and syneresis-resembling phenomena.

5. Conclusions

Within the parameter-span and conditions of our study, the strength of HM pectin with respect to thickening a Ca^{++}-containing yoghurt-beverage-simulating buffer is positively correlated to the [η] that the pectin possesses in a good solvent, and it is positively correlated to the DB of the pectin. Orange pectins tend to be slightly stronger than predicted from their [η], but their strength is over-estimated if M_W is used for predicting in the place of [η]. The calcium sensitive pectin samples are able to build networks, even at concentrations below the threshold for molecular entanglement.

Acknowledgements

The authors are indebted to Laila Pedersen, Mette Enø Hansen, Lene Harboe, Janne Riis Hansen and Annette Nytoft for their extensive analytical work. The data relating to the viscosity in good solvents was part of the PhD study of Heidi L. Nielsen, and the authors thank Dr. Søren Hvidt for supervising this study.

References

Axelos, M.A.V., J.-F. Thibault and J. Lefebvre, 1989. Structure of citrus pectins and viscometric study of their solution properties. Int. J. Biol. Macromol. 11: 186-191.

Braccini, I. and S. Pérez, 2001. Molecular Basis of Ca2+-Induced Gelation in Alginates and Pectins: The Egg-Box Model Revisited. Biomacromolecules 2: 1089-1096.

Christensen, P.E., 1954. Methods of grading pectin in relation to the molecular weight (intrinsic viscosity) of pectin Food Res. 19: 163-172.

Daas, P.J.H., K. Meyer-Hansen, H.A. Schols, G.A. DeRuiter and A.G.J. Voragen, 1999. Investigation of the non-esterified galacturonic acid distribution in pectin with endopolygalacturonaseCarbohydr. Res. 318: 135-145.

Deckers, H.A., C. Olieman, F.M. Rombouts and W. Pilnik, 1986. Calibration and application of high-performance size exclusion columns for molecular weight distribution of pectins. Carbohydr. Polym. 6: 361-378.

Doublier, J.L. and P. Wood, 1995. Rheological properties of aqueous solutions of (1→3)(1→4)-β-D-glucan from oats (Avena sativa L.) Cereal Chem. 72: 335-340.

European Union, 1998. EU-Council Directive of 11th November 1998 (98/86/EC).

Fishman, M.L., P.H. Cooke and D.R. Coffin, 2004. Nanostructure of Native Pectin Sugar Acid Gels Visualized by Atomic Force Microscopy. Biomacromolecules 5: 334-341.

Fishman, M.L., H.K. Chau, F. Kolpak and J. Brady, 2001. Solvent Effects on the Molecular Properties of Pectins. J. Agric. Food Chem. 49: 4494-4501.

Fishman, M.L., H.K. Chau, P.D. Hoagland and A.T. Hotchkiss, 2006. Microwave-assisted extraction of lime pectin. Food Hydrocoll. 20: 1170-1177.

Food Chemicals Codex 5th edition, 2004. Washington D.C. (incl. supplements).

Glahn, P.E., 1982. Hydrocolloid stabilization of protein suspensions at low pH. Prog. Food Nutr. Sci. 6: 171-177.

Glahn, P.E., 1995. Pectin process and composition. EP 0 664 300.

Glahn, P.E. and C. Rolin, 1996. Properties and food uses of pectin fractions: In G.O. Phillips, P.A. Williams and D.J. Wedlock (eds.). Gums and Stabilisers for the Food Industry 8, IRL. Press, Oxford, UK., 393-402.

Grant, G.T., E.R. Morris, D.A. Rees, P.J.C. Smith and D. Thom, 1973. Biological interactions between polysaccharides and divalent cations: the egg-box model. FEBS Lett. 32: 195-198.

Institute of Food Technology (IFT), 1959. Final Report of the Committee, Pectin Standardization, Food Technology 13: 496-500.

Ishii, S., K. Kiho, S. Sugiyama and H. Sugimoto, 1979. Low-methoxyl pectin prepared by pectinesterase from Aspergillus japonicus. J. Food Sci. 44: 611-614.

Kohn, R., I. Furda and Z. Kopec, 1968. Distribution of free carboxyl groups in the pectin molecule after treatment with pectin esterase. Collect. Checz. Chem. Commun. 33: 264-269.

Löfgren, C., P. Walkenström and A.H. Hermanson, 2002. Microstructure and Rheological Behavior of Pure and Mixed Pectin Gels. Biomacromolecules 3: 1144-1153.

Morris, V.J., 2007. Gels. In: P. Belton (ed.). The Chemical Physics of Food, Blackwell Publ. Co., Oxford, UK., p. 151-198.

Neidhart, S., C. Hannak and K. Gierschner, 1996. Investigations of the influence of various cations on the rheological properties of high-esterified pectin gels. In: J. Visser and A.G.J. Voragen (eds.) Pectins and Pectinases, Elsevier Science B.V., The Netherlands, 583-590.

Nielsen H., 2000. Rheological characterization of pectins, Ph.D. thesis, Department of Life Science and Chemistry, Roskilde University.

Oakenful, D. and Scott, A., 1984. Hydrophobic Interaction in the Gelation of High Methoxyl Pectins. J. Food Sci. 49: 1093-1098.

Ren, Y., P.R. Ellis, S.B. Ross-Murphy, Q. Wang and P.J. Wood, 2003. Dilute and semi-dilute solution properties of (1→3), (1→4)-β-D-glucan, the endosperm cell wall polysaccharide of oats (Avena sativa L.) Carbohydr. Polym. 53: 401-408.

Rolin, C., 2002. Commercial pectin preparations. In: G. Seymour and J.P. Knox (eds.), Pectins and their Manipulation, Blackwell Publishing Ltd., Oxford, UK, ISBN 1-84127-228-0. p. 222-241.
Rolin, C., 1994. Calcium sensitivity of high ester citrus pectins. In: G.O. Phillips, P.A. Williams and D.J. Wedlock (eds.), Gums and Stabilisers for the Food Industry 7, IRL. Press, Oxford, UK., 413.
Smith, C.J.B. and E.F. Bryant, 1967. Properties of pectin fractions separated on dethylaminoethyl-cellulose columns J. Food Sci. 32: 197-199.
Speiser, R. and C.R. Eddy, 1946. Effect of Molecular Weight and Method of Deesterification on the Gelling Behavior of Pectin. J. Am. Chem. Soc. 68: 287-293.
United Nations FAO and WHO, 2001. Compendium on Food Additives Specifications / add. 9, FAO Food and Nutrition Paper 52, Rome.
United States Pharmacopoeia and National Formulary, 2004. USP 27/NF 22.

Characterization of sugar beet pectin in relation to emulsion functionality

J.A. Staunstrup, H.L. Nielsen and C. Rolin
CP Kelco ApS, Ved Banen 16, 4623 Lille Skensved, Denmark

Abstract

Sugar beet pectin may be used for stabilizing the citrus oil in carbonated soft drinks. The oil is usually blended with so-called weighting agents, of which there are several possible choices, and prepared as a 10-30% o/w emulsion. For making one litre of soft drink, typically 1 to 3 ml of the emulsion is used as an ingredient. In both systems, the oil should ideally be uniformly suspended and protected against oxidation. The emulsion further needs to be resistant to high-brix conditions that occur temporarily during the soft drink preparation. Only few natural stabilizers like sugar beet pectin and gum arabic can satisfactorily withstand all of these conditions and stabilize the final soft drink for several months. This chapter describes a relationship between the content of acetyl ester groups in sugar beet pectin and its efficacy for stabilizing soft drinks relative to the chosen weighting agent and the sequence of conditions during the soft drink manufacturing.

Keywords: acetylation, stability, soft drink

1. Introduction

When preparing flavour oil emulsions used for soft drinks, the emulsion needs polymeric stabilizers in order to maintain a stable emulsion. Much published work deals with 10 to 30% o/w emulsions with which the oil phase is orange oil and oil-miscible materials that increase the density of the oil phase – so called weighting agents (Tan, 1997; Dluzewska *et al.*, 2004).

These emulsions must be fluid, and it must be possible to dilute them with 500 to 1500 parts of other soft drink ingredients without this causing failure of the emulsion (Tan, 1997). The visual appearance of the oil in the final soft drink should ideally be – and be only – a uniform haze throughout the soft drink. Failure can be, for example, gradual loss of haze, and appearance of macroscopic oil-enriched parts.

The typical production practice in manufacturing hazy orange-flavoured soft drinks comprises blending the emulsion with the sugar and fruit acid for making syrup (Tan, 1997) of about 60-65 brix and pH between 2.5 and 3. The syrup is then later, perhaps several days later, diluted with about five parts of carbonated water, thus becoming the final soft drink. This syrup stage stresses the emulsion for, at least, the reason that the density of the syrup, about 1.28 g/cm^3, is considerably different from the density of even a weighted oil blend (about 0.95 g/cm^3).

The traditional stabilizer for the orange oil emulsion is gum arabic, which in the best cases provides an emulsion that has good fluidity and has excellent stability before as well as after dilution. The problem with gum arabic is lot-to-lot variation and occasional unavailability of the best qualities. For this reason companies that are commercially dependent upon gum arabic have looked for possible replacements like sugar beet pectin and specific qualities of modified starch.

Pectin is a generic term for some materials that occur in plants (Schols and Voragen, 2002) or are commercial water-soluble isolates from plants (Rolin, 2002). According to all definitions of 'pectin' a large part is an $\alpha(1\rightarrow4)$ linked polymer of anhydrogalacturonic acid that is esterified with methanol at some of its 6-C carboxylate groups. With commercial pectin, this galacturonan part (or what is interpreted as being the galacturonan part) must by definition account for at least 65% of the polymer (United Nations FAO and WHO, 2001; EU-Council Directive, 1998; Food Chemicals Codex, 2004). The remainder is for the most part various neutral polysaccharides (Schols and Voragen, 2002) and, depending upon the raw materials and procedures of isolation, other plant materials like protein, for example (McKenna *et al.*, 2006).

The percentage of the carboxylate groups that are methyl-esterified is traditionally termed the degree of methyl esterification, abbreviated DM. In addition to the 6-C methyl esterification, some acetic acid may be esterified to C-2 and C-3 hydroxyls in the galacturonan.

Sugar beet pectin contains a relatively large proportion of neutral carbohydrate as compared to commercial citrus pectin (Rolin, 2002). It has inherently lower DM, and commercial preparations usually have DM < 60. In sugar beet pectin typically about 1 mole per 4 to 5 moles of anhydrogalacturonic acid is acetylated. About 4% of protein is also present (Williams *et al.*, 2005).

Commercial sugar beet pectin is a much better stabilizer of orange oil emulsions than typical commercial citrus pectin (Williams *et al.*, 2005; Dea and Madden, 1986), although a specific quality of citrus pectin with lower than usual molecular weight also has been reported as effective (Akhtar *et al.*, 2002a,b; Mazoyer *et al.*, 1997). It has been suggested that the efficacy of sugar beet pectin is due to the presence of acetate esterification making it more hydrophobic than citrus pectin (Dea and Madden, 1986). Another possible explanation is the presence of protein in sugar beet pectin (Leroux *et al.*, 2003).

In principle, the good stabilizing property of sugar beet pectin also could be owing to one of its other structural characteristics, for example the more compact shape or the higher content of neutral polysaccharides, or, of course, some appropriate combination of the aforementioned characteristics. Introduction of acetate ester groups in citrus pectin has been tried experimentally with the purpose of revealing whether the citrus pectin thereby acquires a stabilizing ability in similarity to sugar beet pectin (Leroux *et al.*, 2003). However, this investigation brought no clear answer. Introduction of acetate ester groups into citrus pectin markedly reduces its solubility, and such modified citrus pectin can become practically insoluble, even when the acetate presence is not as high as that of typical sugar beet pectin (Pippen *et al.*, 1950; Kjeldsen, 2005). Removal of protein from sugar beet pectin clearly decreases its efficacy as stabilizer (Funami *et al.*, 2006). This far, investigation has not been able to point unambiguously to only one of these structural traits as the major contributor to the efficacy as a stabilizer.

The present paper deals with the use of commercial sugar beet pectin and experimental samples produced in accordance with typical commercial pectin production practices for stabilization of orange oil emulsions and soft drinks containing the emulsions. A number of samples with different content of acetate esterification were tested together with orange oil blended with suitable proportions of the weighting agents Hercules Ester Gum and SAIB. The dilution process involved an intermediate syrup stage, which was particularly straining for the pectin, since pectin can be poorly soluble under the condition of high brix in combination with low pH.

2. Materials and methods

2.1. Analytical methods for pectin composition

Degree of methyl esterification (DM) and degree of acetate esterification (DAc) were measured according to the methods of FAO (United Nations FAO and WHO, 2001). The DAc is moles O-acetyl per mole galacturonic acid.

Molecular weight was measured with a Gel Permeation Chromatography (GPC) system from Waters, using well known standards for calibration. Columns; 2 KB 806 M (linear), solvent 0.075M Li-Acetate buffer pH 4.8, data was collected with the GPC software 'Empower'.

Protein content was assessed by measuring nitrogen by Kjeldahl analysis and multiplying the result with 6.25.

2.2 Methods for emulsion and soft drink characterization

The emulsions were evaluated by mean particle size after 1 and 21 days, and by appearance of defects in the top of the emulsion like creaming or occurrence of neckring. These defects were evaluated by visual observation and assigned a rating between 0 (nothing) and 1 (plenty), in some tables recorded as 0 through 100%.

Skin formation was visually observed in the soft drinks after 1, respectively 60, days in the top of the glass bottles, which were stored quiescently at laboratory/room temperature in an upright position. Skin was assigned the ratings '0' for absence, '0.5' for some, and '1' for much.

The size distribution of the suspended particles of emulsions and soft drinks was characterized with a Malvern Mastersizer using 60 seconds of ultrasound before each measurement and using obscuration of 2-4. The parameter settings: refractive index 1.506 and absorption index 0.005. Samples were diluted to the appropriate attenuation in pure water in the apparatus at the measurement. The mean value d(v,0.5) was extracted from the curve.

Viscosity of the emulsions were determined using a Rheometric RS2000 rheometer equipped with a bob-cup geometry, samples were loaded into the rheometer and allowed to rest 5 min. prior to measurement at 25 °C. A flow curve (steady state mode) with increasing stress applied (0.02 -10Pa) with 10points/decade. The viscosity at $10s^{-1}$ was extracted from the curve.

Turbidity measurements were performed on a HACH 2100N Laboratory Turbidimeter. The NTU is measured on soft drink after 60 days. Influence of carbon dioxide was excluded by pouring soft drink 30 times from one beaker to another.

2.3 Pectin samples

All pectin was extracted from sugar beet pulp (leftover from sugar manufacturing) in hot water acidified with nitric acid in accordance with normal industrial procedures [Rolin, 2002]. Analytical data for the samples may be found in Table 1.

2.4 Emulsion recipe

The orange oil was orange oil 8380 NAT, H.N. Fusgaard, Rodovre, Denmark. Orange oil blends were produced by dissolving either Hercules Ester Gum 8BG, Hercules, Delaware, USA, or Sucrose Acetate IsoButyrate (SAIB), Sigma-Aldrich Chemie GmbH, Steinheim, Germany in the oil by gradual addition under gentle agitation. The Oil/Ester Gum mixture was 50:50 by weight and had a density of 0.950 ± 0.003 g/ml at 25 °C. The Oil/SAIB mixture was 50:50 by weight and had a density of 0.970 ± 0.003 g/ml at 25 °C. The emulsions were made according to the descriptions of Table 2 below. The emulsions were homogenized twice, both times in two stages, at 200 bar/50 bar or at 400 bar/50 bar.

Emulsions were diluted with five parts of water and mixed for 10 minutes, and then blended with 84 g of (0.25% Na-benzoate preserved) 65% sucrose solution per g of original emulsion. The syrup was then pH adjusted with citric acid and mixed for 60 minutes. Conditions of the final syrup: 60% Brix, pH 2.8.

Table 1. Analytical data for the sugar beet pectin sample set, DAc (degree of acetylation) DM (degree of esterification), Protein content and Mw (Molecular weight).

Sample	DAc (%)	DM (%)	Protein (% w/w)	Mw (kD)
1	8.5	46.1	5.0	100
2	27.5	61.0	3.9	144
3	13.5	51.4	3.2	156
4	18.5	50.5	4.9	106
5	13.7	49.5	3.7	142
6	22.6	59.4	4.7	162
7	12.4	48.5	4.7	133
8	30.1	59.3	3.9	164
9	31.9	59.6	3.5	155
10	30.9	59.9	3.9	161
11	17.2	58.9	4.2	113
12	18.3	58.8	8.8	119
13	17.0	58.6	7.0	109
14	22.6	59.4	4.7	162
15	13.4	54.7	4.0	153
16	10.4	49.6	3.0	149
17	17.0	49.7	3.8	143
18	12.4	53.8	3.0	136
19	18.7	60.2	6.5	107
20	23.7	45.2	3.1	150
21	12	54.0	5.6	121
22	9.2	49.4	6.3	140
23	4.0	39.6	4.2	64
24	17.2	54.8	6.0	168
25	19.5	59.7	8.1	160
26	21.9	59.3	8.9	164
27	5.1	43.8	4.2	122
28	29.9	62.0	5.5	171
29	2.2	22.1	5.4	32
30	20.3	58.6	4.3	106
31	19.0	58.1	5.2	114

2.5 Soft drink recipe

Model soft drinks were prepared by opening a commercial carbonated spring water (Fürst Bismarck Premium, 0.7 L, glass bottle with resealable screw cap, Fürst Bismarck Quelle, Sachsenwald, Germany), removing 130 g of the contents, adding 130 g of the

Table 2. Standard emulsion recipe.

Order of addition	Ingredients	g/kg
A	Hot deionized water (>80 °C)	913.0 g
	Sugar beet pectin	20.0 g[1]
B	Orange oil blend (see text)	100 g
C	Sodium benzoate (20% sol. w/v)	0.5 ml
D	Citric acid (50% sol. w/v) for pH 3.6	~3 ml
	Total ingredients, approx.	1036.0 g
	Evaporation, approx.	36.0 g
	Yield, approx. ±15 g	1000.0 g

[1]Concentration can vary to achieve an emulsion viscosity between 100-200mPas at a shear rate of $10s^{-1}$ (25 °C).

above syrup, recapping, turning the bottle upside-down ten times, and finally shaking it for 30 min. at 175 rpm.

3. Results

First the chemical composition of the sugar beet pectin is evaluated in a standard emulsion recipe, which is further diluted into soft drink, then follows an evaluation of different weighting agents. Finally a validation set of sugar beet pectin is chosen for determining the best composition of the beet pectin to produce both a stable emulsion and soft drink.

3.1 Evaluation of the chemical composition of the pectin

A broad range of pectin samples (Table 1) was tested in the standard emulsion recipe (Table 2). The emulsions had a pH of 3.2 which is lower than the maximum stability range (3.5-3.7) that was established later on. They all showed a very slight oil film, which is generally seen in all emulsions.

The emulsions were evaluated with respect to droplet size, visual evaluation and the viscosity was measured (Table 3). As seen, the droplet size was stable for most of the emulsions and only very low DAc resulted in increased droplet size.

The formation of neck-ring, on the other hand, seemed to be affected by the presence of DAc, lower than 10 and higher than 30% resulted in the presence of neck-ring as shown in Figure 1.

Table 3. Results of emulsions and soft drinks prepared from the different pectin samples listed in Table 1.

Sample	DAc (%)	Emulsion viscosity (mPas)	Emulsion neckring	d(v.0.5) Day 1 (μm)	d(v.0.5) Day 21 (μm)	Soft drink turbidity (NTU)	Soft drink skin evaluation
1	8.5	21	50%	0.363	0.348	388	30%
2	27.5	56	10%	0.362	0.340	371	100%
3	13.5	64	10%	0.301	0.297	390	30%
4	18.5	22	10%	0.312	0.296	384	50%
5	13.7	54	10%	0.291	0.284	375	30%
6	22.6	82	10%	0.394	0.358	385	100%
7	12.4	53	10%	0.325	0.313	403	30%
8	30.1	65	10%	0.315	0.294	373	100%
9	31.9	54	10%	0.344	0.320	392	100%
10	30.9	67	50%	0.342	0.330	359	100%
11	17.2	49	10%	0.312	0.305	389	100%
12	18.3	31	10%	0.327	0.315	392	50%
13	17.0	35	10%	0.334	0.314	396	80%
14	22.6	84	10%	0.358	0.352	359	100%
15	13.4	75	10%	0.310	0.303	373	80%
16	10.4	71	50%	0.309	0.283	391	50%
17	17.0	46	10%	0.280	0.294	379	80%
18	12.4	135	10%	0.310	0.298	413	30%
19	18.7	30	10%	0.320	0.319	375	100%
20	23.7	85	10%	0.267	0.263	385	80%
21	12.0	41	10%	0.324	0.324	415	30%
22	9.2	61	10%	0.305	0.302	387	50%
23	4.0	12	50%	0.295	0.295	357	0%
24	17.2	64	10%	0.309	0.315	388	100%
25	19.5	46	10%	0.333	0.339	398	100%
26	21.9	38	10%	0.359	0.354	388	80%
27	5.1	48	50%	0.330	0.336	408	0%
28	29.9	65	10%	0.332	0.331	402	100%
29	2.2	4.3	100%	1.489	1.671	182	100%
30	20.3	129	0%	0.331	0.302	375	0%
31	19.0	118	0%	0.322	0.311	378	0%

The emulsion was diluted into soft drink through a syrup stage, to provoke any instability, and after 1 and 60 days the soft drinks were evaluated for the appearance of a skin on top of the soft drink. All soft drinks in this data set were still perfectly turbid after 60 days, and with very little variation in turbidity.

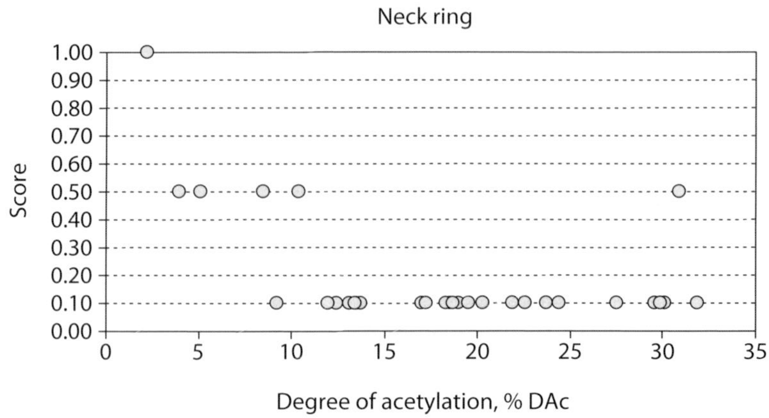

Figure 1. Emulsion evaluation: the occurrence of neck-ring in the emulsion as a function of % DAc of the pectin used for preparing the emulsion.

The data was analysed using the software Modde 8.0 from Umetrics. The Mw and the protein content were not significant at all for the application under investigation. DAc and DM is highly correlated, however the model obtained indicates that % DAc is the most important parameter to explain the visual defects in emulsion and in soft drink, and a good model was obtained only by % DAc.

The model for soft drink has R^2 of 0.75 and Q^2 (goodness of fit) of 0.70. So it is a very good model, especially when bearing in mind that the data is a partly subjective evaluation. The model for Emulsion has R^2 of 0.65 and Q^2 (goodness of fit) of 0.58. So again here the model is good (Staunstrup and Rolin, 2008).

The actual results of the evaluation of neck-ring in emulsions and skin formation in soft drink were plotted versus % DAc in Figure 1 and Figure 2, respectively. The occurrence of neck-ring was reduced as DAc was increased, whereas the formation of skin in soft drink was increased as DAc was increased, as a consequence we propose that the optimal range is 10-15% DAc for optimizing both a stable emulsion and soft drink. The findings have later been verified by testing other pectins in that range, which gave no defects (see 3.3).

3.2 Test of weighting agent: mixing ester gum and SAIB

The average particle size of emulsions where the oil phase was weighted by 50% SAIB, 50% Ester Gum, or 25%SAIB + 25% Ester Gum are shown after 2, 7, and 21 days in Figure 3. The pectin used had a % DAc of 22%.

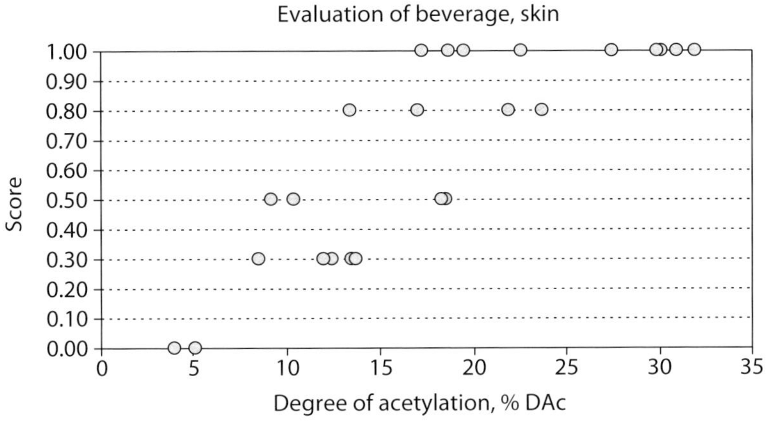

Figure 2. Soft drink skin evaluation as a function of DAc. As seen, the occurrence of skin increases as DAc increases.

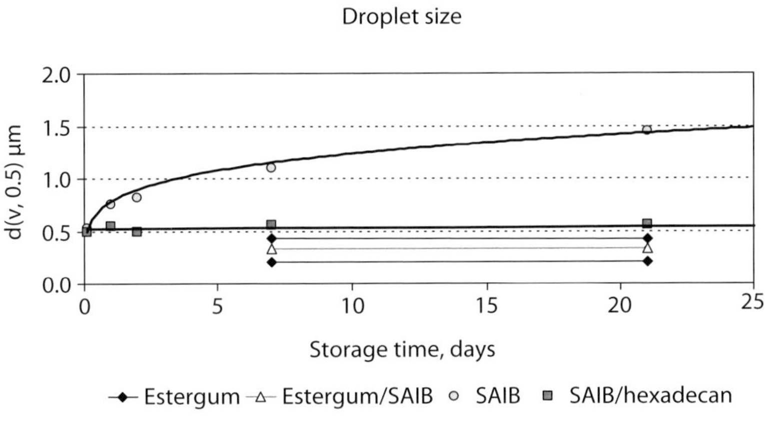

Figure 3. Mean droplet size (d(v,0.5)) of emulsions with SAIB, Ester Gum, or mixtures.

The emulsions based on pure Ester Gum all showed stable droplet size, but high degree of skin formation. The mixtures of Ester Gum and SAIB had stable droplet size and less skin formation.

The instability of the emulsions with SAIB is prevented by mixing SAIB with Ester Gum. The emulsions were diluted through the syrup stage which is described in Section 2.4. Although the emulsions based on pure SAIB showed instability with respect to droplet size, no formation of skin was observed when diluted into soft drink.

3.3 Validation of optimal pectin composition

The major finding from 3.1 was that DAc seems to have an optimal range of 10-15%. Furthermore, the pH was lower than the target. The few samples with viscosities above 100 mPas were generally better than expected. Viscosity is known to be important in emulsion stabilisation according to Stokes law, so it makes sense to increase it by higher pectin concentration and selection of samples.

If the viscosity is too low, the emulsion is not stable, however, highly viscous emulsions are not preferred as they are gel-like. As a consequence, we aim at producing emulsions covering viscosities in the range 100-200 mPas.

To validate the findings in section 3.1 the following setup was made: samples were included that covered the DAc range of 10-15%, the pH in the emulsion was around 3.6, the emulsion viscosity covered the range of 100-200 mPas and only Ester Gum was used. The test range of the validation setup is found in Table 4 and the results of the emulsion and soft drink preparation are shown in Table 5.

In general the samples produced with pectin having a DAc below 20% are showing good performance in the emulsion. The skin formation as a function of DAc for the validation setup is shown in Figure 4 supporting the initial proposal that DAc should be below 20%.

4. Discussion

In the present paper we have demonstrated an apparent relationship between the content of acetate esterification in sugar beet pectin and its efficacy as a stabilizer of 10% o/w orange oil emulsions as well as 0.02% o/w soft drinks. Samples were produced by chemical treatment and as a consequence DM and DAc is highly correlated, therefore it was not possible to separate the effect into an effect by DM and DAc, however the statistical best model was obtained by use of DAc only. We realize that this relationship can only be apparent, since in our experiments it must be considered likely that the structure of the protein part of the pectin, and possibly its attachment to the remainder of the pectin, may have been modified by the same processing conditions which were used for controlling the acetate. Our experiments do not add to, nor detract from, the findings of others (Dea, and Madden, 1986; Leroux *et al.*, 2003; Funami *et al.*, 2006) as concerns the importance of the protein content in comparison to the importance of acetate esterification. On the other hand, there was no correlation between the DAc and the total protein content in our data, and the DAc was the parameter that in our statistical analysis explained the largest part of the variation in the data.

Our suggestion, speculative as it is, is this: an appropriate content of protein as well as an appropriate content of acetate esterification are both necessary, but individually

Table 4. Analytical data for the validation sample set of sugar beet pectin. Emulsion was prepared according to the standard recipe with the following changes; homoganization pressure as well as pectin concentration.

Sample	DAc (%)	DM (%)	Protein (% w/w)	Mw (kD)	Homogenization pressure (bar)	Pectin (%)
V1	8.2	51.9	4.4	140	400	2.2
V2	8.9	56.2	4.3	119	400	2.0
V3	12.4	53.8	3.0	136	200	2.0
V4	12.4	48.5	4.7	133	400	2.0
V5	13.1	51.1	3.6	145	200	2.0
V6	13.7	54.7	4.3	143	400	2.5
V7	13.8	54.7	4.3	143	400	2.0
V8	17.2	58.9	4.2	113	200	2.0
V9	22.7	59.4	4.7	162	200	1.9
V10	22.7	59.4	4.7	162	200	2.1
V11	22.7	59.4	4.7	162	200	2.1
V12	22.7	59.4	4.7	162	200	2.0
V13	22.7	59.4	4.7	162	400	2.0
V14	22.7	59.4	4.7	162	300	2.0
V15	22.7	59.4	4.7	162	300	2.2
V16	22.7	59.4	4.7	162	300	2.5
V17	22.7	59.4	4.7	162	200	2.0
V18	22.7	59.4	4.7	162	200	2.2
V19	22.7	59.4	4.7	162	200	2.5
V20	23.7	45.2	3.1	150	200	2.0
V21	30.9	59.9	3.9	161	200	2.0

not adequate, preconditions for stability. The DAc content must not be too low when stabilizing emulsions, whereas a high DAc results in skin formation in the soft drink – optimal range is 10-15% DAc.

The choice of weighting agent is important. The emulsion viscosity is important also to control, especially low viscosity emulsions are not stable. Emulsion viscosity is adjusted by the pectin concentration. To ensure proper evaluation, soft drinks must be produced following a simulated industrial dilution process.

We suggest that the protein contents of our samples were adequate, as the protein content was not below 3%, and thus we could not study the possible importance of the protein.

All samples used for the soft-drink experiments produced stable 10% o/w emulsions with our recipe. It is our observation that the 10% o/w emulsion in itself can be

Table 5. Results of emulsions and soft drinks prepared from the validation test setup listed in Table 4.

Sample	DAc (%)	Emulsion viscosity (mPas)	Emulsion neckring	d(v.0.5) Day 21 (μm)	Soft drink turbidity (NTU)	Soft drink skin evaluation
V1	8.2	234	25%	0.304	387	25%
V2	8.9	348	0%	0.418	434	25%
V3	12.4	130	13%	0.298	-[1]	0%
V4	12.4	66	38%	0.287	389	25%
V5	13.1	105	13%	0.288	-	0%
V6	13.7	221	0%	0.315	368	25%
V7	13.8	125	25%	0.301	384	25%
V8	17.2	44	25%	0.305	-	25%
V9	22.7	72	13%	0.421	-	50%
V10	22.7	85	13%	0.513	-	75%
V11	22.7	87	13%	0.340	-	75%
V12	22.7	87	13%	0.340	-	75%
V13	22.7	85	50%	0.292	377	75%
V14	22.7	91	25%	0.391	-	50%
V15	22.7	118	38%	0.382	-	50%
V16	22.7	161	25%	0.354	-	50%
V17	22.7	74	25%	0.516	-	75%
V18	22.7	104	50%	0.498	-	50%
V19	22.7	140	38%	0.570	-	75%
V20	23.7	73	25%	0.263	-	50%
V21	30.9	67	38%	0.330	-	100%

[1]data not measured.

satisfactorily stabilized by a broad range of different sugar beet pectin samples. In contrast, the stability of the 0.02% o/w soft drink, and the process of having an intermediate syrup stage, represents greater challenges in the sense that only specific qualities of sugar beet pectin work satisfactorily. In the standard test system the emulsion passes the syrup stage in order to simulate the processes that take place with major manufacturers, Figure 5. The syrup stage is, however, straining, not only because of the high density difference between the oil phase and the aqueous phase, but also because pectin is poorly soluble at the combined condition of high solids and low pH. The defects that could be observed (skin formation) were not seen if emulsions were directly diluted into soft drinks without having an intermediate syrup stage, or if emulsions were diluted into a syrup phase without acid. This is schematically shown in Figure 5.

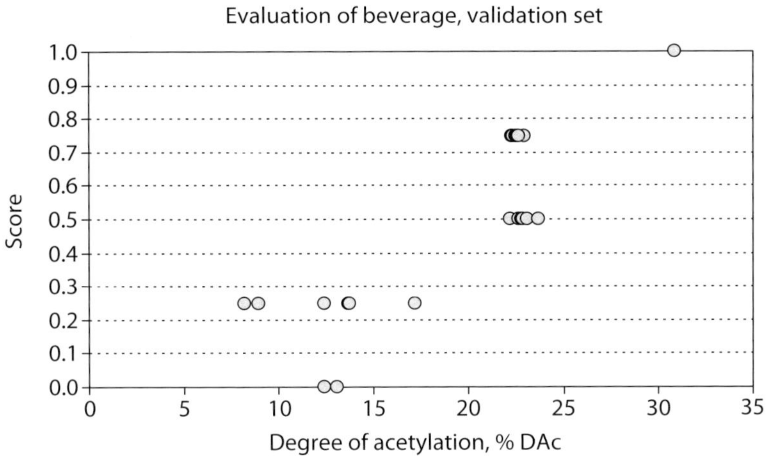

Figure 4. Visual evaluation of soft drink as a function of DAc for the validation setup having an optimized emulsion viscosity and pH.

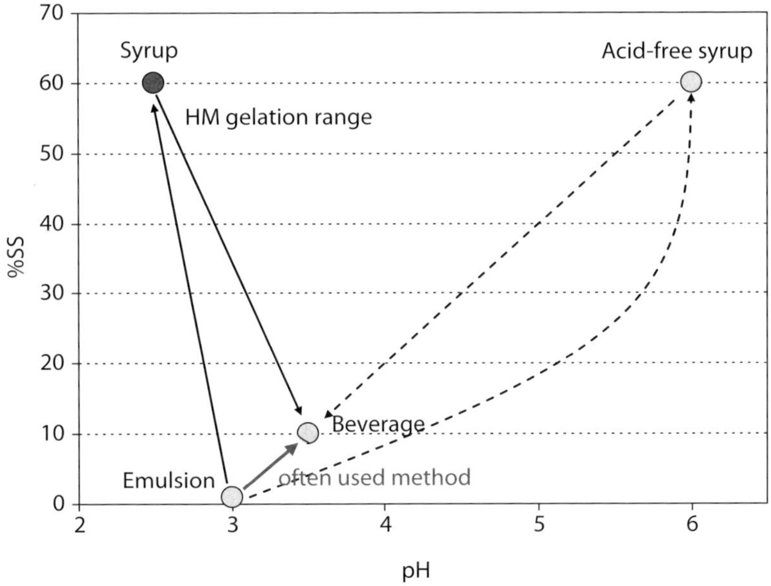

Figure 5. Illustration of the simulating industrial process for diluting emulsions into soft drink. Skin formation in the soft drink is observed, if the emulsion is subjected to conditions of high % SS and low pH (upper left area), which is included in simulating industrial process.

Skin formation in the soft drink is observed if the emulsion is subjected to conditions of high % SS and low pH (upper left area), which are included in simulating industrial process. The defects that could be observed visually were not apparent from particle size distributions, which were stable except with formulations without ester gum. Formulations with only SAIB as weighting agent increased in particle size over time, but there were no visible defects such as skin in the 10% o/w emulsion. In other words, measurements of particle size cannot substitute visual observation.

It is well known that viscosity provides stability in emulsions according to Stokes law, and by slightly increasing the pectin concentration in the emulsion the final emulsion viscosity is increased from around 50-80 mPa·s into the range (100-200 mPa·s) providing improved stability of the emulsion.

Still, we have no specific information about exactly how the intermediate syrup stage, and the fully diluted soft drink stage, impacted our results individually, only we have found a relationship between the pectin acetate esterification and the homogeneity of the ensuing soft drinks.

5. Conclusions

The optimal sugar beet pectin for stabilizing soft drink has lower content of acetate esterification than the pectin, which is optimal for stabilizing 10% o/w emulsion. It cannot be distinguished whether this is due to the acetate esterification itself, or it is due to some other compositional or structural characteristic that is impacted by the same processing conditions as those that are used for controlling the acetate esterification.

The DAc content must not be too low, when stabilizing emulsions, whereas a high DAc results in skin formation in the soft drink; the optimal range is suggested to be 10-15% DAc.

The protein content was higher than 3% in all samples and any clear effect of the protein content could not be determined. It is therefore suggested that the protein contents of the investigated samples were adequate.

The emulsion viscosity is important and especially low viscosity emulsions are not stable. It is recommended to adjust pectin concentration to provide emulsion viscosity in the range 100-200 mPa·s.

To ensure proper evaluation, soft drinks must be produced following a simulated industrial dilution process.

Acknowledgments

The authors are indebted to Majbritt G. Peitersen and Hanne Thulstrup Jensen for their extensive analytical work.

References

Akhtar, M., E. Dickinson, J. Mazoyer and V. Langendorff, 2002a. Emulsifying properties of depolymerised citrus pectin: Role of the protein fraction. Gums. Stabil. Food Ind. 11: 311-317.

Akhtar, M., E. Dickinson, J. Mazoyer and V. Langendorff, 2002b. Emulsion stabilizing properties of depolymerized pectin. Food Hydrocolloids. 16(3): 249-256.

Dea, I.C.M. and J.K. Madden, 1986. Acetylated pectic polysaccharides of sugar beet. Food Hydrocolloids. 1: 71-88.

Dluzewska, E., M. Panasiewicz and K. Leszczynski, 2004. Effect of gum arabic and modified starch on stability of beverage emulsions. Electronic J. Polish Agric. Univ, series Food Science and Technology. 7: 2.

European Union: EU-Council Directive of 11th November 1998 (98/86/EC).

Food Chemicals Codex 5th edition, Washington D.C. 2004 (incl. supplements).

Funami, T., G. Zhang, S. Noda, M. Nakauma, I. Asai, S. Al-Assaf and G.O. Phillips, 2006. Does pectin emulsification mechanism operate via an AGP (Arabinogalactan protein) type fraction? Foods Food Ingred. J. Jpn. 211: 255-263.

Kjeldsen, K.K., MSc thesis B1194, October 2005. The effect of acetylation on the emulsifying properties of citrus pectin. Department of Food Science, The Royal Veterinary and Agricultural University, Denmark.

Leroux, J., V. Langendorff, G. Schick, V. Vaishnav and J. Mazoyer, 2003. Emulsion stabilizing properties of pectin. Food Hydrocolloids 17: 455-462.

Mazoyer, J., J. Leroux and G. Bruneau, 1997. Depolymerised pectins. GB 2 311 024 A, 1-18.

McKenna, C., S. Al-Assaf, G.O. Phillips and T. Funami, 2006. The Protein Component in Pectin- Is It an AGP? Foods Food Ingred. J. Jpn., 211(3): 264-271.

Pippen, E.L., R.M. McCready and H.S. Owens, 1950. Gelation Properties of Partially Acetylated Pectins, J. Am. Chem. Soc, 72(2): 813-816.

Rolin, C., 2002. Commercial pectin preparations. In: G. Seymour and J.P. Knox (eds.), Pectins and their Manipulation, Blackwell Publishing Ltd., Oxford, UK, ISBN 1-84127-228-0. p. 222-241.

Schols, H. and A.G.J. Voragen, 2002. The chemical structure of pectins. In: G. Seymour and J.P. Knox (eds.), Pectins and their Manipulation, Blackwell Publishing Ltd., Oxford, UK, ISBN 1 84127 228 0. p. 1-29.

Staunstrup, J.A. and C. Rolin, 2008. Characterization of sugar beet pectin in relation to emulsion functionality, FFI Journal, 213: 357-364.

Tan C.-T., Beverages Emulsions; Food Emulsions; Marcel Dekker, New York, NY, 1990, 445-478.

United Nations FAO and WHO: Compendium on Food Additives Specifications / add. 9, FAO Food and Nutrition Paper 52, Rome 2001.

Williams, P.A., C. Sayers, C. Vieke, C. Senan, J. Mazoyer and P. Boulenguer, 2005. Elucidation of the emulsification properties of sugar beet pectin. Agric. Food Chem. 53(9): 3592-3597.

Influence of pectin structure on the mechanical properties of flax fibres: a comparison between linseed-winter variety (Oliver) and a fibre-spring variety of flax (Hermes)

S. Alix[1], J. Goimard[2], C. Morvan[1] and C. Baley[2]
[1]Université de Rouen, Laboratory GTV, CNRS, 76821 Mont Saint Aignan Cedex, France
[2]Université de Bretagne Sud, Laboratory L2PIC, Rue St Maudé, 56321 Lorient, France

Abstract

This work evaluates the tensile properties, the structure and the chemical composition of the cellulosic fibres of a linseed-winter variety of flax, Oliver, compared to Hermes, a fibre-flax variety used for textile application and as a reference in our laboratories. The Young's modulus of the elementary fibres of Oliver was evaluated to 38 ± 13 GPa and the breaking strength in the range of 500-1000 MPa, *i.e.* significantly less than those of Hermes. The main discriminating parameters for good tensile properties appeared to be the diameter of the elementary fibres, the content of cellulose, and the total amount of encrusting non cellulosic polysaccharides. The amounts of rhamnogalacturonan I, homogalacturonan and glucomannan that were extracted from the secondary wall of fibres, either with acid or with alkali, might also be of importance for a comprehensive approach of the mechanical properties of cellulosic fibres.

Keywords: cellulose, glucomannan, homogalacturonan, rhamnogalacturonan

1. Introduction

Flax fibres show many advantages as reinforcement for polymers, including good mechanical properties, low density, low-energy cost and biodegradability (Mohanty *et al*., 2000; Wambua *et al*., 2003). However, their properties are highly variable (Baley, 2002) which might originate from 1) genetic factors depending on the choice of the variety (Foster *et al*., 1997; Sharma and Faughey, 1999; Van den Oever *et al*., 2003; Jankauskiene *et al*., 2006), 2) the global environmental conditions (Stamboulis *et al*., 2001; Norton *et al*., 2006). Interestingly, the former papers have pointed out that the fibres of flax cultivated for textile have higher fineness and tensile strength than those from linseed or mixed varieties.

The impact from the environmental conditions might be illustrated when studying the fibre properties along the height of one stem, due to the variation of the meteorological factors over the duration of the plant growth (Fausten, 1992; Charlet *et al*., 2007a,b). Thus, we previously showed that the fibres that differentiated within the medium part of the stems of flax plants grown under optimal weather conditions were the longest, the thinnest, the best filled and presented the highest tensile properties (Charlet *et al.*, 2007a). Although, the mechanical properties of the elementary fibres are essentially due to the orientation of the cellulose microfibrils within the secondary wall (Roland *et al.*, 1995; Wang *et al.*, 2001), our data performed within a collection of pure lines indicated some impact of the fibre composition on the tensile properties (Brutch *et al.*, 2008).

Flax fibres are gathered into bundles located between the phloem and the cortical tissue (Figure 1a). In these so-called cellulosic fibres, which barely have any lignin (Akin *et al.*, 1996; Day *et al.*, 2005), the cellulose microfibrils are embedded in 5-20% of non cellulosic polymers (NCPs) (Morvan *et al.*, 2003). Most papers in the composite research domain have designated NCPs as hemicelluloses, because NCPs were generally extracted from the fibres with alkali. On the other hand, detailed analyses of sugar composition of the NCPs clearly identified two types of NCPs: (1) the hemicelluloses, mainly glucomannans (McDougall, 1993; Van Hazendonk *et al.*, 1996) and (2) pectins, i.e. polysaccharides enriched in galacturonic acid (Goubet *et al.*, 1995; Girault *et al.*, 1997; Mooney *et al.*, 2001). Although glucomannans have never been localized throughout the secondary wall, it is admitted that they act as surfactants of the cellulose microfibrils (Roland *et al.*, 1995), due to the similarity of their backbone with that of cellulose. On the contrary, pectins have been very well localized with ruthenium red or with specific antibodies (Girault *et al.*, 1997; Andeme-Onzighi *et al.*, 2000; His *et al.*, 2001), not only close to the primary wall, but also in the inside part of the fibre (Figure 1b-c). The structure of the pectins, especially the length of the galactan side-chains, was different within each cell-wall domain (Gorshkova and Morvan, 2006). Thus, the external primary wall, which determines the continuity of the fibre, might be more acidic than the secondary wall. Such specific compositions have to be taken into account not only to apprehend the mechanical potential of the multi-lamellar composite structure of these fibres, but also to evaluate the hydratation capacity of the fibres (Gouanve *et al.*, 2006).

The aim of the work was to better characterize a linseed winter variety, Oliver, by comparing (1) the mechanical properties of its elementary fibres, (2) their morphology and (3) their chemical composition, with those of the Hermes variety which has been used as a reference in our laboratories (Charlet *et al.*, 2007a). Hermes, selectioned by the cooperative 'Terre de lin' in Normandy is among the best fibre varieties for its high production, yield, rettability, fineness and tenacity properties of its fibres. Thus, it might be possible to weight the relative importance of the structure and composition of the elementary fibres on their tensile properties.

2. Mechanical properties

Mechanical properties (Young's modulus, ultimate strength and failure strain) of single flax fibres were obtained from tensile tests on a universal MTS type tensile

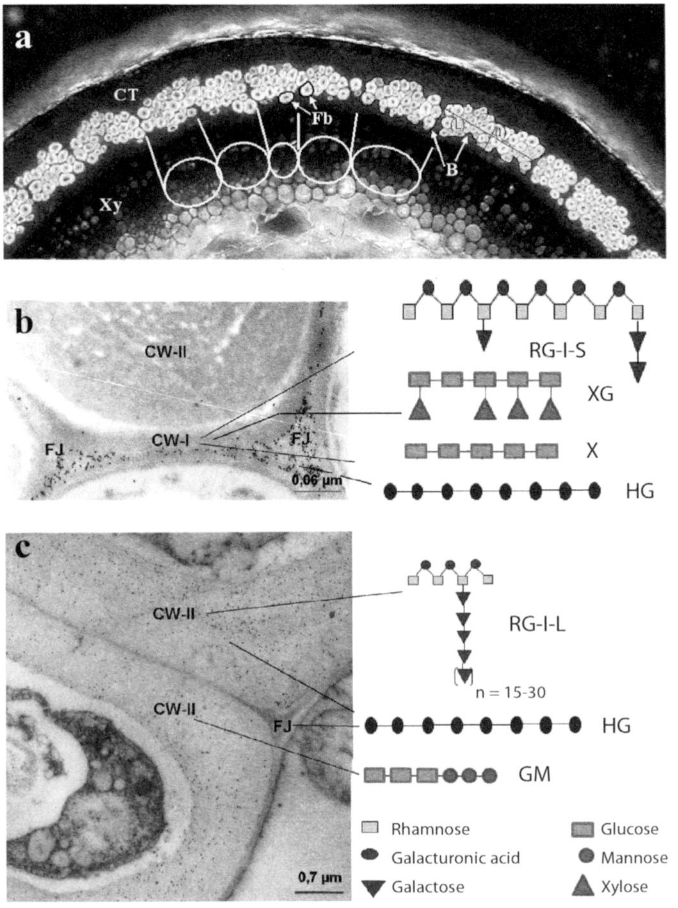

Figure 1. Morphology and structure of flax fibres. a) Transverse section at the middle part of a stem of Oliver. Fibres (Fb); bundle (B); large (L) and small (l) dimensions of B; cortical tissues (CT); xylem (Xy). b) and c) transverse sections observed with a transmission electronic microscope (from Gorshkova and Morvan, 2006). b) The fibre junctions (FJ) were strongly labelled with JIM5 antibody specific to homogalacturonan (HG). The main non cellulosic polymers of the primary wall (CWI) were rhamnogalacturonans I (RGI-S) substituted with short chains of galactose, xyloglucans (XG) and xylans (X). c) The main polymers encrusting the cellulose microfibrils within the secondary wall (CW-II) were glucomannans (GM), RGI-L with long chains of galactans and HG. Note the labelling in CW-II with LM5 antibody raised to long chains of galactans.

testing machine equipped with a 2N capacity load cell at constant temperature (23 °C) and hygrometry (48% relative humidity) with a gauge length of 10 mm (Baley, 2002). The fibre was loaded at a constant crosshead displacement rate of 1 mm/min until rupture. An example of tensile stress-strain curves of a linseed (Oliver) and a fibre flax (Hermes) varieties are shown in Figure 1.

A non-linear region was observed in the earlier stage of the loading that was especially pronounced for the linseed variety. This behaviour might be explained by the sliding of the microfibrils along with their progressive alignment with the fibre axis. These rearrangements would imply an elasto-visco-plastic deformation. Usually in fibre varieties, these microfibrils make an angle of about 10° with the fibre axis, *i.e.* in the tensile direction (Wang *et al.*, 2001), but no value has been reported for linseed varieties yet. The second region of the loading curves was linear, being characteristic of an elastic behaviour. At that level of strain, the microfibrils of cellulose were likely to be the main material deformed. The Young's modulus was calculated as the slope in this region.

Table 1 gathers the values of the tensile parameters of four linseed varieties (Oliver, Everest, Niagara and Alaska). The fibres from Oliver were the stiffest and among the strongest regarding their moderate ultimate strain. Their tensile values were close to those reported for Agatha, a fibre-flax variety well adapted for textile application. Thus, some prospects might exist for Oliver fibres to be used as reinforcing elements in some composite material application. Oliver tensile properties are however weaker than those of Hermes which exhibited mechanical properties comparable to those of E-glass fibres.

Mechanical properties of fibres have been reported to be strongly influenced by their diameter (Baley, 2002); nevertheless, the comparison between Hermes, Agatha and Oliver strongly suggests that other morphological, structural or biochemical parameters might contribute to the tensile properties of Oliver. In the following, the comparison was focussed on Oliver and Hermes varieties.

Table 1. Mean mechanical properties of flax fibres as a function of the variety.

	Diameter[2] (μm)	Young's modulus (GPa)	Strength (MPa)	Ultimate strain (%)
Hermes[1]	20 ± 7	68 ± 36	1,454 ± 835	2.3 ± 0.6
Agatha	22 ± 7	37 ± 10	850 ± 230	2.4 ± 0.6
Glass[1]	17 ± 2	72 ± 2	2,200 ± 100	3.0 ± 0.1
Oliver	23 ± 5	38 ± 13	717 ± 288	2.8 ± 0.8
Everest	24 ± 6	32 ± 19	733 ± 400	3.6 ± 1.0
Niagara	23 ± 5	30 ± 13	603 ± 347	2.9 ± 1.0
Alaska	22 ± 5	23 ± 10	593 ± 241	3.3 ± 0.9

[1]Tensile properties were reported from Charlet *et al.* (2007) and Gay (1997).
[2]The diameter value corresponded to the mean of those of the tested fibres. Average values and the standard error were calculated for at least 30 measurements.

3. Morphology, anatomy and microstructure analyses

To assess morphological parameters, hand cross-sections were cut from the middle part of 2 to 3 average stems per variety, incubated in the dying mixture (carmin 40: 10g/l and methyl green: 1g/l) and observed with a light microscope (Leica DMLB) equipped with a digital camera (Nikon). The number of bundles per transverse section (nB/s), as well as the number of fibres per bundle (nFb/B), were counted (1-5 sections per variety) and the total area of the fibre tissue (aFt) was measured (Figure 2a). A morphological form factor (ffB) was calculated for the bundle as l/L, l and L being the small and large dimensions of the bundle. The equivalent diameter was estimated as Deq = (lf + Lf) /2, lf and Lf being the small and large dimensions of the fibre. A morphological form factor was estimated for the fibre as mf= $(4 \times \pi \times aFb) / P^2$, aFb, and P being the total area and perimeter of the fibre, respectively.

The average diameter of the stems and the area of the fibre tissue (aFt) of Oliver were larger than those of Hermes (Table 2). The number of bundles per section was slightly larger in Oliver. The number of fibre per bundle was the highest in the linseed variety. The stem diameter and fibre yield generally increased with the length within a variety (ecological correlation) (Tikhvinskiy, 1978) and among the different fibre varieties (genotypic correlation) (Sharma and Faughey, 1999; Van den Oever *et al.*, 2003). Oliver was the shortest line but exhibited the largest stem diameter and interestingly its fibre yield (as seen from aFt, nB/s and nFb/B) was the highest.

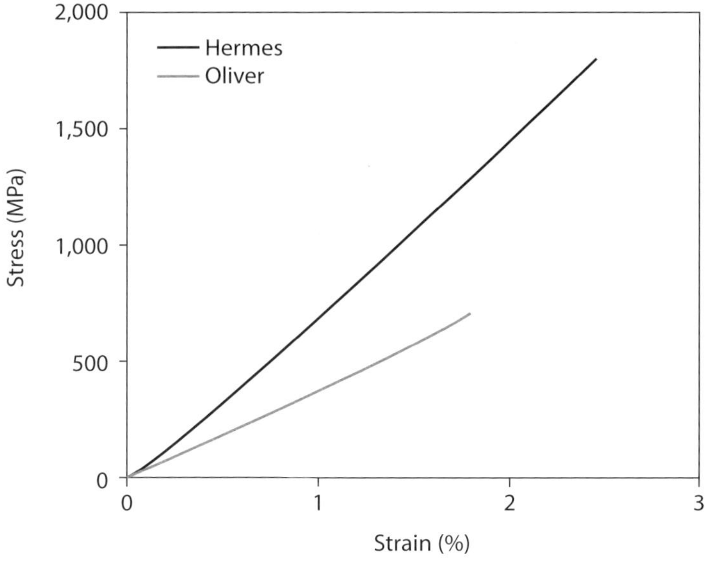

Figure 2. Typical Load-displacement curve for elementary cellulosic fibres of the Oliver variety compared with that of Hermes.

Table 2. Morphology and microstructure parameters as measured by microscopy observations.

	D Stem (mm)	aFt (mm^2)	ffB	nB/s	nFb/B	Deq (μm^2)	mf
Oliver	2.4 ± 0.6	0.46 ± 0.09	0.48 ± 0.15	40 ± 1	30 ± 9	26 ± 5	0.99 ± 0.01
Hermes	1.2 ± 0.6	0.20 ± 0.05	0.65 ± 0.15	38 ± 2	26 ± 9	16 ± 4	0.95 ± 0.05

D, diameter measured in the middle of the stem; aFt, area of the fibre tissue; ffB, form factor of the bundle, nB/s, number of bundles per stem section, nFb/B, number of fibre per bundle; Deq, diameter of the elementary fibres; mf, morphology factor of the elementary fibre. Two to three average stems were chosen per variety.

The mean form factor of the bundle (ffB) of Oliver varied between 0.3 and 0.6 corresponding to an ellipsoid, the best form adapted to spinning (Tikhvinskiy, 1978). On the other hand, this factor ffB was larger for Hermes, some values (15%) being closer to those of a circle (ffB >0.8).

As seen above and importantly for tensile properties, the diameter of Oliver fibres was found significantly larger than that of Hermes. The mean value of the morphological factor of the elementary fibres (mf) of Oliver varied only slightly and was close to unity, meaning that the form of the cells was rather like a circle. The mf values of Hermes were comprised between 0.90 and 1 indicating the presence of two types of forms, circle and hexagon (target value 0.907). It was previously reported that some varieties could be submitted to morphological variation (decrease of Deq and mf) during the late stage of plant growth, due to some cell wall dehydration processing and remodelling of polymer-polymer interactions (reviewed in Morvan *et al.*, 2003). Thus, the lowest values of mf and Deq would be of particular interest if related to the mechanical properties for Hermes.

4. Chemical composition of fibres

The elementary fibres prepared as previously described (Goubet *et al.*, 1995) were then successively treated with 0.02 M Hcl and 1.5 M NaOH/ 100mM $NaBH_4$ (both for 1h at 100 °C + H_2O twice for 1h at 100 °C). The high temperature was needed in order to release the polymers (EH and EOH respectively) which strongly interacted with the cellulose microfibrils. The residues were treated by sulfuric acid (H_2SO_4 for 1h then diluted to 1 M and heated 2h at 110 °C) to dissolve the microfibrils of cellulose and enable the amount of cellulose in the cell walls to be quantified. The insoluble residue was called lignin of Klason. Three independent series were run for each variety. The amounts of total sugars and galacturonic acid of the extracts were obtained by colorimetry (Dubois *et al.*, 1956; Blumenkrantz and Asboe-Hansen, 1973). To determine the relative importance of the pectic and hemicellulosic components, methanolysis and silylation of the EH and EOH samples were performed and the individual sugars were separated by gas chromatography (Goubet *et al.*, 1995).

The main component of elementary fibres consisted of cellulose whose percentage was estimated to 84% for Hermes and 77% for Oliver (Table 3). These values were much higher than those reported in the litterature because the percentage of cellulose was generally given for technical fibres, contaminated by remaining cortical tissues, instead of clean elementary fibres. Nevertheless, whatever the studies, the content of cellulose was larger in fibre-flax than linseed varieties (e.g. Sharma and Faughey, 1999). Conversely, Oliver contained the largest amounts of non cellulosic polysaccharides (NCPs) that encrusted the cellulose microfibrils of the secondary wall. Interestingly, it has been reported that, within a collection of pure lines, the ratio of NCPs to cellulose was negatively linked to the mechanical quality of water-retted fibres (Brutch *et al.*, 2008). Thus, the lowest value measured in Hermes might be another parameter to explain its best tensile performances.

NCPs were successively extracted with HCl (EH) and NaOH (EOH). The residue consisted of cellulose and lignin of Klason (K). Arabinose (Ara), galactose (Gal),

Table 3. Chemical composition of pectins and hemicelluloses released after the chemical extractions of the elementary fibres.

	Hermes (mg/g fibres)		Oliver (mg/g fibres)	
Total NCPs	104 ± 10		147 ± 15	
Cellulose	840 ± 30		773 ± 25	
Lignin K	56 ± 14		80 ± 10	
	Extracts EH (mg/g elementary fibres)		Extract OH (mg/g elementary fibres)	
	Hermes	Oliver	Hermes	Oliver
Ara	2.2 ± 0.2	1.7 ± 0.2	0.5 ± 0.2	0.2 ± 0.1
Gal	20.1 ± 3.0	30.2 ± 3.0	10.2 ± 1.0	12.0 ± 1.0
Rha	3.3 ± 0.3	4.8 ± 0.4	0.8 ± 0.3	0.9 ± 0.1
GalU	8.4 ± 0.5	24.3 ± 3.0	6.5 ± 0.5	7.0 ± 0.5
HG	5.0 ± 1.5	19.5 ± 3.5	5.7 ± 1.0	6.1 ± 1.0
RGI	29.0 ± 4.0	41.5 ± 4.5	12.3 ± 2.0	14.0 ± 2.5
Glc	8.0 ± 1.0	5.0 ± 1.1	17.0 ± 2.5	23.0 ± 2.5
Man	2.0 ± 0.1	4.0 ± 1.0	15.5 ± 2.5	20.5 ± 2.0
Xyl	1.0 ± 0.1	1.0 ± 0.1	5.0 ± 1.0	6.5 ± 1.0
Fuc	<1	<1	<1	<1
GlcU	<1	<1	<1	3.0 ± 2.0
GM	3.3 ± 1.0	6.7 ± 2.0	25.8 ± 3.5	34.2 ± 4.2

rhamnose (Rha) and galacturonic acid (GalUA) were the main sugars contained in pectins. The amount of homogalacturonan (HG) was calculated as GalUA - Rha. The amount of rhamnogalacturonan of type I (RGI) was calculated as 2xRha + Gal + Ara. The sugars specific to hemicellulosic compounds were glucose (Glc), mannose (Man), xylose (Xyl), fucose (Fuc) and glucuronic acid (GlcU). The amount of glucomannan was calculated according to Van Hazendonk *et al.*, 1996 Man = 1.5 Glc.

These NCPs comprised the acidic extract EH which, according to the cell-wall composite-model, consisted of the matrix polymers and the alkali extract EOH which included the polysaccharides the most tightly bound to cellulose. The sugar composition was analysed to evaluate the relative distribution of pectins (P) and hemicelluloses (HC) in each extract EH and EOH. Among the pectins, we distinguished the homogalacturonan (HG) and rhamnogalacturonan (RGI) moities. Among the hemicelluloses we focussed on the glucomannan moities (ratio man/glc = 1.5) as the most representative of the HC family within the elementary fibres (Van Hazendonk *et al.*, 1996). The minor hemicellulosic components were xyloglucans, xylans and other glucans.

4.1 Distribution of the pectins within the encrusting polysaccharides of elementary fibres

The level of pectins was the highest in the extract EH of Oliver. Both the amounts of RGI and HG were significantly higher than in Hermes. Such a tendency was close to that previously reported since Van den Oever *et al.* (2003) found 1% of GalU in Hermes compared to 2-3% in linseed varieties. As previously reported for other fibre-flax varieties (Girault *et al.*, 1997), RGI was also the main pectin present in the secondary wall of the linseed variety, accounting for twice the amount of HG both in EH and EOH. In EOH, the amounts of RGI and HG were similar for both varieties. Moreover, the ratio Gal/Rha which gives an order of the length of the galactan side chains of RGI was about equal in both varieties and higher than that in EH.

Having in mind that long chains of pectic galactans (present in EOH) might strongly interact with cellulose (Zykwinska *et al.*, 2005), we calculated the ratio between the amounts of the RGI in EOH compared to EH. This ratio might be an indicator to estimate the potential of the interphase between some surfacing pectins (in EOH) and matrix pectins (in EH). This ratio was close in Oliver (0.34) and in Hermes (0.42). Those interactions might also exist in flax fibres, like in potato (Zykwinska *et al.*, 2007), but they might be not enough discriminate to participate to the difference of tensile properties between the two varieties. On the other hand, the ratio between the amounts of HG in EOH compared to EH was significantly lower in Oliver (0.3) than in Hermes (1.1). *In vitro*, homogalacturonans do not interact with cellulose (Chanliaud and Gidley, 1999), while, *in vivo,* Roland and Vian (1981) suggested the possibility of ester linkages between HG and cellulose. In the future, we will then have to investigate why these HG were not hydrolyzed with HCl but with NaOH and by which linkages/ or interactions HG might be in relation with cellulose microfibrils.

4.2 Distribution of the hemicelluloses within the encrusting polysaccharides of elementary fibres

Apart from some exceptions, hemicelluloses (HC) are the main components reported to coat the cellulose microfibrils in plant cell walls. Their role is to insure the interface between the microfibrils and the pectic matrix. In the secondary cell wall of flax fibres, the most represented HC was a glucomannan (GM), a polysaccharide with alternated blocks of β-(1,4)-glucan and β-(1,4)-mannan (McDougall, 1993; Van Hazendonk *et al.*, 1996). As expected, GM was more abundant in EOH than in EH in both Oliver and Hermes varieties (Table 3). The amount of GM (accounting for more than two thirds of the total HC) was maximum in Oliver. However, Oliver displayed lower tensile properties than Hermes. To insure its interface role between cellulose and the pectic matrix, the coating polymer HC has to be in sufficient proportion. The ratio between the amount of GM in EOH and the amount of pectins in EH was lower in Oliver (0.56) than in Hermes (0.76). These results suggest that that the content of GM in Oliver might be too low to insure a good interface with the pectic matrix and more importantly to link two different cellulose microfibrils as generally reported in the structural model for primary cell wall (Cosgrove, 2000; Zykwinska *et al.*, 2007).

5. Conclusion

The tensile parameters of Oliver variety were as high as those of Agatha, one fibre flax variety well adapted for textile applications. They were however about half than those measured for Hermes, one of the stiffest fibre flax variety whose use might be optimal for composites reinforced with unidirectional fibres. The differences of tensile properties between Oliver and Hermes might due to structural causes (Deq, mf, orientation of microfibrils). Our data also indicate some influence of the chemistry of the fibres. Firstly, the quantity of cellulose highly contributes to determining the tensile properties of cellulosic fibres. Another distinctive chemical parameter between Hermes and Oliver was the amount of encrusting polymers. As illustrated in Figure 3, a specific parameter might be the ratio between the amounts of coating glucomannans (GM in EOH) and matrix pectins (in EH) that was lower in Oliver than in Hermes. Finally, the interactions of galactans with the microfibrils of cellulose and possibly the presence of homogalacturonan (HG) moieties might be of importance, possibly for the quality of the interface/interphase between the microfibrils of cellulose and the pectic matrix.

The paper points out also some other discriminating morphological parameters between Hermes and Oliver, such as the forms of the bundle and that of the fibres which might influence other mechanical properties such as flexibility or resistance to impact. Moreover the fibre yield of Oliver variety might be among the highest. Altogether, the fibres of Oliver appeared original and their tensile properties might be sufficient to reinforce some polymers, especially natural ones.

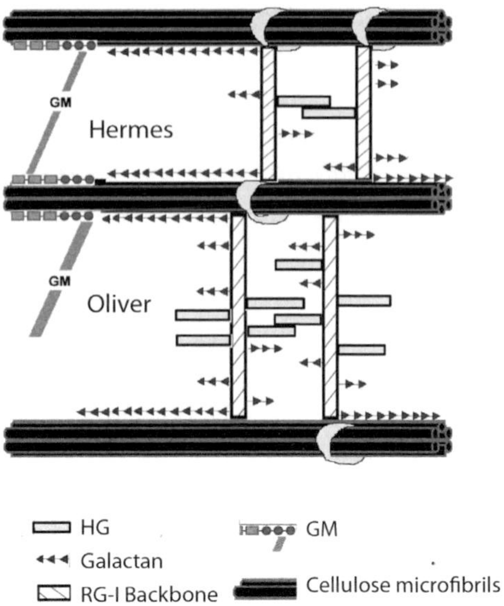

Figure 3. Schematic representation of the composite structure of the secondary wall of flax cellulosic fibres. The model was derived from that of Zykwinska et al. *(2007) established for the primary wall of sugar-beet and potato. The main polymers coating the cellulose microfibrils were glucomannans (GM), the galactan side chains of rhamnogalacturonan (RGI) and possibly homogalacturonan (HG). The main pectins in the matrix consisted of RGI with shorter side chains (see in EH) and to a less extent HG. Note in Oliver the larger amount of the pectic matrix.*

References

Akin, D.E., G.R. Gamble, W.H. Morrison III and L.L Rigsby, 1996. Chemical and structural analysis of fiber and core tissues from flax. J. Sci. Food Agric. 72: 155-165.

Andeme-Onzighi, C., R. Girault, I. His, C. Morvan and A. Driouich, 2000. Immunocytochemical characterization of early-developing flax fiber cells. Protoplasma 213: 235-245.

Baley, C., 2002. Analysis of the flax fibres tensile behaviour and analysis of the tensile stiffness increase. Composites Part A 33: 939-948.

Blumenkrantz N. and G. Asboe-Hansen, 1973. New method for quantitative determination of uronic acids, Anal. Biochem. 54: 484-489.

Brutch, N.B., O. Soret-Morvan, E.A. Porokhovinova, I.Y. Sharov and C. Morvan, 2008. Characters of fibre quality in lines of flax genetic collection. J Natural Fibres 5: 95-126.

Chanliaud, E. and M.J. Gidley, 1999. *In vitro* synthesis and properties of pectin in *Acetobacter xylinus* cellulose composites. Plant J. 20: 25-35.

Charlet, K., C. Baley, C. Morvan, J.P. Jernot, M. Gomina and J. Bréard, 2007a. Characteristics of Hermes flax fibres as a function of their location in the stem and properties of the derived unidirectional composites. Composites part A 38: 1912-1921.

Charlet, K., C. Morvan, J. Bréard, J.P. Jernot and M. Gomina, 2007b. Etude morphologique d'un composite naturel: la fibre de lin. *RCMA*. Hermes/Lavoisier, Paris, 16: 11-24.

Cosgrove, D.J., 2000. Expansive growth of plant cell walls. Plant Physiol. Biochem. 38: 109-124.

Day, A., K. Ruel, G. Neutelings, D. Cronier, H. David, S. Hawkins and B. Chabbert, 2005. Lignification in the flax stem: evidence for an unusual lignin in bast fibers. Planta 222, 234-245.

Dubois, M., K.A. Gilles, J.K. Hamilton, P.A. Rebers and F. Smith, 1956. Colorimetric method for determination of sugars and related substances. Anal. Chem. 28: 350-366.

Fausten, M., 1992. Méthodes de préparation et de qualification des fibres de lin, Journées d'échanges Franco-Allemandes sur le lin, Rouen, France pp. 50-94.

Foster, R., H.S. Pooni and I.J. Mackay, 1997. Quantitative evaluation of *Linum usitatissimum* varieties for dual purpose traits. J. Agric. Sc. 129: 179-185.

Gay, D., 1997. Matériaux Composites. Hermès. Paris 155 p.

Girault, R., F. Bert, C. Rihouey, A. Jauneau, C. Morvan and M. Jarvis, 1997. Galactans and cellulose in flax fibres: putative contributions to the tensile strength. Int. J. Biol. Macromol. 21: 179-188.

Gorshkova, T.A. and C. Morvan, 2006. Secondary cell-wall assembly in flax phloem fibres: role of galactans. Planta 223: 149-158.

Goubet, F., T. Bourlard, R. Girault, C. Alexandre, M.C. Vandevelde and C. Morvan, 1995. Structural features of galactans from flax fibres. Carbohydr. Polymers 27: 221-227.

Gouanvé, F., S. Marais, A. Bessadok, D. Langevin, C. Morvan and M. Métayer, 2006. Study of water sorption in modified flax fibers. J. Applied Polym. Sc. 101: 4281-4289.

His, I., C. Andème-Onzighi, C. Morvan and A. Driouich, 2001. Microscopical analysis of mature flax fibers embedded in London Resin White: immunogold localization of cell-wall matrix polysaccharides. J. Histochem. Cytochem. 49: 1525-1535.

Jankauskiene, Z., K. Bacelis, T. Stuart, R.D. McCall and H.S.S Sharma, 2006. Rapid techniques for assessing fibre quality of flax breeding lines and cultivars using visible and near infrared spectroscopy, and thermal analysis. Ann. Appl. Biol. 149: 91-102.

McDougall, G.J., 1993. Isolation and partial characterisation of the non-cellulosic polysaccharides of flax fibre. Carbohydr. Res. 241: 227-236.

Mohanty, A.K., M. Misra and G. Hinrichsen, 2000. Biofibres, biodegradable polymers and biocomposites: an overview. Macromol. Mater. Eng. 276/277: 1-24.

Mooney, C., T. Stolle-Smits, H. Schols and E. De Jong, 2001. Analysis of retted and non retted flax fibres by chemical and enzymatic means. J. Biotech. 89: 205-216.

Morvan C., C. Andème-Onzighi, R. Girault, D.S. Himmelsbach, A. Driouich and D.E Akin, 2003. Building flax fibres: more than one brick in the walls. Plant Physiol. Biochem. 41: 935-944.

Norton, A.J., S.J. Bennet, M. Hughes, J.P.R.E Dimmock, D. Wright, G. Newman, I.M. Harris and G. Edwards-Jones, 2006. Determining the physical properties of flax fibre for industrial applications: the influence of agronomic practice. Ann. Appl. Biol. 149: 15-25.

Roland, J.-C. and B. Vian, 1981. Use of a purified polygalacturonase for a topochemical study of elongating cell walls at the ultrastructural level. J. Cell Sci. 2: 920-928.

Roland, J.-C., M. Mosiniak and D. Roland, 1995. Dynamique du positionnement de la cellulose dans les parois des fibres textiles du lin (*Linum usitatissimum*). Acta Bot. Gallica 142: 463-484.

Sharma, H.S.S. and G.J. Faughey, 1999. Comparison of subjective and objective methods to assess flax straw cultivars and fibre quality after dew retting. Ann. Appl. Biol. 135: 495-501.

Stamboulis, A., C.A. Baillie and T. Peijs, 2001. Effects of environmental conditions on mechanical and physical properties of flax fibers. Composites Part A 32: 1105-1115.

Tikhvinskiy, S.F., 1978 Improvement of fibre flax quality. Leningrad, Kolos eds 110p.
Van den Oever, M.J.A., N. Bas, L.J.M. Van Soest, C. Melis and J.E.G. Van Dam, 2003. Improved method for fibre content and quality analysis and their application to flax genetic diversity investigations. Ind. Crops Products 18: 213-243.
Van Hazendonk, J.M., E.J.M. Reinerink, P. De Waard and J.E.G. Van Dam, 1996. Structural analysis of acetylated hemicellulose polysaccharides from fibre flax (*Linum usitatissimum* L.). Carbohydr. Res. 291: 141-154.
Wambua, P., J. Ivens and I. Verpoest, 2003. Natural fibres: can they replace glass in fibre feinforced plastics? Comp. Sci. Tech. 63:1259-1264.
Wang, H.H., J.G. Drummont, S.M. Reath, K. Hunt and P.A. Watson, 2001. An improved fibril angle measurement method for wood fibres. Wood Sci. Tech. 34: 493-503.
Zykwinska, A.W., M.C. Ralet, C.D. Garnier and J.F. Thibault, 2005. Evidence for in vitro binding of pectin side chains to cellulose. Plant Physiol. 139: 397-407.
Zykwinska, A.W., J.F. Thibault and M.C. Ralet, 2007. Organization of pectic arabinan and galactan side chains in association with cellulose microfibrils in primary cell wall and models envisaged. J Exp. Bot. 58: 1795-1802.

Part 2

Pectinases

Advances in pectinolytic enzymes, genes and regulation in *Aspergillus*

B.S. Gruben and R.P. de Vries
Microbiology, Utrecht University, Padualaan 8, 3584 CH Utrecht, the Netherlands

Abstract

Aspergillus is the best studied fungus with respect to pectinolytic enzymes and genes. It produces a broad range of pectinolytic activities and has multiple genes for many of these activities. In this chapter the current knowledge on pectinolytic enzymes and genes from *Aspergillus* will be summarized. In addition, the influence of genome sequences on the discovery and characterisation of *Aspergillus* pectinolytic genes will be discussed as well as regulatory aspects of pectinolytic gene expression in *Aspergillus*.

Keywords: *Aspergillus*, novel enzymes, novel genes, genomics, regulation

1. Introduction

Pectinases and pectinolytic genes have been studied for many decades. While examples of characterized pectinases are available for many organisms, some species are at the forefront of this field. Fungal pectinases have been predominantly studied in species of the ascomycete genus *Aspergillus*.

This genus consists of many different species, some of which are opportunistic human pathogens (e.g. *A. fumigatus*), phytopathogens or mycotoxin producers (e.g. *A. flavus*), while others have been used extensively in industry for the production of enzymes and metabolites (e.g. *A. niger, A. oryzae*) or are widely used model organisms for lower eukaryotes (*A. nidulans*). The diversity of topics that are studied in *Aspergillus* have made this the most widely studied fungal group, which is illustrated by the public availability of 9 genome sequences of species of this genus (see 3).

Due to the highly complex structure of pectin, enzymatic degradation and/or modification requires a broad set of activities. These enzymes can be classified into hydrolases, lyases and esterases, each of which can be subdivided into families based on their amino acid sequence (Coutinho and Henrissat, 1999). A database of these families is available at the Carbohydrate Active Enzymes database at www.cazy.org. Some of these families contain only a single enzyme activity, while others contain several activities that often cluster during phylogenetic analysis (De Vries *et al.*, 2005). An overview of the enzyme activities involved in pectin degradation and the CAZy families they are assigned to is given in Table 1. The main chain of homogalacturonan is

Table 1. Fungal enzymatic activities involved in pectin degradation. The enzymes are sorted by class and by substrate.

Enzyme class	Enzyme code	Substrate	CAZy families
Hydrolases			
endopolygalacturonase	PGA	Homogalacturonan	GH28
exopolygalacturonase	PGX	Homogalacturonan	GH28
d-4,5 unsaturated -glucuronyl hydrolase	UGH	Homogalacturonan	GH88
unsaturated rhamnogalacturonan hydrolase	URH	Homogalacturonan	GH105
rhamnogalacturonan galacturonohydrolase / exorhamnogalacturonase	RGX	Rhamnogalacturonan I	GH28
α-rhamnosidase / rhamnogalacturonan rhamnohydrolase	RHA	Rhamnogalacturonan I	GH78
rhamnogalacturonan hydrolase / endorhamnogalacturonase	RHG	Rhamnogalacturonan I	GH28
α-L-arabinofuranosidase	ABF	Rhamnogalacturonan I	GH51,54
endoarabinanase	ABN	Rhamnogalacturonan I	GH43
exoarabinanase	ABX	Rhamnogalacturonan I	GH93
β-1,4-endogalactanase	GAL	Rhamnogalacturonan I	GH53
β-1,6-endogalactanase	GLN	Rhamnogalacturonan I	
β-1,4-D-galactosidase	LAC	Rhamnogalacturonan I	GH2,35
β-1,4-exogalactanase	XFG	Rhamnogalacturonan I	
β-1,6-exogalactanase	XSG	Rhamnogalacturonan I	
β-1,3-exogalactanase	XTG	Rhamnogalacturonan I	
β-1,4-D-xylosidase	BXL	Xylogalacturonan	GH3,43
Lyases			
pectin lyase	PEL	Homogalacturonan	PL1
pectate lyase	PLY	Homogalacturonan	PL1,3,9
rhamnogalacturonan lyase	RGL	Rhamnogalacturonan I	PL4,11
Esterases			
pectin acetyl esterase	PAE	Homogalacturonan	
pectin methyl esterase	PME	Homogalacturonan	CE8
feruloyl esterase	FAE	Rhamnogalacturonan I	CE1
rhamnogalacturonan acetyl esterase	RGAE	Rhamnogalacturonan I	CE12

cleaved by endo- and exopolygalacturonases or pectin and pectate lyases. The activity of these enzymes depends strongly on the degree of methylation and acetylation of pectin, which can be modified by pectin methyl or acetyl esterases. The rhamnogalacturonan main chain is cleaved by rhamnogalacturonan hydrolase, rhamnogalacturonan galacturonohydrolase, α-rhamnosidase and rhamnogalacturonan lyase, with the assistance of rhamnogalacturonan acetyl esterase.

For the degradation of the side chains of rhamnogalacturonan I several activities related to release of L-arabinose and D-galactose are required, while the aromatic residues (mainly ferulic and *p*-coumaric acid) are removed by feruloyl esterases.

In this chapter an update will be presented regarding the pectinolytic enzymes and genes compared to the overview published in 2001 (De Vries and Visser, 2001). In addition, new insights on pectinolytic functions based on genome sequences and regulatory aspects will be discussed.

2. Pectinolytic enzymes from *Aspergillus*

In 2001, a comprehensive overview of enzymes from *Aspergillus* involved in plant cell wall polysaccharide degradation was published (De Vries and Visser, 2001), that also covered the known pectinolytic enzyme activities. In that paper, the physical properties of 17 *Aspergillus* endopolygalacturonases were summarized based on reports on 1 enzyme from *A. alliaceus*, 3 from *A. carbonarius*, 10 from *A. niger* and 3 from *A. oryzae*. In addition, 6 exo-polygalacturonases (3 from *A. allicieus*, 3 from *A. niger*) and 4 endo-rhamnogalacturonan hydrolases from *A. aculeatus* were included. Surprisingly, only a very small number of novel reports on characterized pectin-active enzymes have been published subsequently. In this chapter we summarize the properties of 7 additional endo-polygalacturonases, 2 exo-polygalacturonases, 2 endo-rhamnogalacturonan hydrolases, 2 α-rhamnosidases, 2 pectin lyases and 2 pectin methyl esterases (Table 2). In general, the newly described enzymes acting on the pectin main chain have similar physical properties to those described previously (De Vries and Visser, 2001). The exceptions to this are the acidic pH optimum of *A. kawachii* PGI, which is significantly lower than those reported for other endopolygalacturonases, and the higher molecular mass for *A. niger* exo-PGI (Table 2 and (De Vries and Visser, 2001)).

In addition, a number of novel enzymes acting on the pectin side chains have been described (Table 3). The newly described α-L-arabinofuranosidases and feruloyl esterases have physical characteristics similar to those described previously (De Vries and Visser, 2001). However, the two novel β-galactosidases are significantly different. The *A. japonicus* β-galactosidase has a much lower molecular mass (27 kDa) than that described for any other *Aspergillus* β-galactosidase (De Vries and Visser, 2001), while the *A. nidulans* enzyme is the only β-galactosidase described so far with a neutral to alkaline pH optimum (Table 3).

One of the problems with comparing enzymes from different *Aspergillus* species is that the species designation is not always correct. A typical example of this is *A. awamori*. While this name is commonly used for a number of *Aspergillus* strains, a report in 1997

Table 2. Physical properties of Aspergillus *endo- and exopolygalacturonases, endo-rhamnogalacturonan hydrolases, α-rhamnosidases, pectin lyases and pectin methyl esterases. Enzymes in the table are additions to the previously published overview (De Vries and Visser, 2001).*

Species and enzyme type	Enzyme	Mol mass (kDA)	pH_{opt}	T_{opt}	pI	Reference
Endopolygalacturonases						
A. awaromi	endoPG	41	5	40	6.1	(Nagai *et al.*, 2000)
A. japonicus	PG-I	38	4-5.5	30	5.6	(Semenova *et al.*, 2003)
A. japonicus	PG-II	65	4-5.5	30	3.3	(Semenova *et al.*, 2003)
A. kawachii	PGI	40	2.0-3.0		3.55	(Contreras Esquivel and Voget, 2004)
A. niger	PG-II	61	3.8-4.3	43		(Singh and Appu Rao, 2002)
A. niger	PG-IV	38	3-4.6	45		(Singh and Appu Rao, 2002)
A. niger		40.4	5	36		(Guo *et al.*, 2002)
Exopolygalacturonases						
A. niger	exo-PG1	82	3.4-3.8	60		(Sakamoto *et al.*, 2002)
A. niger	exo-PG2	56	3.4-4.2	60		(Sakamoto *et al.*, 2002)
Endorhamnogalacturonan hydrolase						
A. niger	RhgA	47	3.6		3.7	(Suykerbuyk *et al.*, 1997)
A. niger	RhgB	57	4.1		3.9	(Suykerbuyk *et al.*, 1997)
α-Rhamnosidase						
A. nidulans		102	4.5-6.0	60	5	(Manzanares *et al.*, 2001)
A. terreus		96	4	44	4.6	(Gallego *et al.*, 2001)
Pectin lyases						
A. flavus		38	8	50		(Yadav *et al.*, 2008)
A. japonicus	PL	50			3.8	(Semenova *et al.*, 2003)
Pectin methyl esterases						
A. japonicus	PEI	46			3.8	(Semenova *et al.*, 2003)
A. japonicus	PEII	47			3.8	(Semenova *et al.*, 2003)

(Parenicová *et al.*, 1997) demonstrated that most *A. awamori* strains are in fact either *A. niger*, *A. tubingensis* or *A. foetidus*. This demonstrates that new taxonomical insights are not being incorporated in other fields, preventing accurate species comparisons. Accurate species (re-)identification using modern techniques will not only facilitate a better understanding of differences between closely related species, but will also assist in gene mining for novel enzymes from different *Aspergillus* species.

Table 3. Physical properties of Aspergillus *α-L-arabinofuranosidases, β-galactosidases and feruloyl esterases. Enzymes in the table are additions to the previously published overview (De Vries and Visser, 2001).*

Species and enzyme type	Enzyme	Mol mass (kDA)	pH_{opt}	T_{opt}	pI	Reference
α-L-Arabinofuranosidases						
A. awamori	I	81	4	60	3.3	(Kaneko *et al.*, 1998)
A. awamori	II	62	4	60	3.6	(Kaneko *et al.*, 1998)
A. kawachii	AbfA	80	4	55		(Koseki *et al.*, 2003)
A. kawachii	AbfB	62	4	55		(Koseki *et al.*, 2003)
β-Galactosidase						
A. japonicus		27	5.2	45	4.6	(Saad, 2004)
A. nidulans		450[a]	7.5	30		(Diaz *et al.*, 1996)
Feruloyl esterases						
A. nidulans	FAE	56	7	45	4.66	(Shin and Chen, 2007)
A. niger	FaeB	74	6.0			(De Vries *et al.*, 2002b)

[a] Multimeric complex.

3. Pectinolytic genes and genomics

The 2001 review (De Vries and Visser, 2001) also provided an overview of *Aspergillus* genes encoding plant polysaccharide degrading enzymes. The availability of novel molecular biology technologies and genome sequences has strongly increased the number of putative genes encoding pectinolytic enzymes. However, only a small number of papers report the functional characterisation of these genes. Analysis of the genome sequences of *Aspergillus* species for the presence of genes encoding putative pectinolytic enzymes has shown that only part of the potential for pectin degradation has been characterised. In a recent paper, the genome sequence of *A. niger* was described and the contents of the genome with respect to plant polysaccharide degradation was compared to those of *A. nidulans*, *A. oryzae*, and *A. fumigatus* (Table 4) (Pel *et al.*, 2007). This study identified functions for which genes had not been described before for *Aspergillus* (exorhamnogalacturonan hydrolase, exoarabinanase, rhamnogalacturonyl hydrolase, d-4,5-unsaturated glucuronyl hydrolase) as well as additional genes encoding previously identified functions, and demonstrated significant differences in the pectinolytic gene set of these four Aspergilli. *A. niger* contains the largest set of GH28 genes of the four Aspergilli, but significantly less pectate lyases (Pel *et al.*, 2007). This difference correlates with the strong acidifying nature of *A. niger* and the fact that pectate lyases have low activity at acidic pH. A closer examination of the family GH28 from these 4 species using ClustalX demonstrated an even higher level of differentiation (Figure 1). Only 1 endopolygalacturonase, 2 exopolygalacturonases, 1 endorhamnogalacturonan

Table 4. Comparison of the number of putative pectinolytic genes in four Aspergilli. The putative genes are sorted by CAZy family. Based on (Pel et al., *2007).*

Enzyme	CAZy family	*A. niger*	*A. nidulans*	*A. fumigatus*	*A. oryzae*
Hydrolases					
β-galactosidase	GH2	0	3	2	1
endopolygalacturonase	GH28	7	3	4	4
endorhamnogalacturonan hydrolase	GH28	6	1	2	6
exopolygalacturonase	GH28	4	3	3	2
exorhamnogalacturonan hydrolase	GH28	3	1	2	4
xylogalacturonan hydrolase	GH28	1	1	1	2
β-galactosidase	GH35	5	4	5	6
α-L-arabinofuranosidase	GH43	2	3	3	3
α-L-arabinofuranosidase	GH51	3	2	2	3
endo- β-1,4-galactanase	GH53	2	1	1	1
α-L-arabinofuranosidase	GH54	1	1	1	1
α-rhamnosidase	GH78	8	8	6	8
d-4,5-unsaturated glucuronyl hydrolase	GH88	1	2	2	3
exoarabinanase	GH93	0	2	3	3
Rhamnogalacturonyl hydrolase	GH105	2	3	2	2
Lyases					
pectate lyase	PL1	1	3	2	4
pectin lyase	PL1	5	5	4	8
pectate lyase	PL3	0	5	3	3
rhamnogalacturonan lyase	PL4	2	4	3	4
pectate lyase	PL9	0	1	1	1
rhamnogalacturonan lyase	PL11	0	1	0	0
Esterases					
feruloyl esterase	CE1	1	1	0	2
pectin methyl esterase	CE8	3	3	4	5
rhamnogalacturonan acetyl esterase	CE12	2	2	2	4

hydrolase and 1 xylogalacturonan hydrolase have orthologues in all 4 species, while other have orthologues in only 1 or 2 of the other species, or have another ORF from the same species as closest homologue (Figure 1). This demonstrates that this family has been subject to both gene duplication and gene loss, which has likely fine-tuned the pectinolytic system of the different species to suit their natural biotope.

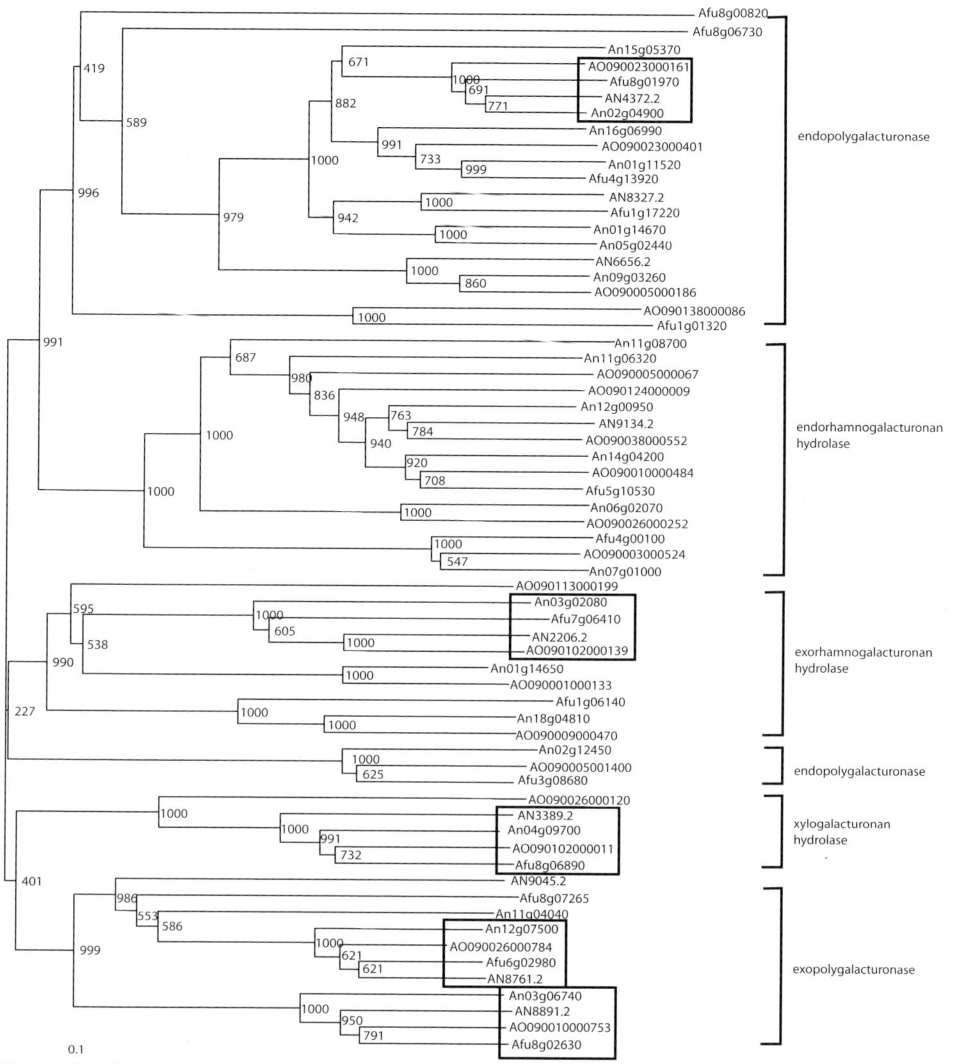

Figure 1. Neighbour Joining tree with 1000 bootstraps of putative GH28 glycosyl hydrolases from A. niger, A. nidulans, A. fumigatus *and* A. oryzae. *Genes with orthologues in all four species are boxed.* A. niger, A. nidulans, A. fumigatus *and* A. oryzae *ORFs are depicted as AnXXgXXXX, ANXXXX.2, AfuXgXXXXX and AOXXXXXXXXXXXX, respectively, with X being any number between 0 and 9.*

A. niger does not contain a putative exoarabinase encoding gene, which suggests a different strategy for arabinan degradation than the three other Aspergilli. However, the second genome sequence for *A. niger* performed by the Joint Genome Institute does contain a gene for this function (Coutinho *et al.*, 2009). This could indicate a difference between these two *A. niger* strains, but could also be explained by the presence of this gene in one of the remaining gaps in the CBS 513.88 sequence or an annotation error.

The genes that were not included in detail in this analysis were feruloyl esterases, but a recent paper included a genomic analysis of feruloyl esterases in 31 fungal genomes (Benoit *et al.*, 2008). This demonstrated that feruloyl esterase A from *A. niger* is only present in the Aspergilli and not in any of the other genomes tested in hat study, while *A. niger faeB* is present in most filamentous ascomycetes (Benoit *et al.*, 2008). Additional candidate feruloyl esterases were identified in *A.oryzae* (4), *A. niger* (2), *A. terreus* (4) and *A. fumigatus* (2). However, these putative genes were not conserved between the species, suggesting a high level of differentiation between the Aspergilli with respect to feruloyl esterases.

Biochemical analysis has only been performed for a few of the newly discovered genes, such as the exorhamnogalacturonan hydrolase from *A. niger* (Martens-Uzunova *et al.*, 2006). As the function for the other novel genes has been assigned based on amino acid homology, these may not all be correct and the corresponding enzymes may have different functions. Addition of some of these newly identified enzymes to the pectinolytic enzyme cocktails will likely result in new or improved applications.

4. Regulation of pectinolytic gene expression in *Aspergillus*

Regulation of pectinolytic gene expression has not been studied in detail and most of the available literature limits the data to evidence for induction by pectin or pectin derived components (e.g. polygalacturonic acid, D-galacturonic acid, L-arabinose) (De Vries, 2003). Only a single study has been published so far that describes a detailed expression analysis of pectinolytic genes (De Vries *et al.*, 2002a). In this study, the expression of 26 genes encoding enzymes involved in the degradation of pectin were studied on 16 carbon sources, at three time points and in a wild type strain as well as a strain with a mutation in the major carbon catabolite repression system CreA (Dowzer and Kelly, 1991). This study demonstrated that all of the genes except for two feruloyl esterases are expressed in the presence of D-galacturonic acid, polygalacturonic acid and/or sugar beet pectin, suggesting the presence of a general D-galacturonic acid responsive transcriptional activator. However, they did not all have identical expression profiles on all substrates, but could be divided into distinct groups with similar expression profiles (Figure 2) (De Vries *et al.*, 2002a). Some of the differences between the groups are based on the time point of the samples, indicating temporal regulation of pectinolytic gene expression. This may reflect the need for the different enzymes during initial or later stages of pectin degradation. The subgroup with the earliest expression profile consists of PelA, PelD, PgaA, PgaX and PmeA, suggesting that these enzymes are important for the initial steps of pectin degradation. As PgaX releases D-galacturonic acid, the early expression of this gene may also ensure further induction of the total pectinolytic gene system.

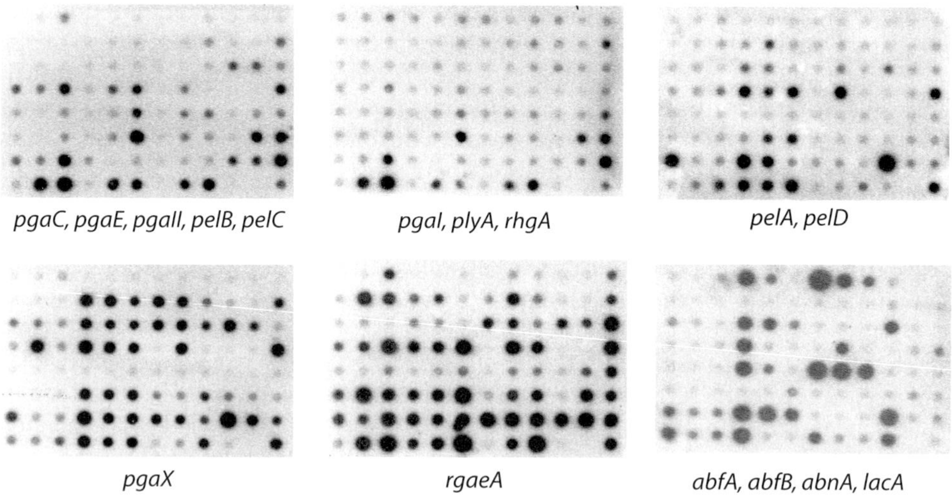

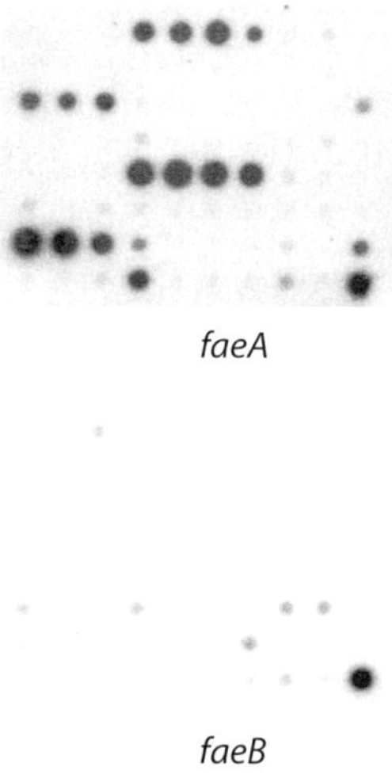

	1	2	3	4	5	6	7	8	9	10	11	12
	13	14	15	16	17	18	19	20	21	22	23	24
N402	25	26	27	28	29	30	31	32	33	34	35	36
	37	38	39	40	41	42	43	44	45	46	47	48
	49	50	51	52	53	54	55	56	57	58	59	60
NW200	61	62	63	64	65	66	67	68	69	70	71	72
	73	74	75	76	77	78	79	80	81	82	83	84
	85	86	87	88	89	90	91	92	93	94	95	96

1-3: 25 mM D-glucose (glc); 2, 4 and 8 h
4-6: 25 mM D-xylose (xyl); 2, 4 and 8 h
7-9: 25 mM L-arabinose (ara); 2, 4 and 8 h
10-12: 25 mM D-galactose (gal); 2, 4 and 8 h
13-15: 25 mM L-rhamnose (rham); 2, 4 and 8 h
16-18: 25 mM D-galacturonic acid(gal.a.); 2, 4 and 8 h
19-21: 25 mM D-glucuronic acid (gluc.a.); 2, 4 and 8 h
22-24: 25 mM glc + 25 mM gal.a.; 2, 4 and 8 h
25-27: 25 mM xyl + 25 mM gal.a.; 2, 4, and 8 h
28-30: 25 mM ara + 25 mM gal.a.; 2, 4 and 8 h
31-33: 25 mM rham + 25 mM gal.a.; 2, 4 and 8 h
34-36: 1% polygalacturonate (PG); 2, 8 and 24 h
37-39: 1% PG + 25 mM glc; 2, 8 and 24 h
40-42: 1% sugar beet pectin (SBP); 2, 8 and 24 h
43-45: 1% SBP + 25 mM glc; 2, 8 and 24 h
46-48: 1% SBP + 0.03% ferulic acid; 2, 8 and 24 h
49-96: same as 1-48

Figure 2. Grouping of pectinolytic genes by expression profiling. A representative profile is given for each of the subgroups. Data previously presented in (De Vries et al., *2002a).*

In addition, subgroups of genes were identified that were expressed in response to other monomeric inducers (De Vries *et al.*, 2002a). Genes encoding L-arabinose releasing functions (α-L-arabinofuranosidase, *abfA* and *abfB*, and endoarabinanase, *abnA*) and D-galactose releasing functions (β-galactosidase, *lacA*) were also expressed in the presence of L-arabinose, suggesting the presencc of an L-arabinose responsive transcriptional activator. This was confirmed by the isolation of UV mutants from *A. niger* that had reduced expression of genes encoding L-arabinose releasing enzymes as well as genes of the L-arabinose catabolic pathway (De Groot *et al.*, 2003). Recently, the gene encoding this transcriptional activator (*araR*) was identified and analysis of strains in which this gene was disrupted confirmed that AraR controls L-arabinose release from xylan and pectin (arabinan) as well as the L-arabinose catabolic pathway (Battaglia *et al.*, unpublished results).

The rhamnogalacturonan acetyl esterase encoding gene (*rgaeA*) was induced by L-rhamnose, while the feruloyl esterase encoding genes (*faeA*, *faeB*) were induced by ferulic acid and the β-galactosidase encoding gene (*lacA*) was induced by D-galactose, suggesting the presence of three additional transcriptional activators (De Vries *et al.*, 2002a). No other publications have reported L-rhamnose-induced gene expression in *Aspergillus*, but induction of feruloyl esterases by ferulic acid has been described in detail for *A. niger faeA* and *faeB* (De Vries and Visser, 1999; De Vries *et al.*, 2002b) and galactose-induction of *lacA* has been described before ((De Vries *et al.*, 1999). Recently, the D-galactose responsive transcriptional activator (GalR) has been identified in *A. nidulans* (Nikolaev *et al.*, unpublished results).

Finally, all pectinolytic genes, except for *pgaX*, *pgaA*, *pgaB*, *pgaD* and *rglA*, had higher expression levels in the CreA derepressed strain than in the wild type (De Vries *et al.*, 2002a), indicating an important role for the carbon catabolite repressor protein CreA in pectinolytic gene regulation as is the case for most genes involved in plant polysaccharide degradation (De Vries, 2003). This system ensures that production of polysaccharide degrading enzymes does not occur when sufficient amounts of easy metabolizable monomeric carbon sources are available, and therefore prevents wasting energy on the production of enzymes that are not essential at that time.

Based on the currently available data, the regulation of pectinolytic gene expression involved at least 5 transcriptional activators and 1 transcriptional repressor (Figure 3).

5. Concluding remarks

Since the last Pectins and Pectinases Conference in 2001 (Voragen *et al.*, 2003), a relatively small number of functionally characterized pectinolytic enzymes and/ or genes have been reported for *Aspergillus*. One could argue that already most of them were identified and it is therefore not surprising. However, recent availability of genome sequences for several Aspergilli demonstrated that there is a significantly higher number of putative pectinolytic genes than was known in literature. Functional characterization of some of these novel functions has been initiated, but many are still unstudied and biochemical characterisation may still show novel specificities. In

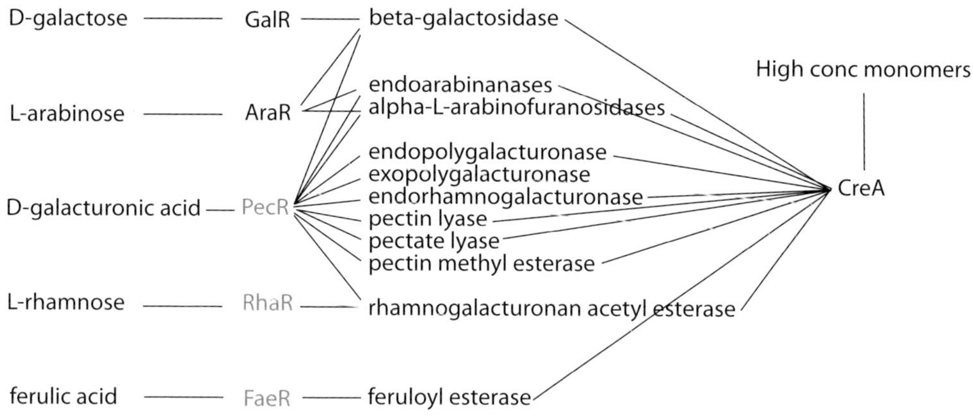

Figure 3. Inducers and regulators involved in the production of pectinases in Aspergillus niger. *Regulators in grey are not yet identified.*

addition, the genome sequences have shown that the Aspergilli are among the richest fungi with respect to the number of ORFs encoding putative plant polysaccharide degrading enzymes.

Genome sequences have also opened doors to novel strategies for identifying regulators involved in pectinolytic gene expression. This has already resulted in the identification of the L-arabinose-responsive (AraR) and D-galactose-responsive (GalR) transcriptional activators in *A. niger* and *A. nidulans*, respectively.

Further mining of genomes from various Aspergilli as well as other fungi will not only strongly increase the pectinolytic enzymes available for industrial applications, but will also provide new insights in fungal biodiversity, utilization of natural carbon sources and natural biotope characteristics.

References

Benoit, I., E.G. Danchin, R.J. Bleichrodt and R.P. de Vries, 2008. Biotechnological applications and potential of fungal feruloyl esterases based on prevalence, classification and biochemical diversity. Biotechnol Lett 30: 387-396.

Contreras Esquivel, J.C. and C.E. Voget, 2004. Purification and partial characterization of an acidic polygalacturonase from *Aspergillus kawachii*. J Biotechnol 110: 21-28.

Coutinho, P.M. and B. Henrissat, 1999. Carbohydrate-active enzymes: an integrated database approach. In: H.J. Gilbert, G.J. Davies, B. Henrissat and B. Svensson (eds.) Rec. Adv. Carbohydr. Bioeng. The Royal Society of Chemistry. Cambridge. pp.3-14.

Coutinho, P.M., M.R. Andersen, K. Kolenova, P.A. vanKuyk, I. Benoit, B.S. Gruben, B. Trejo-Aguilar, H. Visser, P. van Solingen, T. Pakula, B. Seiboth, E. Battaglia, G. Aguilar-Osorio, J.F. Jong, R.A. Ohm, M. Aguilar, B. Henrissat, J. Nielsen, H. Stålbrand and R.P. de Vries, 2009. Post-genomic insights into the plant polysaccharide degradation potential of *Aspergillus nidulans* and comparison to *Aspergillus niger* and *Aspergillus oryzae*. Fung Genet Biol 46: S161-S169.

De Groot, M.J.L., P.J.I. van de Vondervoort, R.P. de Vries, P.A. vanKuyk, G.J.G. Ruijter and J. Visser, 2003. Isolation and characterization of two specific regulatory *Aspergillus niger* mutants shows antagonistic regulation of arabinan and xylan metabolism. Microbiol 149: 1183-1191.

De Vries, R.P., 2003. Regulation of *Aspergillus* genes encoding plant cell wall polysaccharide degrading enzymes; relevance for industrial production. Appl Microbiol Biotechnol 61: 10-20.

De Vries, R.P. and J. Visser, 1999. Regulation of the feruloyl esterase (*faeA*) gene from *Aspergillus niger*. Appl Environ Microbiol 65: 5500-5503.

De Vries, R.P. and J. Visser, 2001. *Aspergillus* enzymes involved in degradation of plant cell wall polysaccharides. Microb Mol Biol Rev 65: 497-522.

De Vries, R.P., H.C. van den Broeck, E. Dekkers, P. Manzanares, L.H. de Graaff and J. Visser, 1999. Differential expression of three α-galactosidase genes and a single β-galactosidase gene from *Apergillus niger*. Appl Environ Microbiol 65: 2453-2460.

De Vries, R.P., J. Jansen, G. Aguilar, L. Parenicová, J.A.E. Benen, V. Joosten, F. Wulfert and J. Visser, 2002a. Expression profiling of pectinolytic genes from *Aspergillus niger*. FEBS Lett 530: 41-47.

De Vries, R.P., H.C.M. Kester, P.A. vanKuyk and J. Visser, 2002b. The *Aspergillus niger faeB* gene encodes a second feruloyl esterase involved in pectin and xylan degradation, and is specifically induced on aromatic compounds. Biochem J 363: 377-386.

De Vries, R.P., C. van Grieken, P.A. vanKuyk and H.A.B. Wösten, 2005. The value of genome sequences in the rapid identification of novel genes encoding specific plant cell wall degrading enzymes. Curr Genom 6: 157-187.

Diaz, M., A.M. Pedregosa, J.R. de Lucas, S. Torralba, I.F. Monistrol and F. Laborda, 1996. Purification and properties of beta-galactosidase from Aspergillus nidulans. Microbiologia 12: 585-592.

Dowzer, C.E.A. and J.M. Kelly, 1991. Analysis of the *cre*A gene, a regulator of carbon catabolite repression in *Aspergillus nidulans*. Mol Cell Biol 11: 5701-5709.

Gallego, M.V., B. Pinan-Lucarré, D. Ramon and S. Valles, 2001. Purification and Characterization of an α-L-Rhamnosidase from *Aspergillus terreus* of Interest in Winemaking. J Food Sci 66: 204-209.

Guo, C.-T., W.-M. Xue, T.-B. Chen, W.-H. Deng and P.-F. Rao, 2002. Purification and partial characterization of an endo-polygalacturonase from *Aspergillus niger*. J Food Biochem 26: 253-265.

Kaneko, S., M. Arimoto, M. Ohba, H. Kobayashi, T. Ishii and I. Kusakabe, 1998. Purification and substrate specificities of two alpha-L-arabinofuranosidases from Aspergillus awamori IFO 4033. Appl Environ Microbiol 64: 4021-4027.

Koseki, T., M. Okuda, S. Sudoh, Y. Kizaki, K. Iwano, I. Aramaki and H. Matsuzawa, 2003. Role of two alpha-L-arabinofuranosidases in arabinoxylan degradation and characteristics of the encoding genes from shochu koji molds, *Aspergillus kawachii* and *Aspergillus awamori*. J Biosci Bioeng 96: 232-241.

Manzanares, P., H.C. van den Broeck, L.H. de Graaff and J. Visser, 2001. Purification and characterization of two different alpha-L- rhamnosidases, RhaA and RhaB, from *Aspergillus aculeatus*. Appl Environ Microbiol 67: 2230-2234.

Martens-Uzunova, E.S., J.S. Zandleven, J.A.E. Benen, H. Awad, H.J. Kools, G. Beldman, A.G.J. Voragen, J.A. van den Berg and P.J. Schaap, 2006. A new group of exo-acting family 28 glycoside hydrolases of *Aspergillus niger* that are involved in pectin degradation. Biochem J 400: 43-52.

Nagai, M., T. Katsuragi, T. Terashita, K. Yoshikawa and T. Sakai, 2000. Purification and characterization of an endo-polygalacturonase. From Aspergillus awamori. Biosci Biotechnol Biochem 64: 1729-1732.

Parenicová, L., J.A.E. Benen, R.A. Samson and J. Visser, 1997. Evaluation of RFLP analysis of the classification of selected black aspergilli. Mycol Res 101: 810-814.

Pel, H.J., J.H. de Winde, D.B. Archer, P.S. Dyer, G. Hofmann, P.J. Schaap, G. Turner, R.P. de Vries, R. Albang, K. Albermann, M.R. Andersen, J.D. Bendtsen, J.A.E. Benen, M. van den Berg, S. Breestraat, M.X. Caddick, R. Contreras, M. Cornell, P.M. Coutinho, E. Danchin, G,J,, A.J.M. Debets, P. Dekker, P.W.M. van Dijck, A. van Dijk, L. Dijkhuizen, A.J.M. Driessen, C. d'Enfert, S. Geysens, C. Goosen, G.S.P. Groot, P.W.J. de Groot, T. Guillemette, B. Henrissat, M. Herweijer, J.P.T.W. van den Hombergh, C.A.M.J.J. van den Hondel, R.T.J.M. van der Heijden, R.M. van der Kaaij, F.M. Klis, H.J. Kools, C.P. Kubicek, P.A. vanKuyk, J. Lauber, X. Lu, M.J.E.C. van der Maarel, R. Meulenberg, H. Menke, A.M. Mortimer, J. Nielsen, S.G. Oliver, M. Olsthoorn, K. Pal, N.N.M.E. van Peij, A.F.J. Ram, U. Rinas, J.A. Roubos, C.M.J. Sagt, M. Schmoll, J. Sun, D. Ussery, J. Varga, W. Vervecken, P.J.I. van de Vondervoort, H. Wedler, H.A.B. Wösten, A.-P. Zeng, A.J.J. van Ooyen, J. Visser and H. Stam, 2007. Genome sequence of *Aspergillus niger* strain CBS 513.88: a versatile cell factory. Nature Biotechnol 25: 221-231.

Saad, R.R., 2004. Purification and some properties of β-galactosidase from *Aspergillus japonicus*. Annals Microbiol 54: 299-306.

Sakamoto, T., E. Bonnin, B. Quemener and J.F. Thibault, 2002. Purification and characterisation of two exo-polygalacturonases from Aspergillus niger able to degrade xylogalacturonan and acetylated homogalacturonan. Biochim Biophys Acta 1572: 10-18.

Semenova, M.V., S.G. Grishutin, A.V. Gusakov, O.N. Okunev and A.P. Sinitsyn, 2003. Isolation and properties of pectinases from the fungus Aspergillus japonicus. Biochemistry (Mosc) 68: 559-569.

Shin, H.D. and R.R. Chen, 2007. A type B feruloyl esterase from Aspergillus nidulans with broad pH applicability. Appl Microbiol Biotechnol 73: 1323-1330.

Singh, S.A. and A.G. Appu Rao, 2002. A simple fractionation protocol for, and a comprehensive study of the molecular properties of, two major endopolygalacturonases from Aspergillus niger. Biotechnol Appl Biochem 35: 115-123.

Suykerbuyk, M.E.G., H.C.M. Kester, P.J. Schaap, H. Stam, W. Musters and J. Visser, 1997. Cloning and characterization of two rhamnogalacturonan hydrolase genes from *Aspergillus niger*. Appl Environ Microbiol 63: 2507-2515.

Voragen, F., H. Schols and R. Visser (eds.), 2003. Advances in pectin and pectinase research. Kluwer Academic Publishers, Dordrecht, the Netherlands.

Yadav, S., P.K. Yadav, D. Yadav and K.D.S. Yadav, 2008. Purification and characterization of an alkaline pectin lyase from *Aspergillus flavus*. Process Biochem 43: 547-552.

Purification and cloning of a rhamnogalacturonase tolerant to an acetylated rhamnogalacturonan

J. Normand[1], P. Delavault[2], J.-F. Thibault[1] and E. Bonnin[1]
[1]Unité de Recherche Biopolymères, Interactions, Assemblages, INRA Nantes, BP 71627, 44316 Nantes Cedex 03, France
[2]Laboratoire de Biologie et Pathologie Végétale, Université de Nantes, 44322 Nantes, France

Abstract

A rhamnogalacturonase acting on acetylated substrate was detected in the commercial preparation Driselase®, an enzymatic mixture derived from the basidiomycete *Irpex lacteus*. The enzyme was purified by hydrophobic interaction chromatography, gel filtration and preparative isoelectric focusing, resulting in the isolation of different rhamnogalacturonan hydrolases having various isoelectric points. SDS-PAGE analysis of the fractions revealed a similar molar mass around 55 kDa and mass spectrometry analysis after trypsin cleavage did not permit to find difference in the aminoacid sequence of the different isoforms. In order to produce a recombinant rhamnogalacturonase using a molecular approach to produce high amount of protein, PCR primers were designed based on a sequence alignment of six known rhamnogalacturonases from ascomycetes. Using 5' and 3' RACE experiments, a full-length cDNA was isolated and the corresponding putative aminoacid sequence shared significant identities not only with ascomycete rhamnogalacturonases but also with the purified rhamnogalacturonase from *Irpex lacteus*.

Keywords: rhamnogalacturonase; acetylated rhamnogalacturonan; *Irpex lacteus*

1. Introduction

Pectins, a major component of plant cell wall, are composed of two major regions: homogalacturonans (HG) which can be acetylated and/or methylated, and rhamnogalacturonans (RG) which carry neutral sugars side chains and can also be substituted by acetyl groups. Acetylation can occur on *O*-2 and/or *O*-3 position of the galacturonic acid (GalA) (Ishii, 1995; Voragen *et al.*, 1995). The acetylation degree of pectins, which is defined as the number of acetyl groups present on 100 GalA, depends on the plant source. Pectins isolated from apple and citrus are quite not acetylated (Axelos and Thibault, 1991) whereas pectins from sugar beet and potato tuber are highly acetylated (Voragen *et al.*, 1995). In most of the plants, acetyl substituents are more present in RG than in HG (Ishii, 1995; Schols and Voragen, 1996). The presence and the distribution of these non-sugar substituents influence the association properties of pectins.

Pectin structure can be studied by enzymatic tools. The enzymatic degradation of the pectic backbone includes rhamnogalacturonases and polygalacturonases. Most of the known rhamnogalacturonan-hydrolases have been purified from ascomycete organisms (Schols *et al.*, 1990; Suykerbuyk *et al.*, 1997). They are sensitive to the substrate acetylation and their activity can be hampered towards pectins of certain plants having a high degree of acetylation (Schols *et al.*, 1990; Schols and Voragen, 1994; Kofod *et al.*, 1994). An enzyme able to act on acetylated rhamnogalacturonan will be a powerful tool to study the structure of acetylated pectin and the acetylation pattern.

Irpex lacteus is a wood-decaying basidiomycete that produces numerous cell wall polysaccharide degrading enzymes. Driselase®, a commercial preparation derived from this microorganism presents xylanase, cellulase and polygalacturonase as major activities. By using acetylated rhamnogalacturonans extracted from sugar beet as the substrate, it was shown that it contains a rhamnogalacturonase activity tolerant to acetylated substrate. Purification and molecular strategies have been used to get the relevant protein, the characteristics of this enzyme were studied and a construction was made to clone the protein in a yeast expression system.

2. Material and methods

2.1 Enzymatic assays and substrates

Acetylated RG was prepared by enzymatic degradation of sugar beet pectin (Teknisk SBP Danisco). This preparation was derived from the degradation of sugar beet pectin (Ralet *et al.*, 2005). Sugar beet pectin was dissolved in 50 mM succinate pH 4.5 buffer. Degradation by endopolygalacturonase I from *Aspergillus niger* (purified in the lab from Pectinex AR, Novozymes, Sakamoto *et al.*, 2002), β-galactosidase from *Aspergillus oryzae* (Sigma) and other enzymes cloned from *Aspergillus aculeatus* in *Aspergillus oryzae*: PME (Novozyme-PPJ4300), galactanase (purified in the lab from Novozyme-PPJ4381 by Sepharose DVS according to Lahaye *et al.*, 1991), arabinanase (purified in the lab by Superose 12 from Novozyme-PPJ4381), and arabinofuranosidase (Novozyme-PPJ4494), was performed at 40 °C for 24 h. Degradation products were first applied on a DEAE Sepharose CL-6B column (30 × 2.6 cm; GE-Helathcare), eluted with 50 mM succinate buffer pH 4.5 at a flow rate of 1 ml/min in a linear gradient of NaCl: 0-0.5 M NaCl in 50 mM succinate buffer pH 4.5, 1600mL. 12 ml fractions were collected and analysed for their GalA and neutral sugars content (Thibault, 1979; Tollier and Robin, 1979). The fractions of interest were then pooled and dialyzed against water and 50 mM succinate buffer pH 4.5. The resulting sample was then applied on size exclusion chromatography on Sephacryl-S 200 column (60 × 1.6 cm; GE Healthcare) equilibrated in 50 mM succinate buffer pH 4.5. The sample was eluted at a flow rate of 1 ml/min. 2 ml fractions were collected and analysed for their GalA and neutral sugars content. Fractions of interest were pooled, concentrated to 2.5 mg/ml and conditioned in 50 mM succinate buffer pH 5.0. The acetylated rhamnogalacturonan had a DAc of 45 and a GalA/Rha ratio of 1.5.

Polygalacturonic acid and carboxymethylcellulose (CMC) are from commercial origin (Sigma), others substrates are from the laboratory collection, non acetylated RG (Bonnin *et al.*, 2001), xylan (Bobin-Dubigeon *et al.*, 1997), arabinan (Lahaye and Thibault, 1990), galactan (Lahaye *et al.*, 1991).

Enzymatic assays for depolymerases activities were performed by measuring the release of reducing groups in a reaction mixture containing substrates (5 mg/ml for all substrates except for acetylated RG: 2.5 mg/ml) in 50 mM sodium acetate buffer, pH 5.0, and enzyme sample incubated at 40 °C. Reducing sugars were measured by the method of Nelson (Nelson, 1944; Sturgeon, 1990) using monomer as the standard except for RG using GalA. Enzymatic activities were expressed in nkatal (nkat), one nkat being defined as the amount of enzyme that catalyses the release of 1 nmol of reducing end per second in the conditions described above.

2.2 Enzyme purification

Driselase® from *Irpex lacteus* (Sigma-D9515-25G) was solubilised at 200 mg/ml in 50 mM acetate buffer, pH 5.0 and then centrifuged. The supernatant was used as the starting material. This supernatant was dialysed against 50 mM Tris-HCl, 1.25 M ammonium sulfate, pH 6.5. The dialysate was loaded on a Phenyl-Sepharose CL6B column (55 ml; GE Healthcare) equilibrated with the dialysis buffer. The bound proteins were eluted in 50 mM Tris-HCl buffer, pH 6.5 as buffer A and 50 mM Tris-HCl, 1.25 M ammonium sulfate, pH 6.5 as buffer B: 2 column volumes at 100% buffer B, 1 column volume from 100% to 50% buffer B, 3 column volumes from 50% to 25% buffer B, 1 column volume from 25% to 0% buffer B and at least 2 column volumes at 0% buffer B. The elution was realised at a flow rate of 1.5 ml/min and monitored by absorption at 280 nm. The fractions of interest according to the elution profile were pooled.

After concentration the sample was conditioned in 20 mM acetate buffer, pH 5.0. The enzyme was applied on two Superdex 75 columns (120 ml; GE Healthcare) mounted in series and equilibrated with the same buffer than the sample. The elution was carried out with the same buffer at a flow rate of 0.3 ml/min. The pool containing the activity was dialysed against cold deionized water to be focused in a preparative Rotofor cell (BioRad) in the presence of carrier ampholytes Bio-Lyte 6/8 (BioRad). 20 fractions were recovered and pooled according to the activity profile.

2.3 Enzyme analysis

Protein homogeneity was evaluated by SDS-PAGE with a continuous 10-20% polyacrylamide gel. Protein bands were excised and digested by porcin trypsin (Promega). The peptides obtained were analysed by liquid chromatography-tandem mass spectrometry (nanoLC-MS/MS) (Steen and Mann, 2004), *de novo* sequenced and then homology matched against known peptides in the Swiss-Prot protein database.

To study the pH, the enzyme reaction was performed at various pHs using 50 mM sodium acetate buffer (pH 3.5-6) for non acetylated RG and 50 mM succinate buffer for acetylated RG at 40 °C. To study temperature optimum, the enzyme reaction was performed at various temperatures (20-60 °C) in 50 mM sodium acetate buffer, pH 5.0 for non acetylated RG and 50 mM succinate buffer pH 5 for acetylated RG.

2.4 Cloning of full-length rhamnogalacturonan hydrolase gene

Irpex lacteus strain KY2902 (LGC-Promochem; ATCC 44426) was cultivated in YM medium containing sugar beet pectin (1%) at 26 °C. Each day the enzymatic activity on acetylated RG and not acetylated RG was measured in the culture medium. After 11 days of culture the total RNA was extracted. PCR primers were constructed after blasting the sequences of four RG-hydrolases from *Aspergillus aculeatus* (GenBank accession no. L35499), *Aspergillus niger* (GenBank accession no. X94220 and X94221), and *Botryotinia fuckeliana* (GenBank accession no. U62397), several regions with high similarities were identified. The partial cDNA fragment of *I. lacteus* RG-hydrolase was amplified by RT-PCR. The primers used were RGH3(5'-GGHYTGGAYGGKATTGATGTCTGG-3') and RGH6 (5'-GTNGGDATGGTNGGDATNGGDAT-3'), which were designed according to the amino acid sequences GLDGIDVW and IPIPTIPT respectively. After DNA sequencing, a partial cDNA sequence information was identified. To obtain the full length cDNA, 5' and 3' rapid amplications of cDNA ends (RACE) were performed using the GeneRacer™ kit (Invitrogen). The primers used in 3'RACE were GeneRacer™ 3'(5'-GCTGTCAACGATACGCTACGTAACG-3') and irpex 3RACE (5'-CGGTGCAGAACGTGAACTTGGAGAA-3'). The primers used in 5'RACE were GeneRacer™ 5'(5'-CCAACTCCCACGCAA-3') and irpex 5RACE (5'-CGGTGCTCTGGCTACTCCAATACTGA-3'). By aligning the sequences of the 3' and 5' RACE products, the full-length cDNA sequence of a RG-hydrolase was deduced and obtained through RT-PCR using the specific primers (forward primer 5'-AAAGAAGTTCCACCCACTTCAAAGGA-3' and reverse primer 5'-TACATTAAATCGTGTCTCATGACGTC-3'). The full-length product was purified and cloned in the pGEM-T easy vector (Promega). Proper construction of the vector containing full-length *I. lacteus* RG-hydrolase cDNA was confirmed by restriction digestion and DNA sequencing and was named pIRPEXFL.

2.5 Construction of the expression plasmid and transformation of P. pastoris

The cDNA encoding the mature *I. lacteus* RH-hydrolase gene was PCR amplified using the specific primers (irpexEcorI: 5'-CCGGAATTCCAGCTGTCTGGGTCCGTGGG-3' and irpexXbaI: 5'-GCTCTAGATTAGGCGCTATCAATAGCAGC-3') and the recombinant pIRPEXFL as a template. Primer irpexEcorI introduced an EcoRI site, and primer irpexXbaI contained an XbaI site (underlined in the above sequence).

After digestion with EcoRI and XbaI, the PCR product was inserted into the vector pPICZαA (Invitrogen). Proper construction was confirmed by restriction digestion and DNA sequencing and was designated as pPICZαA-RGH.

The recombinant plasmid pPICZαA-RGH was linearized with PmeI (New England BioLabs, USA), then transformation of *P. pastoris* and cultivation in shaken flasks were performed according to the EasySelect Pichia Expression Kit (Invitrogen).

3. Results and discussion

3.1 Enzyme purification

The Driselase® preparation was first characterized. Various activities of degradation of plant cell wall polysaccharides were measured (Table 1). A rhamnogalacturonase activity on RG acetylated to 45% was detected, but no acetylesterase activity. The activity ratio naRG *vs* acRG is 26, showing that *Irpex lacteus* RG-hydrolase is rather tolerant to substrate acetylation.

To purify this new rhamnogalacturonase from Driselase®, a hydrophobic interaction chromatography was first performed, which made it possible to eliminate xylanase and cellulase activities. The pool resulting from this step and containing the rhamnogalacturonase still exhibited polygalacturonase and arabinanase activities. A gel filtration was then performed which resulted in two equal major protein peaks. The second one corresponded to the rhamnogalacturonase activity making possible to remove contaminating proteins present in the first peak (Figure 1). A preparative isoelectric focusing was performed as the last purification step separating RG activity in five peaks and shoulders, so resulting in the isolation of different RG-hydrolases having various isoelectric points: 6.2, 6.7, 7.0, 7.2, and 7.7 (Figure 2). Those pI are quite high in comparison with the pI of the RGase A from *Aspergillus aculeatus*

Table 1. Cell wall polysaccharides degrading enzymes in Driselase®.

Substrates	Activity (nkat/ml)
acRG	33
naRG	853
PG	1,133
Xylan	1,210
CMC	994
Galactan	82
Arabinan	144
Acetylesterase	0

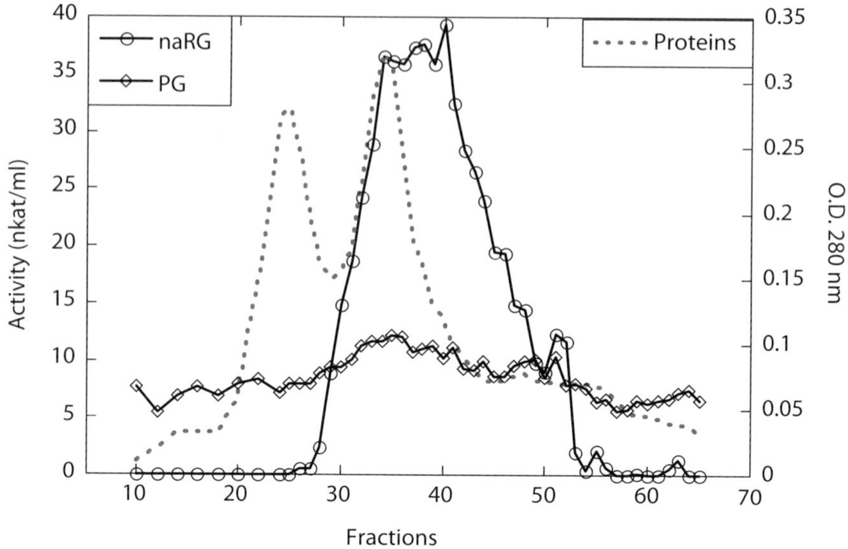

Figure 1. Elution profile after Superdex chromatography.

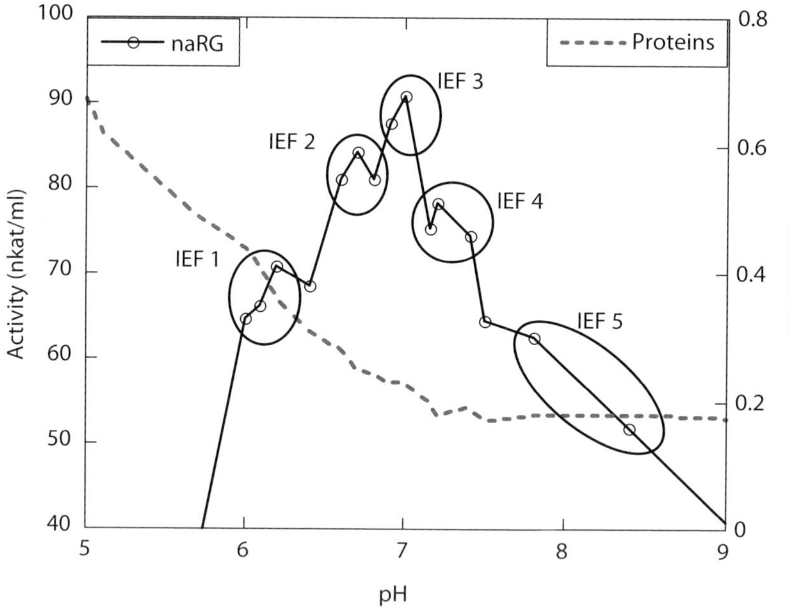

Figure 2. Elution profile after isoelectrofocalisation.

which is of 4.5 (Kofod *et al.*, 1994). The purification procedure for RG-hydrolase is summarized in Table 2.

3.2 Enzyme analysis

For each fraction issued from IEF fractionation, the ratio activity naRG/acRG was between 11 and 14, indicating no difference between those fractions. SDS-PAGE analysis (Figure 3) showed 3 bands in fractions IEF 1 and IEF 2, with a Mw of 31, 43 and 55 kDa. There are 2 bands in the fraction 3 where the band at 43 kDa Mw is lacking. There is a single band of 55 kDa in the fractions 4 and 5. Trypsic digestion and peptide analysis by tandem mass (Steen and Mann, 2004) spectrometry and *de novo* sequencing allowed to identify the 43 kDa band as an *Irpex lacteus* peptidase. Four peptides from the 31 kDa protein were homologous to peptides from fungi proteases.

Table 2. Purification of the RG-hydrolase from Irpex lacteus

Sample	Activity (nkat)	Specific activity (nkat/mg)	Yield in activity (%)	Purification rate
Crude preparation	1,402	3		
HIC	351	19	25.0	6.3
Superdex 75	160	26	11.4	8.7
IEF 3	22	149	1.6	49.7
4	18	137	1.3	45.7
5	15	124	1.1	41.3

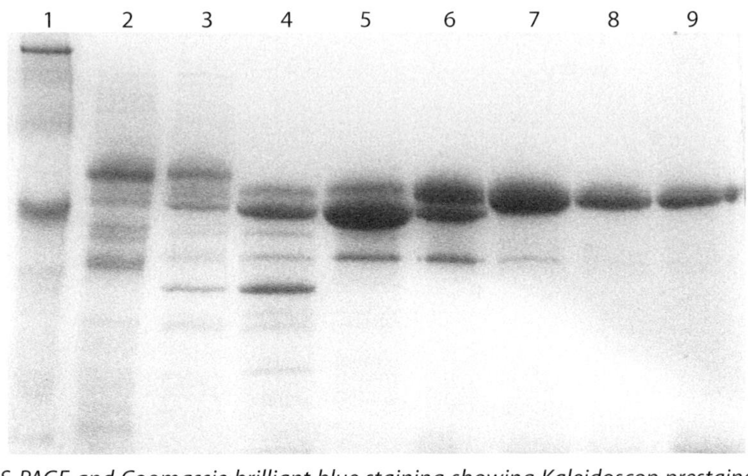

Figure 3. SDS-PAGE and Coomassie brilliant blue staining showing Kaleidoscop prestained molecular weight markers (lane 1), Driselase® mixture (lane 2), HIC pool (lane 3), Superdex pool (lane 4), IEF fraction 1 (lane 5), IEF fraction 2 (lane 6), IEF fraction 3 (lane 7), IEF fraction 4 (lane 8), IEF fraction 5 (lane 9).

The 55 kDa protein showed high peptides identities with the known RG-hydrolases from ascomycetes. In the 15 peptides identified, 11 were identified as RG-hydrolases peptides, with some being contiguous in the sequence. It allowed to identify it as a RG-hydrolase. The peptides analysis gave no difference between the different 55 kDa proteins with different isoelectric points. As there was no difference between IEF fractions, we decided to work further on the IEF 4 fraction as it presented a single band on the SDS-PAGE and it contained the highest amount of protein.

Influence of pH and temperature were examined towards non acetylated RG as well as acetylated RG (Figure 4). The optimum activity occurred between pH 4.5 and 5.0 and the highest activity was obtained for a temperature range from 40 to 50 °C. These results were not dependent on the substrate and were close to those obtained for RG-hydrolases from *Aspergillus aculeatus* and *Aspergillus niger* (Kofod *et al.*, 1994; Suykerbuyk *et al.*, 1997).

3.3 Cloning and expression of a recombinant RG-hydrolase

Sequences of four known RG-hydrolases from *Aspergillus aculeatus*, *Aspergillus niger* and *Botryotinia fuckeliana* were aligned allowing identification of several regions with high similarities. The primer set, RGH3 and RGH6, amplified a 700-bp cDNA product from RNA isolated after 11 days culture where depolymerase activities on RG substrates was maximum. Using 5' and 3' RACE experiments, a full-length 1437-bp cDNA was isolated, containing an open reading frame of 1329-bp. The deduced amino acid sequence corresponds to a 421 amino acid protein with a 22 amino acid signal sequence. The 15 peptides that have been identified from the purified rhamnogalacturonase from *Irpex lacteus* were present in the deduced protein sequence (Figure 5). This sequence was searched against protein databank and shared

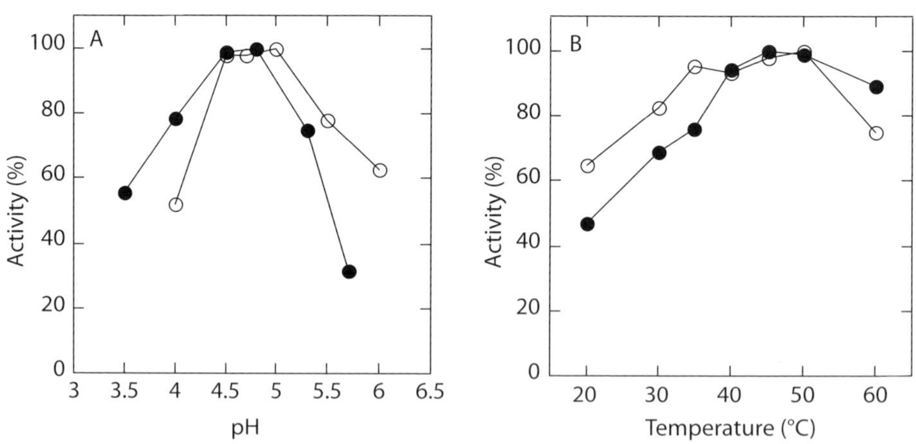

Figure 4. A, Optimum pH of RG-hydrolase, 100% = enzyme activity at optimum pH; B, Optimum temperature of RG-hydrolase, 100% = enzyme activity at optimum temperature.

QLSGSVGPTSSLSSKQGTICNVLNYGGSVGSSDIGPAIGKAFSDCVTKATNGATLYVP
PGNYNMQTWQTLNHGTKWAFQLDGVITRTSTTGGNMIVIQNANDFEFFSSTGKGAI
QGNGYQCRNAGPRLIRVVTSNNWSLHDIIMVDSPEFHLVIQDGSNGEVYNTVIRGGN
LGGSDGIDVWGTNYWIHDIEVTNRDECVTVKSPANHIQVEQIWCNQSGGSAIGSLGA
NTTIQNVLYRNVYTNGGNQIFMIKSNGGSGTVQNVNLENFIARNTAYGLDIDQYWSS
QSTAPGNGVQLKDITFSNWDGFITDGARRAPIQVLCADGAPCTDININNVNLWAANN
QATNKCRSAYGTGACLKSGSGGSYSQVTKTISKPPAFTTPATMSGDLSDGFPTNSPIPI
PTIPPSFFPGTQPLKALAGK

Figure 5. Protein sequence of the Irpex lacteus *RG-hydrolase. In underlined, the sequences encountered from mass spectrometry.*

53% identity with the RG-hydrolases A from *Aspergillus aculeatus* and *Aspergillus niger* (Figure 6). Predicted isoeletric point was 6.2 and predicted Mw was 44.6 kDa. Compared to the 55 kDa native protein, the difference could signify that the protein is glycosylated. The cDNA was cloned in a *Pichia pastoris* pPICZαA expression vector for further production of the recombinant protein.

```
Aspergillusaculeatus      QLSGSVGPLTSASTKGATKTCNILSYGAVADNSTDVGPAITSAWAACK----SGGLVYIP
Aspergillusniger          QLSGSVGPLTSAHSKAATKTCNVLDYGAVADNSTDIGSALSEAWDACS----DGGLIYIP
Irpexlacteus              QLSGSVGPTSSLSSKQGT-ICNVLNYGGSVG-SSDIGPAIGKAFSDCVTKATNGATLYVP
                          ******** :*  :* .*  **:*.**. .. *:*:*.*: .*:  *     .*. :*:*

Aspergillusaculeatus      SGNYALNTWVTLTGGSATAIQLDGIIYRTGTASGNMIAVTDTTDFELFSSTSKGAVQGFG
Aspergillusniger          PGDYAMDTWVSLSGGKATAIILDGTIYRTGTDGGNMILVENSSDFELYSNSSSGAVQGFG
Irpexlacteus              PGNYNMQTWQTLNHGTKWAFQLDGVITRTSTTGGNMIVIQNANDFEFFSSTGKGAIQGNG
                          .*:* ::** :*. *.   *: *** * **.* .**** : ::.***::*.:..**:** *

Aspergillusaculeatus      YVYHAEG-TYGARILRLTDVTHFSVHDVILVDAPAFHFTMDTCSDGEVYNMAIRGGNEGG
Aspergillusniger          YVYHREGDLDGPRILRLQDVSNFAVHDIILVDAPAFHFVMDDCSDGEVYNMAIRGGNSGG
Irpexlacteus              YQCRNAG----PRLIRVVTSNNWSLHDIIMVDSPEFHLVIQDGSNGEVYNTVIRGGNLGG
                          *  :  *    .*::*:   .::::**:*:**:* **:.::  *:***** .***** **

Aspergillusaculeatus      LDGIDVWGSNIWVHDVEVTNKDECVTVKSPANNILVESIYCNWSGGCAMGSLGADTDVTD
Aspergillusniger          LDGIDVWGSNIWVHDVEVTNKDECVTVKGPANNILVESIYCNWSGGCAMGSLGADTDITD
Irpexlacteus              SDGIDVWGTNYWIHDIEVTNRDECVTVKSPANHIQVEQIWCNQSGGSAIGSLGANTTIQN
                          *******:*  *:**:****:*******.***:* **.*:** ***.*:*****:* : :

Aspergillusaculeatus      IVYRNVYTWSSNQMYMIKSNGGSGTVSNVLLENFIGHGNAYSLDIDGYWSSMTAVAGDGV
Aspergillusniger          ILYRNVYTWSSNQMYMIKSNGGSGTVNNTLLENFIGRGNRYSLDVDSYWSSMTAVDGDGV
Irpexlacteus              VLYRNVYTNGGNQIFMIKSNGGSGTVQNVNLENFIARNTAYGLDIDQYWSSQSTAPGNGV
                          ::****** ..**::***********.*. *****.:.. *.**:* **** ::. *:**

Aspergillusaculeatus      QLNNITVKNWKGTEANGATRPPIRVVCSDTAPCTDLTLEDIAIWTESGSSELYLCRSAYG
Aspergillusniger          QLSNITFKNWKGTEADGAERGPIKVVCSDTAPCTDITIEDFAMWTESGDEQTYTCESAYG
Irpexlacteus              QLKDITFSNWDGFITDGARRAPIQVLCADGAPCTDININNVNLWAAN-NQATNKCRSAYG
                          **.:**..**.*  ::** * **:*:*:* *****:.:::. :*: . ..     *.****

Aspergillusaculeatus      SGYCLKDSSSHTSYTTTSTVTAAPSGYSATTMAADLATAFGLTASIPIPTIPTSFYPGLT
Aspergillusniger          DGFCLEDSDSTTSYTTTQTVTTAPSGYSATTMAADLTTDFGTTASIPIPTIPTSFYPGLT
Irpexlacteus              TGACLKSGSGGSYSQVTKTISKPPAFTTPATMSGDLSDGFPTNSPIPIPTIPPSFFPGTQ
                           * **:.... :   .*.*:: .*:  :.:**:.**:  *  .:.*******.**:**

Aspergillusaculeatus      PYSALAG-----
Aspergillusniger          AISPLASAATTA
Irpexlacteus              PLKALAGK----
                          . ..**.
```

Figure 6. CLUSTAL-W alignment of RG-hydrolases sequences from Aspergillus aculeatus *(RGase A; UniProt accession no Q00001),* Aspergillus niger *(rhgA; UniProt accession no P87160) and* Irpex lacteus.

4. Conclusion

We purified a rhamnogalacturonase from the basidiomycete *Irpex lacteus* that can act on highly acetylated substrate. This is a 55 kDa enzyme with a pI of 7.3. It exhibits high identities with RG-hydrolases from ascomycetes. Its optimum pH is 4.5-5 and its optimum temperature is 40-50 °C. At the same time, we isolated a 1437-bp cDNA containing an open reading frame of 1329-bp that corresponded to a 421 amino acid protein. The predicted molecular mass of the deduced protein is 44.6 kDa. The cDNA has been inserted in a *Pichia pastoris* expression vector for production a recombinant RG-hydrolase.

Such an enzyme could be a powerful new tool to be used with other pectinolytic enzymes to study the acetylation pattern and structure of RG.

References

Axelos, M.A.V. and J.-F. Thibault, 1991. Influence of the substituents carboxyl and of the rhamnose content on the solution properties and flexibility of pectins. International Journal of Biological Macromolecules 13: 77-82.

Bobin-Dubigeon, C., C. Hoebler, V. Lognoné, C. Dagorn-Scaviner, S. Mabeau, J.-L. Barry and M. Lahaye, 1997. Chemical composition, physico-chemical properties, enzymatic inhibition and fermentative characteristics of dietary fibres from edible seaweeds. Science des aliments 17: 619-639.

Bonnin, E., M. Brunel, Y. Gouy, L. Lesage-Meessen, M. Asther and J.-F. Thibault, 2001. *Aspergillus niger* I-1472 and *Pycnoporus cinnabarinus* MUCL39533, selected for the biotransformation of ferulic acid to vanillin, are also able to produce cell wall polysaccharide-degrading enzymes and feruloyl esterases. Enzyme and Microbial Technology 28: 70-80.

Ishii, T., 1995. Pectic Polysaccharides From Bamboo Shoot Cell-Walls. Mokuzai Gakkaishi 41(7): 669-676.

Kofod, L.V., S. Kauppinen, S. Christgau, L.N. Andersen, H.P. Heldt-Hansen, K. Dörreich, and H. Dalbøge. 1994. Cloning and characterization of two structurally and functionally divergent rhamnogalacturonases from *Aspergillus aculeatus*. Journal of Biological Chemistry 269: 29182-29189.

Lahaye, M. and J.-F. Thibault, 1990. Purification of arabinanases and galactanases from *Aspergillus niger*. Presented at the 3rd -Workshop on Plant Polysaccharides, Structure, and Function, Le Croisic, France.

Lahaye, M., J. Vigouroux and J.-F. Thibault, 1991. Endo--1,4-D-galactanase from *Aspergillus niger* var. *aculeatus*. Purification and some properties. Carbohydrate Polymers 15: 431-444.

Nelson, N., 1944. A photometric adaptation of the Somogyi method for determination of glucose. Journal of Biological Chemistry 153: 375-380.

Ralet, M.C., J.C. Cabrera, E. Bonnin, B. Quémener, P. Hellìn and J.-F. Thibault, 2005. Mapping sugar beet pectin acetylation pattern. Phytochemistry 66(15): 1832-1843.

Sakamoto, T., E. Bonnin, B. Quémener and J.-F. Thibault, 2002. Purification and characterisation of two exo-polygalacturonases from *Aspergillus niger* able to degrade xylogalacturonan and acetylated homogalacturonan. Biochimica and Biophysica Acta 1572 (1): 10-18.

Schols, H.A. and A.G.J. Voragen, 1994. Occurrence of pectic hairy regions in various plant cell wall materials and their degradability by rhamnogalacturonase. Carbohydrate Research 256: 83-95.

Schols, H.A. and A.G.J. Voragen, 1996. Complex pectins: structure elucidation using enzymes. In: J. Visser and A.G.J. Voragen (eds.) Progress in biotechnology 14, Pectins and Pectinases. Elsevier, Amsterdam, NL, pp. 3-19.

Schols, H.A., C.C.J.M. Geraeds, M.F. Searle-van Leeuwen, F.J.M. Kormelink and A.G.J. Voragen, 1990. Rhamnogalacturonase: a novel enzyme that degrades the hairy regions of pectins. Carbohydrate Research 206: 105115.

Steen, H. and M. Mann, 2004. The abc's (and xyz's) of peptide sequencing. Nature Reviews Molecular Cell Biology 5(9): 699-711.

Sturgeon, R.J., 1990. Monosaccharides. In: P.M. Dey and J.B. Harbone (eds.) Methods in Plant Biochemistry. Academic press, London, UK, Vol. 2, pp 1-37.

Suykerbuyk, M.E.G., H.C.M. Kester, P.J. Schaap, H. Stam, W. Musters and J. Visser, 1997. Cloning and characterization of two rhamnogalacturonases from *Aspergillus niger*. Applied and Environmental Microbiology 63: 2507-2515.

Thibault, J.-F., 1979. Automatisation du dosage des substances pectiques par la méthode au méta-hydroxydiphenyl. Lebensmittel Wissenchaft Und Technologie 12: 247-251.

Tollier, M.T. and J.P. Robin, 1979. Adaptation de la méthode à l'orcinol sulfurique au dosage automatique des oses neuters et totaux. Annales de Technologie Agricole 28: 1-15.

Voragen, A.G.J., W. Pilnik, J.-F. Thibault, M.A.V. Axelos and C.M.G.C. Renard, 1995. Pectins. In: A.M. Stephen (ed.) Food polysaccharides and their applications. Marcel Dekker, New York, USA, pp. 287-339.

Behavior of pectin methylesterases in pectic gels

A. Slavov[1], M.-J. Crépeau[2], S. Durand[2], C. Garnier[2], J.-F. Thibault[2] and E. Bonnin[2]
[1]University of Food Technologies, Dept.Organic Chemistry, 4002 Plovdiv, Bulgaria
[2]INRA, UR1268 Biopolymères Interactions Assemblages, 44300 Nantes, France

Abstract

Pectic gel has been used as a simplified solid matrix to study the behaviour of two pectin methylesterases (PMEs) - one from *Aspergillus aculeatus* and one from orange. Gel was obtained during demethoxylation of a high methoxy lime pectin with a initial degree of methylation of 71 by PME action in the presence of calcium ions. Orange PME action led to moderate decrease of the degree of methylation and rheological measurements showed the formation of strong gels even when the degree of methylation was around 70. On the opposite, fungal PME treatment led to fast decrease of the degree of methylation and, this decrease induced a gel appearance only at the highest activities or for prolonged time, when DM decreased below 20. Degree of blockiness determination showed the presence of block structures of free galacturonic acid residues after treatment by orange PME whereas it was characterized with a rather random pattern after treatment by fungal PME. To explain these results, behaviour of the two PMEs was compared in gel (*i.e.* in the presence of calcium) and solution (i.e. in the absence of calcium). Fungal PME treated-pectins have similar DM in the two conditions. However in the gel, pectins showed higher degree of blockiness, suggesting a less random mode of action of the enzyme. For the plant PME treated-pectins calcium affected not only the degree of blockiness but also the degree of methylation. Indeed in presence of calcium, plant PME treatment led to a lower decrease of degree of methylation but this was accompanied with a more blockwise mode of action since the degree of blockiness was higher in the presence of Ca than in the absence. Possible explanations for calcium-dependent mode of action of PME are changes in visco-elastic properties of the medium, gel formation or influence of Ca on the enzyme-substrate interactions.

Keywords: pectin methylesterases, calcium pectic gels, degree of methylation, degree of blockiness

1. Introduction

Pectin is a complex cell wall heteropolysaccharide, which has wide application as gelling agent in food and non-food industry. Galacturonic acid (GalA) is its main building block and can be partially methylesterified. The degree of methylation (DM = number of methyl groups for each 100 GalA) governs the gelation conditions (Ralet *et al.*, 2002): High Methoxyl pectins (HM, DM>50) gel in the presence of sucrose at low pH values whereas Low Methoxyl pectins (LM, DM<50) gel in the presence of calcium ions in a wider range of pH.

Pectin methylesterases (PME, E.C. 3.1.1.11, CE family 8) hydrolyse the ester bond between methanol and GalA. They have a marked effect in the gelation process as these enzymes release free carboxyl groups able to interact with calcium ions. Their mode of action in soluble medium is well documented (Markovic *et al.*, 1983, Grasdalen *et al.*, 1996, van Alebeek *et al.*, 2003) and it is well known that fungal PMEs cause random distribution of the free GalA whereas plant PMEs lead to blocky structure (Limberg *et al.*, 2000).

In the present work, the pectic gel has been used as a solid matrix to study the behaviour of two PMEs – a fungal one from *Aspergillus aculeatus* and a plant one from orange – during the gel formation and to compare these observations with their modes of action in homogenous soluble medium. The aim is to outline how the de-esterification process affects the gelation kinetics and to investigate if the gel structure affects the mode of action of the two enzymes. The gel is considered as simulating a food matrix as well as a simplified cell wall.

2. Materials and methods

2.1 Materials

P71 is a lime pectin kindly provided by Cargill Texturizing Solution (Redon, France) containing 882.0 mg GalA /g whose 71% is methylesterified.

The orange PME (*O*-PME, Sigma P5400, L'Isle d'Abbeau, France) and the *Aspergillus aculeatus* PME (*Aa*-PME, UniProt Q12535, kindly provided by Novozymes A/S, Copenhagen, Denmark) were both solubilized at 5mg/ml in 2-[N-Morpholino] Ethane-Sulphonic acid (MES) buffer (10 mM, pH 6) and dialyzed overnight at 4 °C against the same buffer. The absence of polygalacturonase and pectin acetylesterase was checked in both enzymes, by the viscometry measurement and determination of the remaining degree of acetylation, respectively.

The endopolygalacturonase (*An*PGII, E.C. 3.2.1.15, UniProt P26214, provided by Novozymes) was from *Aspergillus niger* and was purified in the laboratory as described (Sakamoto *et al.*, 2003).

2.2 Incubation of pectin with PMEs

1.5 ml of a 2% pectin solution in 50 mM MES buffer (pH 6) was preheated to 50 °C and mixed with 1.5 ml of 6 mM $CaCl_2$ in 50 mM MES buffer, pH 6 (or 1.5 ml of 50 mM MES buffer for the experiments made without Ca). The PME solution at the required activity was added and the mixture was incubated at 30 °C for the specified time.

Visual gelation of the system was checked at regular interval by carefully tilting the tube, and was observed until the mixture did not flow any more. Enzymatic reaction was stopped by boiling 5 min and equal volume of 2.5% EDTA was added. Samples were dialyzed with 2.5% EDTA followed by an extensive dialysis with deionized water before to recover pectin using freeze-drying.

2.3 DM determination

Pectin DM was determined by quantification of methanol released by alkaline de-esterification (0.5 M NaOH) for 1h at 4 °C in presence of $CuSO_4$. Reverse phase HPLC was carried out with a C18 Superspher column (Merck) using 4 mM H_2SO_4 as solvent at a flow rate of 0.7 ml/min at 25 °C and with isopropanol as internal standard (Levigne *et al.*, 2002). DM was calculated as the molar ratio of methanol to GalA.

2.4 Enzymatic fingerprinting

Pectins were dissolved at 1 mg/ml in 50 mM acetate buffer pH 4. 0.21 nkat of *An*PGII were added at 24 and at 48 hours and allowed to incubate for 72 hours at 40 °C. Degradation products were analyzed by HPAEC at pH 13 according to Ralet *et al.* (2005) and allowed to calculate the degree of blockiness (DB) of the de-esterified pectin. The DB is calculated from the amount of non methyl-esterified monomer, dimer, trimer released by the enzyme and is expressed as a percentage of the amount of non methyl-esterified GalA present in the pectin (Daas *et al.*, 1999).

2.5 Rheological measurements

The enzyme was added at the necessary activity to a mixture of 3 mM $CaCl_2$ and 1% HM pectin in 50 mM MES buffer, pH 6 at 50 °C and approximately 1.5 ml was transferred onto the rheometer plate preheated at 50 °C.

Rheological measurements were performed at 30 °C using a controlled-stress rheometer (AR2000, TA instruments, Guyancourt, France) equipped with a Peltier temperature device and with a cone-plane geometry (20 mm diameter, 4 ° angle, gap between cone and plane 113 μm). G' (storage) and G" (loss) moduli were recorded to characterize the viscoelastic properties of the systems. Time sweep test was performed at a frequency of 1 rad/s and at a strain amplitude of 1%, followed by a frequency sweep test at the same deformation rate. It was checked that the strain applied was well within the linearity domain of viscoelasticity.

3. Results and discussion

3.1 Time course of enzymatic de-esterification

Time course of de-esterification of P71 by both PMEs at different activities was followed for 24h in the presence of calcium (Figure 1). De-esterification was much faster by *Aa*-PME than by *O*-PME. With *Aa*-PME, a gel was visible after 165 min and 320 min with 7.11 and 3.56 nkat, respectively and is accompanied with a very low residual DM below 20. On the contrary with *O*-PME, a gel was visible from the lowest activity and after a very short incubation time (120 min at 0.24 nkat, 25 min at 1.19 nkat). In these conditions, gels were observed with very high DM close to 70 that is rather unusual for a calcium gel, although is was already described by Ström *et al.*, 2007 for pectin de-esterified by using PME before to the addition of calcium.

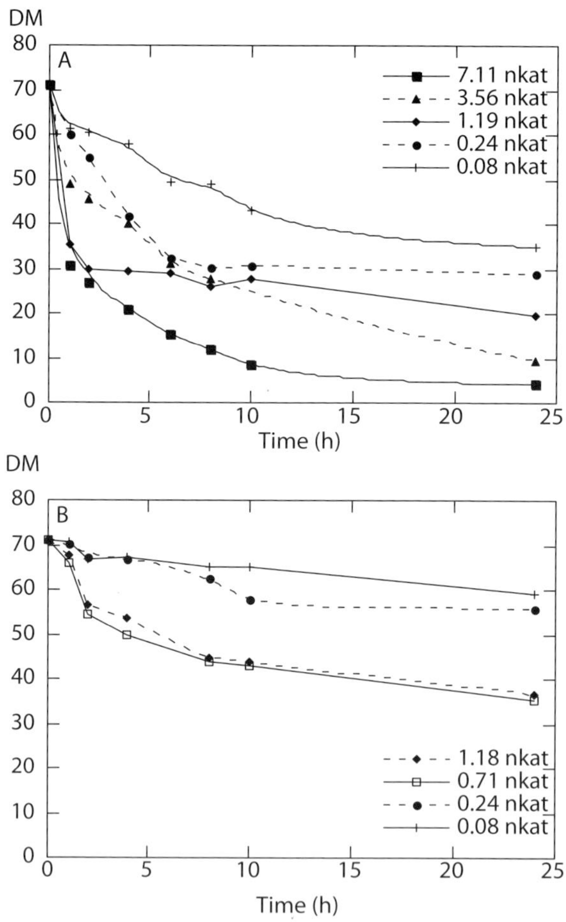

Figure 1. DM kinetics of P71 with different activities of each enzyme. A: Aa-PME; *B:* O-PME.

It is the reason why the time course of gelation was investigated in the presence of the same amounts of each enzyme. G' and G" evolution with time upon addition of 1.19 nkat of PMEs are shown in Figure 2. A very fast increase of G' from the beginning of the incubation was observed in presence of *O*-PME while *Aa*-PME led to a slow evolution of the system. However, G' was over G" for both systems, showing a structuration of the system at this frequency. *O*-PME-system reached a plateau around 2000 Pa in around 6h, whereas *Aa*-PME system did not reached a plateau after 24h incubation and led to a final G' value around 10 Pa.

Analysis of the gel structure was investigated though mechanical spectra after 24h of incubation, as shown for 1.19 nkat of each enzyme on Figure 3. *Aa*-PME action led to a system with G' over G" on nearly all the frequency range, but with G' depending on the frequency at high frequency. However, G' reached a plateau at low frequency, showing a structure of weak gels, with a G' value around 10 Pa. On the contrary, *O*-PME-deesterified system showed a strong gel behavior, G' being well over G", not frequency dependent and around 2000 Pa.

O-PME provoked a slow DM decrease and a very rapid gel evolution to obtain strong gel, while the opposite effect was observed with *Aa*-PME. In relation with what is known about the PMEs modes of action in homogenous soluble medium, it could be thought that the gel structure is determined by the distribution of de-esterified GalA appearing during the enzymatic demethoxylation.

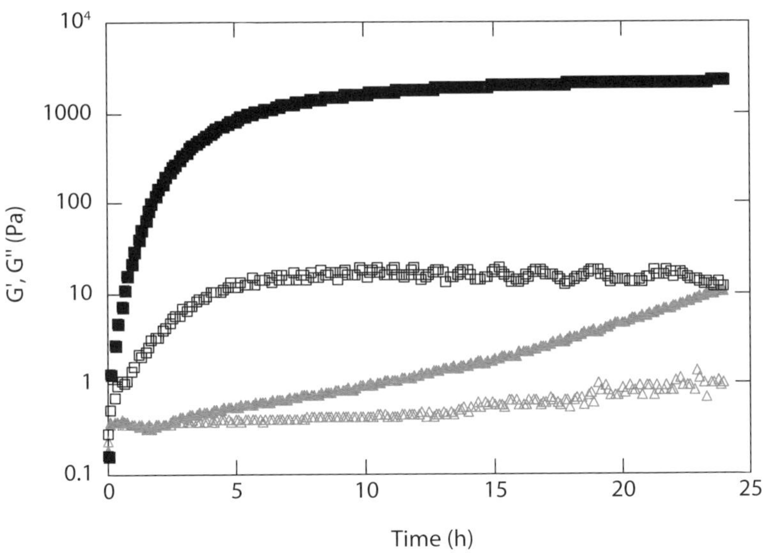

Figure 2. Gel evolution as a function of time during in situ *de-esterification of P71 reacting with 1.19 nkat of each enzyme in the presence of calcium. Full markers, G'; empty markers, G"; gray triangles,* Aa-PME*; black squares,* O-PME*.*

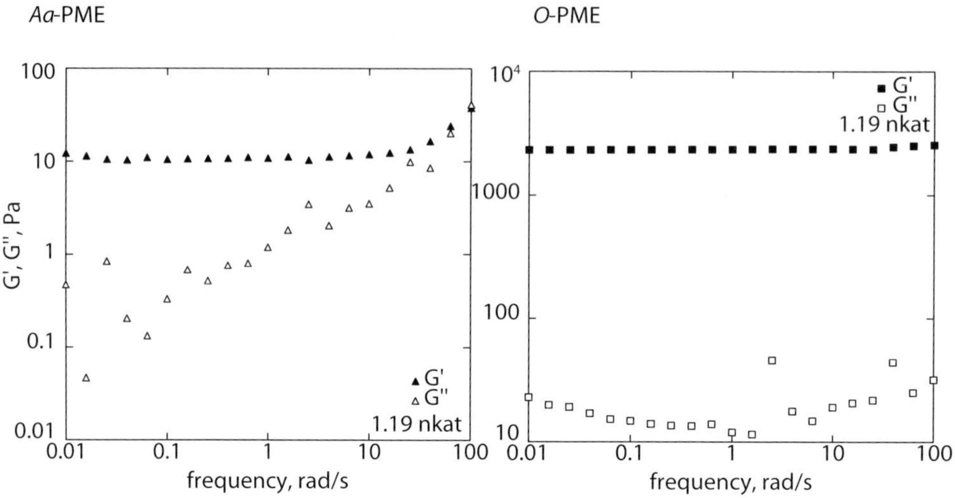

Figure 3. Mechanical spectra of gels obtained by in situ *de-esterification of P71 reacting with 1.19 nkat of each enzyme.*

3.2 Distribution of free GalA on the de-esterified pectins

To investigate this point, enzymatic fingerprinting was performed with the help of *An*PGII, that allows to quantify the more or less blocky structure of the pectin and to calculate the degree of blockiness from the amount of non methyl-esterified monomer (1^0), dimer (2^0), trimer (3^0) released by the enzyme and is expressed as a percentage of the amount of non methyl-esterified GalA present in the pectin. Thus, the DB will be higher for blockwise distribution than for random distribution (Daas *et al.*, 1999).

The effects of *Aa*-PME and *O*-PME, both at 1.19 nkat, on DB were compared and tentatively related to DM and gel formation (Figure 4). A very rapid decrease of DM with *Aa*-PME was accompanied with a slow increase of DB and G', whereas with *O*-PME, the DM decreased slowly since DB and G' increased very fast. This behaviour was markedly observed in the two first hours of the reaction. During these experiments, a gel was visible after 22h with *Aa*-PME, when DM was 20 and DB was 30.5, and corresponded to G' around 7 Pa. With *O*-PME, a visible gel was observed after 20 min when DM was 70 and DB was 17, and corresponded to G' around 18 Pa.

Once again, we highlight here the specificity of *O*-PME to create block structures and the role of these blocks in the gel formation. As previously mentioned by Löfgren *et al.* (2005), the kinetic behaviour is a key point to understand the gel formation with HM pectins. In their work, the macromolecular aggregation of the pectin network occurs during the first hours after the addition of calcium and sucrose, and the further G' increase can be related to internal strengthening without any visible effects on the gel microstructure.

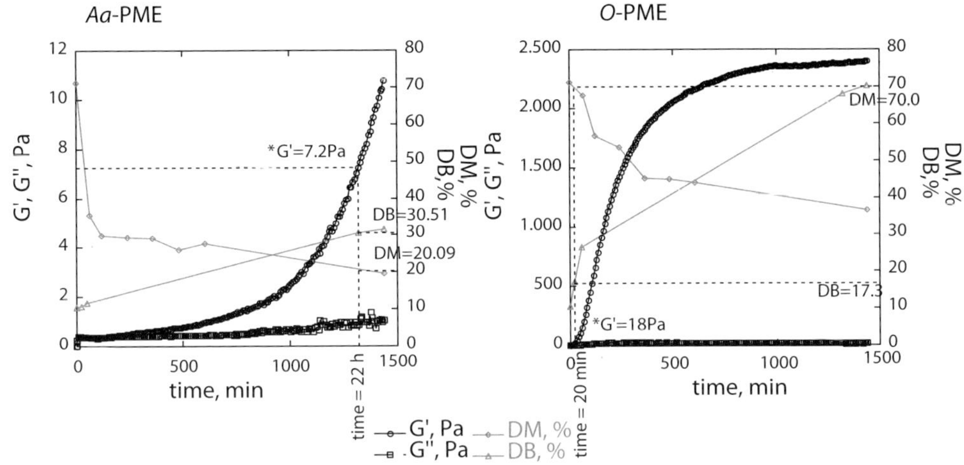

Figure 4. DM (lozenges), DB (triangles) and gel evolution (circles: G', squares: G'') of systems containing P71 and Aa-PME *and* O-PME *at 1.19 nkat. Visual gelation is shown with dotted lines.*

However, it is noticeable that *O*-PME-de-esterified system gelled although pectin DM was 70 and DB value lower than in *Aa*-PME system. DM and DB are both involved in the gelation process but none of them is directly related with gel strength. This can not be explained with what is known about pectin gelation in the presence of calcium and this could suggest that either calcium ions and/or viscoelastic properties of the matrix play a role on the enzyme mode of action.

3.3 Role of calcium and viscoelastic contraints on enzyme mode of action

The activity of both enzymes was measured on P71 in the presence and in the absence of calcium. The specific activity of *Aa*-PME was 195.2 nkat/mg and 167.6 nkat/mg, in the presence of 6 mM and in the absence of calcium, respectively. In the same conditions, the specific activity of *O*-PME was 124.1 nkat/mg and 67.1 nkat/mg, respectively. That shows that *Aa*-PME activity was slightly higher than that of *O*-PME, and that both enzymes showed higher activity in the presence of Ca, which is in accordance with the literature (Chirstgau *et al.*, 1996; Micheli, 2001).

P71 was incubated with *Aa*-PME and *O*-PME at 0.24 or 1.19 nkat for different times, either in the presence or in the absence of calcium. For each point, DM and DB were measured, as well as the gel strength for the samples incubated in the presence of calcium. When DB is plotted as a function of DM, the two enzymes gave a quite linear relation between the two parameters (Figure 5). The effect of calcium was the same for both enzymes, showing a higher DB in presence of calcium for a given DM.

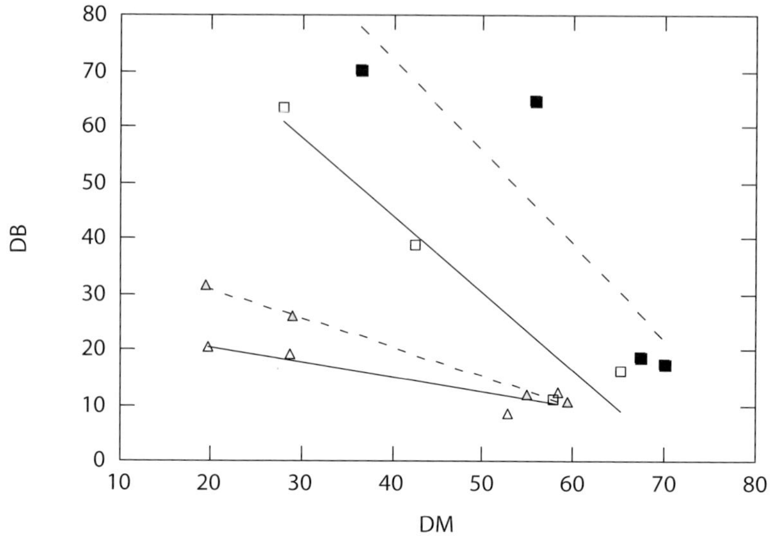

Figure 5. DB and DM evolution during incubation of P71 with Aa-*PME (triangles) or* O-*PME (squares) in the absence (full lines) or in the presence (dotted lines) of calcium.*

However, calcium seems to be much more effective on *O*-PME, as for example a DM of 40 corresponded to a DB of 44 without calcium and more than 70 with calcium. With *Aa*-PME, a given DM of 30 corresponded to a DB of 20 without calcium and 26 with calcium. Finally, in the presence of calcium the processivity of both enzymes was enhanced but much more on *O*-PME that has already a processive mode of action without calcium, than on *Aa*-PME, that is not processive without calcium. Indeed, DB of pectins demethylated by (plant or fungal) PMEs in gel and in solution differs. An effect of Ca^{2+} and Zn^{2+} was shown for *Erwinia chrysanthemi* (Laurent *et al.*, 2000) and the authors attributed this effect to the gel formation. It can be thought that strong gel structure hinders the PME diffusion and thus diminishes the enzyme activity. However, porosity measurements were performed on gels made with pectin having similar DM but different methyl distribution (Willats *et al.*, 2001) and it was shown that plant PME treated pectin led to higher gel porosity than fungal PME-treated pectin. Thus, it would be difficult to conclude about the respective influences of either calcium ions or viscoelastic properties of the matrix on the mode of action of the enzymes as calcium ions induces obviously changes in the textural properties of the system.

4. Conclusions

To summarize our results, the fungal PME treatment was shown to induce a fast demethylation and a moderate increase of DB, whereas the plant PME treatment induced a moderate decrease of DM and a rapid increase of DB. Along these treatments, strong gels were obtained either with *O*-PME even at high DM, or with *Aa*-PME with a higher amount of enzyme.

On the other hand, our results show that DB of pectins demethylated by – plant or fungal – PMEs either in gel or in solution differs, even if they did not explain if this observation has to be attributed to changes in visco-elastic properties of the medium, changes in PME properties or changes in PME-pectin complexes.

In the near future, it would be powerful to compare the viscoelastic properties of calcium gels prepared by mixing pectin and PME *in situ* or with pre-deesterified pectins and to investigate the diffusion of the enzymes inside the calcium-pectin gel. These findings would have important implications for the action of PMEs inside the cell wall.

References

Christgau, S., L.V. Kofod, T. Halkier, L.N. Anderson, M. Hockauf, K. Dörreich, H. Dalboege and S. Kauppinen, 1996. Pectin methylesterase from Aspergillus aculeatus: expression cloning in yeast and characterization of the recombinant enzyme. Biochemical Journal, 319, 705-712.

Daas, P.J.H., K. Meyer-Hansen, H.A. Schols, G.A. De Ruiter and A.G.J. Voragen, 1999. Investigation of the non-esterified galacturonic acid distribution in pectin with endopolygalacturonase. Carbohydrate Research, 318, 135-145.

Grasdalen, H., A.K. Andersen and B. Larsen, 1996. NMR spectroscopy studies of the action pattern of tomato pectinesterase: generation of block structure in pectin by a multiple-attack mechanism. Carbohydrate Research 289: 105-114.

Laurent, F., A. Kotoujansky and Y. Bertheau, 2000. Overproduction in *Esherichia coli* of the pectin methylesterase A from *Erwinia chrysanthemi* 3937: One-step purification, biochemical characterization, and production of polyclonal antibodies. Canadian Journal of Microbiology 46: 474-480.

Levigne, S., M. Thomas, M.-C.Ralet, B. Quéméner and J.-F. Thibault, 2002. Determination of the degrees of methylation and acetylation of pectins using a C18 column and internal standards. Food Hydrocolloids 16: 547-550.

Limberg, G., R. Körner, H.C Buchholt, T.M.I.E. Christensen, P. Roepstorff and J.D. Mikkelsen, 2000. Analysis of different de-esterification mechanisms for pectin by enzymatic fingerprinting using endopectin lyase and endopolygalacturonase II from A. niger. Carbohydrate Research 327: 293-307.

Löfgren, C., S.E. Guillotin, H. Evenbratt, H.A. Schols and A.-M. Hermansson, 2005. Effects of calcium, pH, and blockiness on kinetic rheological behavior and microstructure of HM pectin gels. Biomacromolecules 6: 646-652.

Markovic, O., E. Machova and A. Slezarik, 1983. The action of tomato and *Aspergillus foetidus* pectinesterases on oligomeric substrates esterified with diazomethane. Carbohydrate Research 116: 105-111.

Micheli, F., 2001. Pectin methylesterases: cell wall enzymes with important roles in plant physiology. Trends in Plant Science, 6(9), 414-419.

Ralet, M.-C., E. Bonnin and J.-F. Thibault, 2002. Pectins. In: A. Steinbüchel (ed.) Biopolymers. Wiley-VCH Verlag Gmbh, Weinheim, pp 345-380.

Ralet, M.-C., J.C. Cabrera, E. Bonnin, B. Quémener, M.P. Hellin and J.-F. Thibault, 2005. Mapping sugar beet pectin acetylation pattern. Phytochemistry 66: 1832-1843.

Sakamoto, T., E. Bonnin and J.-F. Thibault, 2003. A new approach for studying interaction of the polygalacturonase-inhibiting proteins with pectins. Biochimica Biophyica. Acta 1621: 280-284.

Ström, A., P. Ribelles, L. Lundin, I. Norton, E.R. Morris and M.A.K. Williams, 2007. Influence of pectin fine structure on the mechanical properties of Calcium-pectin and acid-pectin gels. Biomacromolecules 8: 2668-2674.

Van Alebeek, G.-J.W.M., K. van Scherpenzeel, G. Beldman, H.A. Schols and A.G.J. Voragen, 2003. Partially esterified oligogalacturonides are the preferred substrates for pectin methylesterase of *Aspergillus niger*. Biochemistry Journal 372: 211-218.

Willats, W.G.T., C. Orfila, G. Limberg, H.C. Buchholt, G.-J.W.M. van Alebeek, A.G.J. Voragen, S.E. Marcus, T.M.I.E. Christensen, J.D. Mikkelsen, B.S. Murray and J.P. Knox 2001. Modulation of the degree and pattern of methyl-esterification of pectic homogalacturonan in plant cell walls. Implications for pectin methyl esterase action, matrix properties, and cell adhesion. Journal of Biological Chemistry, 276: 19404-19413.

Part 3

Pectins in plant cell walls

Pectins, cell wall biology and the elucidation of functions

Y. Verhertbruggen and J.P. Knox
Centre for Plant Sciences, Faculty of Biological Sciences, University of Leeds, Leeds LS2 9JT, United Kingdom

Abstract

Our knowledge of how the structural and biochemical complexities of pectic polysaccharides are integrated with cell wall microstructures and with cell and tissue biology is limited. Monoclonal antibodies to defined structural elements of pectic polysaccharides are essential tools to map pectic structures in relation to cellular and developmental processes and in relation to taxonomic diversity as a prelude to the elucidation of functions. Restricted patterns of cell wall epitope occurrence within individual cell walls or across sections of plant materials are well documented and absence of an antibody binding has generally been taken as absence of the polymer. Two emerging strands of work indicate that patterns of cell wall epitope occurrence must be interpreted with caution. Firstly, recent work has indicated that cell wall hemicellulose epitopes can be effectively masked by pectin. Secondly, a new monoclonal antibody to 1,5-arabinan, LM13, has binding properties distinct from the previously isolated and established anti-1,5-arabinan monoclonal antibody LM6. This pair of antibodies indicates that the structure of 1,5-arabinan *in muro* is highly regulated and specific cell-related occurrences of the epitopes can vary between organs and between species.

Keywords: immunocytochemistry, monoclonal antibody, rhamnogalacturonan-I, arabinan

1. Introduction

Pectins are structurally complex polymers. In broad terms, components of pectins can be classified as distinct polysaccharides: the major ones being homogalacturonan (HG), rhamnogalacturonan-I (RGI), rhamnogalacturonan-II and xylogalacturonan. Understanding the biological significance and functional roles of these individual polysaccharides within the large ramifying complex pectic networks that occur in primary cell walls remains a challenge. This is due, not least, to the fact that it is difficult to place the structural complexity of pectins in cell and developmental contexts. How do the distinct pectic polysaccharides (and variants of these polysaccharides that can arise in cell walls due to sets of modifying enzymes) occur in the context of individual cell walls or in terms of the functionally and structurally diverse sets of cells walls that comprise most plant organs?

One of the best ways to map the occurrence of pectin structures on to cell walls within tissues and organs is with the use of antibody probes in conjunction with immunocytochemistry procedures. Glycan-directed antibodies usually bind to three-to-five sugar residues and therefore, generally, have the capacity to bind to a diagnostic structural feature of a pectic polysaccharide or to specific structural variants of a pectic polysaccharide. Such probes can detect the appropriate structural element or epitope in sections of plant materials and in some cases at the surface of intact cells or organs (Knox, 1997; Willats *et al.*, 2001a;b). The sets of available monoclonal antibody probes that are directed to pectic polysaccharides have been discussed elsewhere (Willats *et al.*, 2001b; 2006; Verhertbruggen and Knox, 2007) and are part of extensive panels of cell wall-directed molecular probes that are being use to explore cell wall biology, to document the heterogeneity of cell walls at both cell and tissue levels and also to explore cell wall functions in response to genetic, environmental and biotic impacts.

2. Results and discussion

2.1 Cell wall biology: understanding the functions of cell and tissue patterns of cell wall polymers

Pectins are components of all primary cell walls. As molecular probes for cell wall polymers have been applied to intact cell walls in sections of plant materials it has become apparent that cell walls are diverse and that not all cell walls are comprised of equivalent polymers. In particular, components of the pectic polymers appear to be highly developmentally regulated (Willats *et al.*, 2001b; Knox, 2008a). Such a view has significant implications for the consideration of current models of cell wall architectures which often try to project a unitary model of primary cell wall structure and function (Carpita and Gibeaut, 1993; Somerville *et al.*, 2004). For example, primary cell walls are often considered to be comprised of cellulose microfibrils, tethered by evenly distributed hemicelluloses (or cross-linking glycans) and the cellulose-hemicellulose network is embedded in a relatively evenly distributed matrix of pectic polysaccharides. Immunocytochemical approaches with molecular probes indicate that cellulose is evenly distributed within cell walls although variant substructures relating to microfibril crystallinity as evidenced by the binding by a range of carbohydrate-binding modules indicate variations within cell walls and between cell types (Blake *et al.*, 2006). Diverse hemicelluloses and their structural variants may not always be evenly distributed within primary cell walls and in the case of pectins the methyl-esterification status of HG and the arabinan, galactan and arabinogalactan motifs found in the side chains of RGI appear to be extensively developmentally and taxonomically regulated (Willats *et al.*, 2001b; 2003).

Restricted patterns of detection of a cell wall epitope within a cell wall or an organ can result from:

- a restricted occurrence of the polymer (antigen),

- subtle variations in the structure or conformation of the polymer resulting in a particular epitope being present or absent when the polymer is present, or
- the masking of an epitope by the intimate proxmity of another cell wall polymer.

It may not be easy to distinguish between these possibilities. It has often been assumed, including by the current authors, that a section through a plant organ, cutting across cell walls would expose all polymers present and allow detection by antibodies. However, recent work has indicated that, in pea and tobacco stems at least, the access of a monoclonal antibody to xyloglucan is prevented by the presence of pectic HG (Marcus *et al.*, 2008). These observations indicate that in certain cell walls the xyloglucan epitope is masked by the presence of pectic HG. An example of such xyloglucan epitope masking by pectic HG is shown in Figure 1. This set of observations raises important questions relating to how extensive the masking of glycan epitopes in cell walls is. It is now established that xyloglucan is associated with pectin and the two sets of polymers are covalently attached (e.g. Popper and Fry, 2008) so the identified masking may be a specific consequence of the covalent interaction and intimacy of these two sets of polymers in primary cell walls. Alternatively, epitope masking may be a widespread phenomenon with a range of cell wall polysaccharide epitopes masked by pectic HG and also masking by other cell wall polysaccharides.

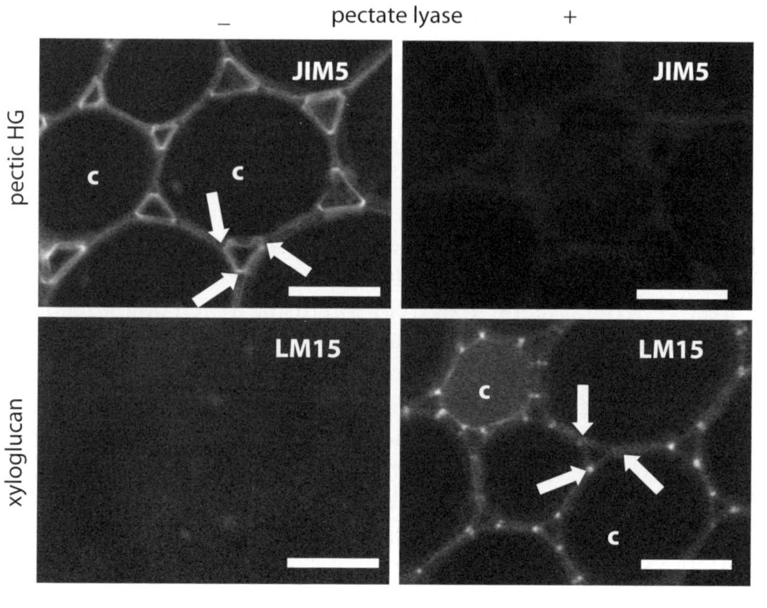

Figure 1. Pectic HG masks xyloglucan epitopes in tobacco stem cell walls. Indirect immunofluorescence detection of the JIM5 pectic HG and the LM15 xyloglucan epitopes in transverse sections of tobacco stem pith parenchyma when sections had been subject to an enzymatic pre-treatment to remove pectic HG. Xyloglucan epitopes, revealed by pectate lyase treatment are abundant in cell walls at the corners of intercellular spaces. c = cell interior. Arrows indicate corners of triangular intercellular spaces. Bar = 20 μm.

For example, the xyloglucan epitopes revealed in tobacco stem pith parenchyma by pectic HG removal are not evenly distributed but are most abundant at cell junctions at intercellular spaces. Are other cell wall polymers involved in masking subsets of xyloglucan in the adhered cell walls in this system or are the xyloglucans distributed unevenly and in a manner responsive to needs relating to cell adhesion rather than control of cell enlargement? These possibilities must be borne in mind when patterns of cell wall polysaccharide occurrence are interpreted.

2.2 A novel antibody to rhamnogalacturonan-I and 1,5-arabinans

Pectins are multifunctional. Of the major polysaccharides that comprise the pectic matrix or network of primary cell walls, RGI is the most obscure in terms of both structure and function. It is comprised of a heterogeneous set of polysaccharides that have a rhamnogalacturonan backbone where rhamnose residues, alternating with galacturonic acid residues, can act as points of attachments for complex and varied sets of arabinans, galactan and arabinogalactans in which β-1,4-linked galactan and α-1,5-linked arabinan are major structural features (Ridley *et al.*, 2001). It is also a matter of debate and remains an open question as to how the rhamnogalacturonan backbone is linked to HG within the pectic networks that ramify within primary cell wall matrices (Vincken *et al.*, 2003; Coenen *et al.*, 2007; Verhertbruggen and Knox, 2007). In terms of function, this subset of the pectic polysaccharide family is variously implicated in the generation/maintenance of the mechanical properties of cells and organs but mechanistic understanding is scant (e.g. McCartney *et al.*, 2000; Jones *et al.*, 2003; 2005).

To determine the cell biological context of RGI molecular probes capable of the specific detection of structural features of this set of polymers within cell walls are required. This is not easy or straightforward as defined oligosaccharide fragments of the range of structural motifs of this set of polymers (for preparation of immunogens and antibody characterization) are not readily available. In the 1990s two rat monoclonal antibodies were generated using neoglycoproteins in which oligosaccharides isolated from 1,4-galactan and 1,5-arabinan side chains were attached to protein to act as immunogens (Jones *et al.*, 1997; Willats *et al.*, 1998). These antibodies, known as LM5 and LM6, have been widely used to detect these structural features of the side chains of RGI in a wide range of plant tissues. Such studies have been useful in strengthening, in our understanding, the links between RGI and aspects of cell and tissue mechanical properties (e.g. McCartney *et al.*, 2000; Jones *et al.*, 2003). Until now, these two antibody probes have allowed the major *in situ* molecular view of 'pectic galactan' and 'pectic arabinan' in cell walls in a range of systems.

Arabinans, although often occurring as a sub-domain within RGI, are a structurally diverse set of polymers within themselves. Linear chains of 1,5-arabinan can be branched with α-1,2- and α-1,3-linked arabinosyl residues and may also be substituted with galactan chains (Beldman *et al.*, 1997; Ridley *et al.*, 2001; Øbro *et al.*, 2004). Arabinans appear to be highly flexible polymers that may fill space in cell wall matrices (Ha *et al.*, 2005) but can also associate with cellulose microfibrils (Zykwinska *et al.*, 2005). Arabinan polymers are proposed to contribute to the maintenance of appropriate cell wall flexibilities and are implicated in guard cell wall function and also in cell adhesion events in plants (Orfila *et al.*, 2001; Iwai *et al.*, 2001; Jones *et al.*, 2003; Peña and Carpita, 2004; Devaux *et al.*, 2005), but how variations in arabinan structures relate to cell wall structures and properties and thus cell functions is unclear.

In a recent screen for novel monoclonal antibodies the hierarchical clustering of antibody binding to a microarrayed collection of defined cell wall polysaccharides led to the identification of a novel arabinan-binding rat monoclonal antibody (Moller *et al.*, 2008). This monoclonal antibody, designated LM13, binds strongly to 1,5-arabinans, the binding is highly sensitive to arabinanases and its binding is optimally inhibited by arabinooligosaccharides with a degree of polymerization greater than seven (Moller *et al.*, 2008). LM13 has some similar but some distinct properties in terms of binding to arabinans to the previously generated and characterized LM6 (Willats *et al.*, 1998). In summary, it appears that LM6 can bind to a short epitope of 1,5-arabinan comprising two to four arabinosyl residues and LM13 requires a longer stretch of 1,5-linked arabinosyl residues (accounting for its enhanced sensitivity to arabinanases). The requirement for a longer region of arabinan for antibody recognition may indicate that the LM13 arabinan epitope has conformational elements i.e. that the arabinan takes on distinct conformations when the polymer is over a certain length or the branches have a particular distribution. Overall binding characteristics suggest that LM6 can bind more readily to a highly branched arabinan and LM13 only to less branched arabinans where five or more unsubstituted residues are available for antibody recognition. Possible epitope structures and how this pair of antibodies could possibly discriminate between arabinan structures are shown schematically in Figure 2.

It is of considerable interest that the monoclonal antibodies LM6 and LM13 do not always bind in an identical way to plant materials. Most strikingly, immunocytochemistry with transverse sections of the *Arabidopsis thaliana* inflorescence stem demonstrates that LM6 binds to all cell walls whereas LM13 binds to epidermal cell walls only as shown in Figure 2. The hypothesis developed from this observation with this pair of arabinan-directed antibodies is that arabinans in epidermal cell walls are less branched than the pectic arabinans in the cell walls of cortical and vascular cells in this organ. The functional basis of this is not known as it is not clear what attributes a more branched or un-branched arabinan would impart to a cell wall. It is clear that the shoot epidermis has a special role in restraining cell expansion in shoot organs (Salvaldi-Goldstein *et al.*, 2007) and the specific location of a structurally-distinct arabinan may reflect this. This observation of a restricted occurrence of the LM13 arabinan epitope is not a case of masking in terms of occlusion by pectic HG but appears to be a subtle structural alteration to a set of polymers across the cell walls of the organ that results in a specific epitope appearing in only certain cell layers at a certain stage of development. We have also observed that the LM13 epitope is not detected in an equivalent way in epidermal cell walls along the developing/growing inflorescence

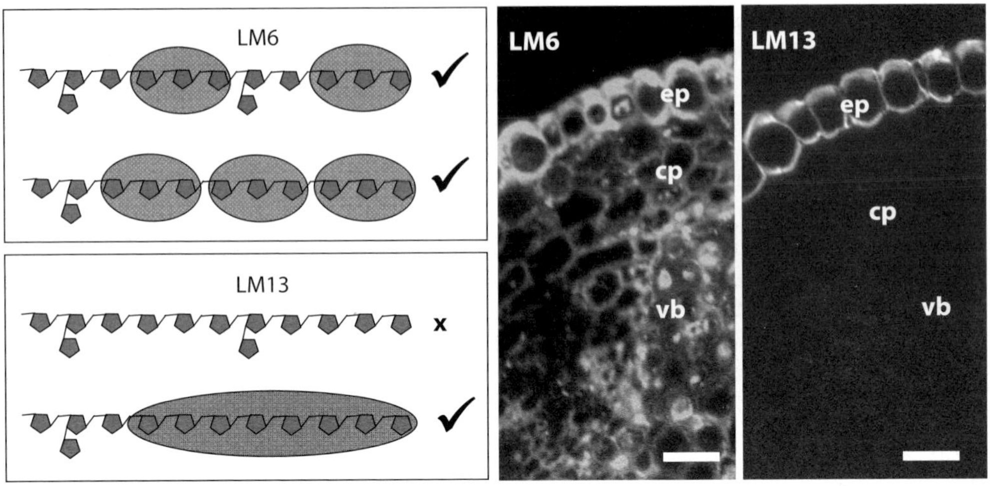

Figure 2. Comparative binding of arabinan-directed monoclonal antibodies LM6 and LM13. Schematic views of the potential for LM6 and LM13 binding to branched and unbranched arabinans. Potential antibody binding is indicated by ellipses over arabinosyl residues. Immunofluorescence detection of the LM6 and LM13 arabinan epitopes in transverse sections of inflorescence stems of Arabidopsis thaliana. *LM6 binds to all cell walls. LM13 binding is restricted to the epidermis. ep = epidermis, cp = cortical parenchyma, vb = vascular bundle. Bar = 20 μm.*

stem. The epitope appears in a gradient down the stem with a higher abundance at stem apices declining towards stem bases. The varying arabinan branching patterns (that we propose lead to diverse functional arabinans) are presumably due to the action of genes/enzymes involved in arabinan synthesis (Harholt *et al.*, 2006) or the *in muro* action of enzymes such as arabinofuranosidases with the potential to trim branches to form more linear arabinans (Chávez Montes *et al.*, 2008). The epidermis is the load bearing cell layer of shoots and distinct mechanical properties may be required to control the growth rate or maintain organ cohesion or flexibility. The role of cell wall arabinan in this cell layer requires further study into the growth and mechanical properties of the *A. thaliana* inflorescence stem and plants with reduced arabinan such as *arad1* with a disrupted putative arabinosyltransferase and 50% reduction in arabinan (Harholt *et al.*, 2006) may be useful in this regard.

It is also of interest that the specific epidermal location of the LM13 arabinan epitope is restricted to the inflorescence stem in *Arabidopsis thaliana* seedlings and does not occur in leaves or in the root. In stems of tobacco the LM13 arabinan epitope occurs in cell walls of all tissue systems, in pea stems in subsets of cells in cortical and vascular tissues and in maize leaves it is restricted to stomatal guard cell walls and some xylem vessels. Such observations on the varied cell- and tissue-based occurrence of the same cell wall epitope in different species lead to questions about the taxonomic aspects of cell wall polymers and, indeed, of the relationships of pectins and other cell wall polysaccharides to evolution and speciation events (Knox, 2008b).

2.3 Molecular imaging of plant cell walls: the challenging way ahead

Pectins do not function in isolation within plant cell walls. The diverse polysaccharide sub-domains of pectins may interact with other cell wall polymers and factors in diverse ways. Untangling the mechanisms of the functional roles of specific pectic structural elements will not be easy. A concerted effort to explore the basis of cell wall heterogeneity as revealed by molecular probes such as monoclonal antibodies will require careful use of an extended set of probes in conjunction with enzymatic dissection of cell wall structures. Such an approach may also provide insight into any functional links between pectins and other polymers within walls in relation to cell wall architectures. For example, does the cell junction abundance of the pectate-lyase revealed xyloglucan (seen in Figure 1) relate to a role for xyloglucan in maintaining cell adhesion or intercellular space in this tissue through association with specific pectin structures, identified by the LM7 HG monoclonal antibody (Willats *et al.*, 2001c), that are known to occur specifically at these regions of cell walls?

A further major aspect of studies aimed at the imaging of cell wall structures is defining what set of polymers/antigens are being detected by a particular probe in a specific immunocytochemical experiment. A glycan epitope may be common to biochemically distinct sets of polymers. This can be a particularly important issue for arabinans as the LM6 arabinan epitope is associated with pectins (RGI) but also with the class of plant proteoglycans known as arabinogalactan-proteins (AGPs) as seen in the moss *Physcomitrella patens* (Lee *et al.*, 2005). AGPs are proposed to act in a range of plant processes but mechanisms of action are unknown (Seifert and Roberts, 2007). The arabinogalactan (AG) structures of AGPs (known as type 2 AG and based on a 1,3-1,6-galactan core) form the major component of these molecules and similar structures have been identified to be associated with pectin (e.g. Renard *et al.*, 1991). The structural and mechanistic details of the relationships between pectins and AGPs are not clear. Can type 2 AG structural features be common to the two distinct sets macromolecules or do AGPs that become attached to pectins during cell processes account for the type 2 AG in pectins? AGPs have been shown to be covalently attached to pectins in carrot cell walls (Immerzeel *et al.*, 2006). It is possible that in a single section of plant material (or even the same cell wall) an antibody such as LM6 may bind to a range of both pectic and AGP molecules. Use of antibody probes in immunocytochemistry procedures with sections of plant materials in conjunction with some form of chromatographic separation of macromolecules isolated from the same material will answer these questions and may, for some epitopes at least, be an important complementary approach to accompany immunolocation experiments.

As we delve further into the molecular architectures and the developmental and physiological aspects of cell walls our appreciation of the complexity of both pectic polysaccharides and cell walls increases. As sets of molecular probes advance in range and scope we should be able to approach an integrated view and detailed knowledge of the structural complexity of sets cell wall polysaccharides such as arabinans (Beldman *et al.*, 1997) in relation to cell wall contexts, cell biological events and cell/tissue functions across growing and developing plant organs.

References

Beldman, G., H.A. Schols, S.M. Pitson, M.J.F. Searle-van Leeuwen and A.G.J. Voragen, 1997. Arabinans and arabinan-degrading enzymes. In: Advances in Macromolecular Carbohydrate Research. Volume 1, pp 1-64. JAI Press, London.

Blake, A.W., L. McCartney, J.E. Flint, D.N. Bolam, A.B. Boraston, H.J. Gilbert and J.P. Knox, 2006. Understanding the biological rationale for the diversity of cellulose-directed carbohydrate-binding modules in prokaryotic enzymes. Journal of Biological Chemistry 281: 29321-29329.

Carpita, N.C. and D.M. Gibeaut, 1993. Structural models of primary cell walls in flowering plants: consistency of molecular structure with the physical properties of the walls during growth. Plant Journal 3: 1-30.

Chávez Montes, R.A., P. Ranocha, Y. Martinez, Z. Minic, L. Jouanin, M. Marquis, L. Saulnier L, L.M. Fulton, C.S. Cobbett, F. Bitton, J.P. Renou, A. Jauneau and D. Goffner, 2008. Cell wall modifications in arabidopsis plants with altered α-L-arabinofuranosidase activity. Plant Physiology 147: 63-77.

Coenen, G.J., E.J. Bakx, R.P. Verhoef, H.A. Schols and A.G.J. Voragen, 2007. Identification of the connecting linkage between homo- or xylogalacturonan and rhamnogalacturonan type I. Carbohydrate Polymers 70: 224-235.

Devaux, M.F., A. Barakat, P. Robert, B. Bouchet, F. Guillon, B. Navez and M. Lahaye, 2005. Mechanical breakdown and cell wall structure of mealy tomato pericarp tissue. Postharvest Biology and Technology 37: 209-221.

Ha, M.-A., R.J. Viëtor, G.D. Jardine, D.C. Apperley and M.C. Jarvis, 2005. Conformation and mobility of the arabinan and galactan side-chains of pectin. Phytochemistry 66: 1817-1824.

Harholt. J., J.K. Jensen, S.O. Sørensen, C. Orfila, M. Pauly and H.V. Scheller, 2006. Arabinan deficient 1 is a putative arabinosyltransferase involved in biosynthesis of pectic arabinan in arabidopsis. Plant Physiology 140: 49-58.

Immerzeel, P., M.M. Eppink, S.C. de Vries, H.A. Schols and A.G.J. Voragen, 2006. Carrot arabinogalactan-proteins are interlinked with pectins. Physiologia Plantarum 128: 18-28.

Iwai, H., T. Ishii and S. Satoh, 2001. Absence of arabinan in the side chains of the pectic polysaccharides strongly associated with cell walls of *Nicotiana plumbaginifolia* non-organogenic callus with loosely attached constituent cells. Planta 213: 907-915.

Jones, L., G.B. Seymour and J.P. Knox, 1997. Localization of pectic galactan in tomato cell walls using a monoclonal antibody specific to (1→4)-β-D-galactan. Plant Physiology 113: 1405-1412.

Jones, L., J.L. Milne, D. Ashford and S.J. McQueen-Mason, 2003. Cell wall arabinan is essential for guard cell function. Proc Natl Acad Sci U S A. 100: 11783-11788.

Jones, L., J.L. Milne, D. Ashford, M.C. McCann and S.J. McQueen-Mason, 2005. A conserved functional role of pectic polymers in stomatal guard cells from a range of plant species. Planta 221: 255-264.

Knox, J.P., 1997. The use of antibodies to study the architecture and developmental regulation of plant cell walls. International Review of Cytology 171: 79-120.

Knox, J.P., 2008a. Revealing the structural and functional diversity of plant cell walls. Current Opinion in Plant Biology 11: 308-313.

Knox, J.P., 2008b. Mapping the walls of the kingdom: the view from the horsetails. New Phytologist 179: 1-3.

Lee, K.J.D., Y. Sakata, S.-L. Mau, F. Pettolino, A. Bacic, R.S. Quatrano, C.D. Knight and J.P. Knox, 2005. Arabinogalactan-proteins are required for apical cell extension in the moss *Physcomitrella patens*. Plant Cell 17: 3051-3065

Marcus, S.E., Y. Verhertbruggen, C. Hervé, J.J. Ordaz-Ortiz, V. Farkas, H.L. Pedersen, W.G.T. Willats and J.P. Knox, 2008. Pectic homogalacturonan masks abundant sets of xyloglucan epitopes in plant cell walls. BMC Plant Biology 8: 60.

McCartney L., A.P. Ormerod, M.J. Gidley and J.P. Knox, 2000. Temporal and spatial regulation of pectic (1→4)-β-D-galactan in cell walls of developing pea cotyledons: implications for mechanical properties. Plant Journal 22: 105-113.

Moller, I., S.E. Marcus, A. Haeger, Y. Verhertbruggen, R. Verhoef, H. Schols, J.D. Mikklesen, J.P. Knox and W. Willats, 2008. High-throughput screening of monoclonal antibodies against plant cell wall glycans by hierarchial clustering of their carbohydrate microarray binding profiles. Glycoconjugate Journal 25: 37-48.

Øbro, J., J. Harholt, H.V. Scheller and C. Orfila, 2004. Rhamnogalacturonan-I in *Solanum tuberosum* tubers contains complex arabinogalactan structures. Phytochemistry 65: 1429-38.

Orfila, C., G.B. Seymour, W.G.T. Willats, I.M. Huxham, M.C. Jarvis, C.J. Dover, A.J. Thompson and J.P. Knox, 2001. Altered middle lamella homogalacturo nan and disrupted deposition of (1→5)-α-L-arabinan in the pericarp of *Cnr*, a ripening mutant of tomato. Plant Physiology 126: 210-221.

Peña, M.J. and N.C. Carpita, 2004. Loss of highly branched arabinans and debranching of rhamnogalacturonan I accompany loss of firm texture and cell separation during prolonged storage of apple. Plant Physiology 135: 1305-1313.

Popper, Z.A. and S.C. Fry, 2008. Xyloglucan-pectin linkages are formed intra-protoplasmically, contribute to wall assembly, and remain stable in the cell wall. Planta 227: 781-794.

Renard, C.M.G.C., A.G.J. Voragen, J.-F., Thibault and W. Pilnik, 1991. Studies on apple protopectin V: structural studies on enzymatically extracted pectins. Carbohydrate Polymers 16: 137-154.

Ridley, B.L., M.A. O'Neill and D. Mohnen, 2001. Pectins: structure, biosynthesis, and oligogalacturonide-related signaling. Phytochemistry 57: 929-967.

Savaldi-Goldstein, S., C. Peto and J. Chory, 2007. The epidermis both drives and restricts plant shoot growth. Nature 446: 199-202.

Seifert, G.J. and K. Roberts, 2007. The biology of arabinogalactan proteins. Annual Review of Plant Biology 58:137-161.

Somerville, C., S. Bauer, G. Brininstool, M. Facette, T. Hamann, J. Milne, E. Osborne, A. Paradez, S. Persson, T. Raab, S. Vorwerk and H. Youngs, 2004. Towards a systems approach to understanding plant cell walls. Science 306: 2206-2211.

Verhertbruggen Y. and J.P. Knox, 2007. Pectic polysaccharides and expanding cell walls. In: J.-P. Verbelen and K. Vissenberg (eds.) Plant Cell Monographs 5, The Expanding Cell. Springer-Verlag Berlin, pp. 139-158.

Vincken, J.-P., H.A. Schols, R.J.F.J. Oomen, M.C. McCann, P. Ulvskov, A.G.J. Voragen, and R.G.F. Visser, 2003. If homogalacturonan were a side chain of rhamnogalacturonan-I. Implications for cell wall architecture. Plant Physiology 132: 1781-1789.

Willats, W.G.T., S.E. Marcus and J.P. Knox, 1998. Generation of monoclonal antibody specific to (1→5)-α-L-arabinan. Carbohydrate Research. 308: 149-152.

Willats, W.G.T., L. McCartney and J.P. Knox, 2001a. *In-situ* analysis of pectic polysaccharides in seed mucilage and at the root surface of *Arabidopsis thaliana*. Planta 213: 37-44.

Willats, W.G.T., L. McCartney, W. Mackie and J.P. Knox JP, 2001b. Pectin: cell biology and prospects for functional analysis. Plant Molecular Biology 47: 9-27.

Willats, W.G.T., C. Orfila, G. Limberg, H.C. Buchholt, G.-J.W.M. van Alebeek, A.G.J. Voragen, S.E. Marcus, T.M.I.E. Christensen, J.D. Mikkelsen, B.S. Murray and J.P. Knox, 2001c. Modulation of the degree and pattern of methyl-esterification of pectic homogalacturonan in plant cell walls: implications for pectin methyl esterase action, matrix properties and cell adhesion. Journal of Biological Chemistry 276: 19404-19413.

Willats, W.G.T., L. McCartney and J.P. Knox, 2003. Pectin cell biology: complexity in context. In: F. Voragen, H. Schols and R. Visser (eds.) Advances in Pectin and Pectinase Research. Dordrecht, Kluwer Academic Publishers, pp. 147-157.

Willats, W.G.T, J.P. Knox and J.D. Mikkelsen, 2006. Pectin: new insights into an old polymer are starting to gel. Trends in Food Science & Technology 17: 97-104.

Zykwinska, A., M.-C. Ralet, C. Garnier, J.-F. Thibault, 2005. Evidence for *in vitro* binding of pectic side chains to cellulose. Plant Physiology 139: 397-407.

Pectins in secondary cell walls: modifications during cell wall assembly and maturation

T.A. Gorshkova, P.V. Mikshina, N.N. Ibragimova, N.E. Mokshina, T.E. Chernova, O.P. Gurjanov and S.B. Chemikosova
Kazan Institute of Biochemistry and Biophysics, Kazan Scientific Center, Russian Academy of Sciences, P.O. Box 30, 420111 Kazan, Russia

Abstract

Pectins are the major components of middle lamellae and primary cell wall. Though in relatively small proportion, they are also present in secondary cell walls and can play there an important role, which received little attention. This is especially true for secondary cell wall of gelatinous type, which is characteristic for many plant fibers. Pectic molecules can be so tightly bound to the secondary cell wall cellulose, that they cannot be extracted even after treatment with concentrated alkali, which confirms their role as an important player in cell wall assembly. Developing flax fibres, while making their secondary cell wall, contain a particular tissue- and stage-specific high-molecular mass pectic galactan, which can be collected from homogenization buffer, and is not fixed yet within the cell wall. The galactan comes out from the specific Golgi-derived vesicles and can be separated from the other polymers by gel filtration due to its considerable molecular weight (900-2000 kDa), providing a rare opportunity to characterize individual polysaccharides before they are incorporated into the cell wall. Moreover, the fiber-specificity of the galactan gives possibility to trace the metabolism of certain polymers in certain cell type, while analyzing whole stem portions. The set of approaches permitted to analyze the composition of several pools: the galactan from Golgi vesicles, the galactan from various fractions of the cell wall and the galactose mono- and oligomers. Label redistribution in pulse-chase experiments indicated that disassembly of the nascent galactan into smaller blocks occurred before or during the interaction with cellulose. A significant amount of free galactose is present specifically in fibers and only at the stage of secondary cell wall formation. The activity of beta-galactosidase, which corresponded in time and place with the content of free galactose was measured. The changes in the galactan structure are discussed from the perspective of its role in cell wall assembly and maturation.

Keywords: secondary cell wall, gelatinous cell wall, pectic galactan, plant fibers

1. Pectins can be the major non-cellulosic polysaccharides in secondary cell walls

Pectins – polysaccharides, containing (1→4)-α-D-galacturonic acid – are known to be among the main components of the middle lamellae and primary cell wall. However, they can also be present in the secondary cell wall, though in relatively small proportions. This is especially true for secondary cell walls of the gelatinous type, which are characteristic for many plant fibers.

The gelatinous type of secondary cell wall is getting to be recognized as a specific type of plant cell wall, which has very peculiar characteristics in terms of composition, structure and way of formation (Mellerowicz *et al.*, 2001; Pilate *et al.*, 2004; Gorshkova and Morvan, 2006; Meloche *et al.*, 2007). The usually considered type of plant cell secondary wall contains three major polymers - xylan, cellulose and lignin in roughly equal proportion. It has a helicoidal orientation of cellulose microfibrils and is usually 1-2 μm thick. The completely different, 'gelatinous', type of secondary cell wall has cellulose as the overwhelming component – it comprises over 80% of cell wall mass. Xylan and lignin are virtually absent and the main matrix polysaccharides are galactose-containing ones. Cell walls of gelatinous type are often extraordinary thick (around 10 μm) and the most striking feature is the axial orientation of all cellulose microfibrils within the huge S2 layer (Roland *et al.*, 1995; Prodhan *et al.*, 1995; Gorshkova and Morvan, 2006). Two types of secondary cell wall can be present within one stem, like in many fibrous crops or tension wood.

Gelatinous secondary cell walls in fibers of several plants contain pectic galactan. It was indicated by monosaccharide analysis of isolated fibers (Mooney *et al.*, 2001; Morvan *et al.*, 2003) and proved by immunolabelling with LM5 antibody (His *et al.*, 2001; Gorshkova *et al.*, 2004; Meloche *et al.*, 2007; Arend, 2008) and biochemical isolation of the polymer or its fragments (Gorshkova and Morvan, 2006; Gurjanov *et al.*, 2008; Chernova, Gurjanov, Gorshkova, unpublished data). Immunolabelling of flax stem cross- and longitudinal sections with LM5 antibody, which binds tetramers of (1→4)-β-D-galactose (Jones *et al.*, 1997), revealed that the galactan is located throughout the whole thickness of secondary cell wall (His *et al.*, 2001; Gorshkova *et al.*, 2004). At fiber maturity, galactose accounts for 3% (Mooney *et al.*, 2001) to 7% (Girault *et al.*, 1997) of dry mass, making 40-80% of non-cellulosic polysaccharides. The minor component of gelatinous secondary cell wall could be polygalacturonic acid (Morvan *et al.*, 2003; Gurjanov *et al.*, 2008).

2. Pectins can be tightly bound to cellulose

The bulk of pectins are usually obtained from a cell wall by treatment with chelating agents, while further extraction may be obtained by alkali. However, some pectic polysaccharides remain un-extracted even after treatment with concentrated alkali, being tightly bound to cellulose. For instance, galacturonic acid (without rhamnose), was found in tomato fruit cell wall residue after alkaline extraction, indicating the presence of polygalacturonans (Rose *et al.*, 1998). Galacturonic acid together with rhamnose and galactose are present in alkali-insoluble residue of flax fiber (Mooney *et al.*, 2001), suggesting that pectic galactan can be tightly bound to cellulose of secondary cell wall.

Polymers, which cannot be extracted by alkali, are thought to be entrapped by cellulose microfibrils during their crystallization as was suggested for primary cell wall xyloglucan (Hayashi, 1989; Pauly *et al.*, 1999; Cosgrove, 1997). Polysaccharides, strongly associated with cellulose can play an important role in the formation of cell wall supramolecular structure, but as a rule, they are characterized only by acid hydrolysis, which permits just to determine monosaccharide composition and it is safe to state that not much is known about the molecular masses and structures of these polysaccharides in native form.

To obtain the polysaccharides, tightly bound to cellulose, in polymeric form, cellulose of the primary cell wall can be digested with cellulase. However, in secondary cell walls of the gelatinous type the cellulose microfibrils are characterized by a larger size of the crystalline unit and a lower surface/volume ratio than in most other plant sources (Šturcová *et al.*, 2004; Muller *et al.*, 2006). Cellulose molecules within the crystal unit are inaccessible to the surrounding solvent and are resistant towards hydrolysis; only one third of flax fiber cell wall could be degraded by cellulase (Girault *et al.*, 2000). To isolate matrix polysaccharides, tightly associated with cellulose, from the cell walls of such types we designed a special protocol (Gurjanov *et al.*, 2008). Residue, remaining after cell wall treatment with chelators and alkali, was dissolved in solution of lithium chloride in N,N-dimethylacetamide. Cellulose was precipitated by water; such treatment turned cellulose I into cellulose II and made it digestible with cellulase, giving the possibility to separate matrix polysaccharides, which remained in polymeric form.

The obtained non-celluosic polymers were fractionated by gel permeation chromatography on Sepharose 4B-CL column (Figure 1a). The major fraction (molecular mass 100-400 kDa), which accounted for around 80% of all non-cellulosic polymers in alkali-insoluble residue, was composed of galactose with the ratio between two other significant monomers, GalA and Rha, being 1.1-1.4. This fraction was heavily labeled with LM5 antibody and was not stained by Yariv reagent. Both ^{1}H and ^{13}C NMR spectra confirmed that the polymer was mainly composed of β-(1→4)-linked galactose. The ratio between t-Gal and 1,4-Gal indicated the presence of long β-(1→4)-galactan chains. A small amount of similar polysaccharide (and also of RGI with only short side chains of galactose) can be extracted by chelators, hot water or DMSO (Davis *et al.*, 1990; McDoughall, 1993; Van Hazendonk *et al.*, 1996). It remains to be elucidated whether the nuances in structure, differences in function or time of deposition distinguish these galactan subfractions.

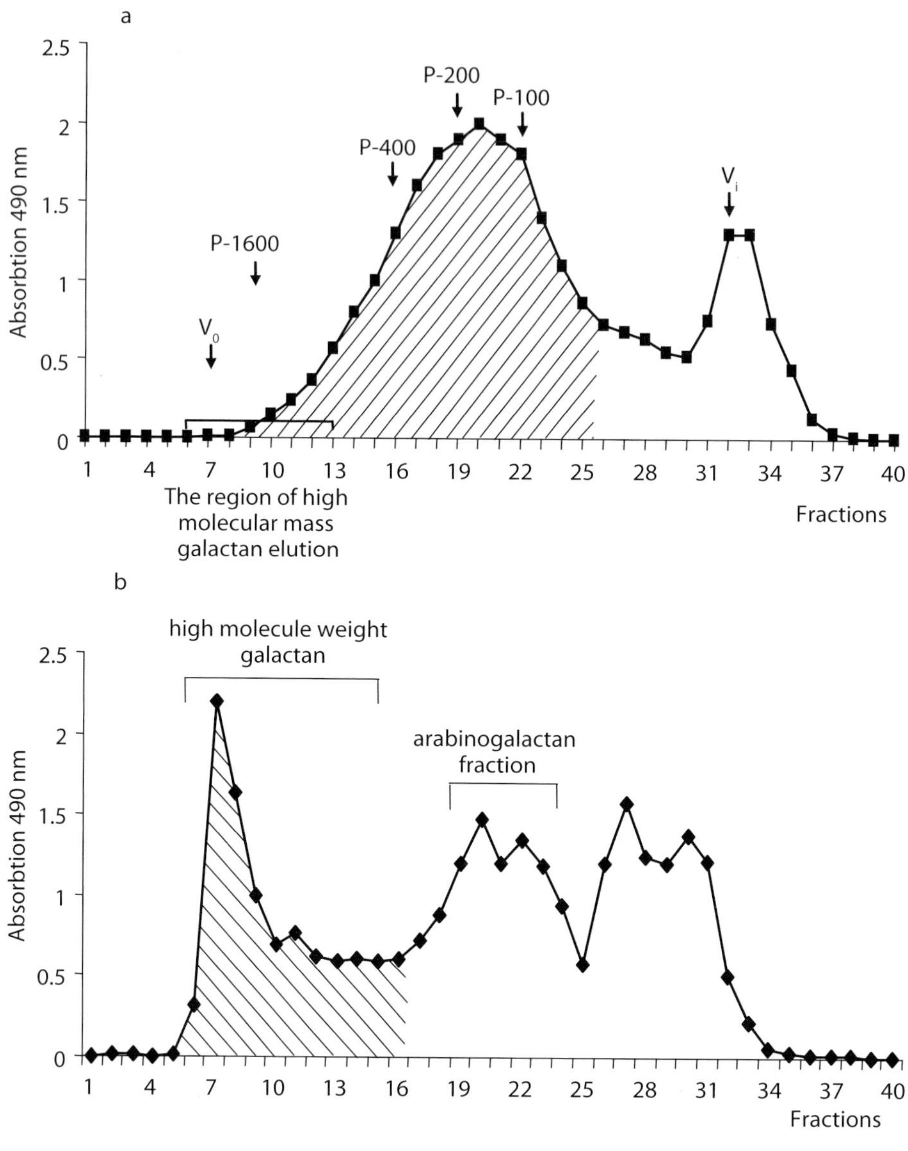

Figure 1. Elution profile of fractions of flax polysaccharides on Sepharose 4B-CL column (Gurjanov et al.*, 2008, with modifications). a) Polymers, tightly bound to cellulose in mature fibers. Calibration was performed using pullulan standards with molecular masses 1600, 400, 200, and 100 kDa. b) Polymers, collected from homogenization buffer of fiber-enriched peels of developing flax plants. Striped part – region of pectic galactan elution.*

Hemp phloem fibers, both primary and secondary ones, which also form secondary cell wall of gelatinous type, contain similar galactan with RGI backbone (though with smaller Gal/Rha ratio), which is also tightly associated with cellulose (Chernova, Gurjanov, Gorshkova, unpublished data). Polygalacturonic acid can be among the minor components of matrix polymers, tightly bound to cellulose in flax phloem fiber (Gurjanov *et al.*, 2008). Thus, pectic molecules can be so tightly bound to the secondary cell wall cellulose, that remain unextracted even after treatment with concentrated alkali, making to consider them as the players in secondary cell wall assembly. This important role of pectins has received little attention yet.

3. Nascent pectic galactan from flax phloem fibers

Developing flax fibres, while making a secondary cell wall, contain a particular tissue- and stage-specific high-molecular mass pectic galactan, which can be collected from homogenization buffer, being not fixed yet within cell wall (Gorshkova *et al.*, 1996; 2004). The galactan comes out from the specific Golgi-derived vesicles (Gorshkova *et al.*, 2005; Salnikov *et al.*, 2008) and can be separated from the other polymers by gel filtration due to its considerable molecular weight (900-2000 kDa) (Figure 1b). A similar type of galactan can be collected from developing primary and secondary fibers of hemp, while forming secondary cell wall of the gelatinous type (Chernova, Gurjanov, Gorshkova, unpublished data).

The possibility of rather easy isolation of nascent polymer provides for a rare opportunity to characterize individual polysaccharides before they are incorporated into the cell wall, avoiding severe treatments (with acids, alkali or enzymes), which are usually required for extraction of polymers from the cell wall and which can modify their structure. Moreover, fiber-specificity of the nascent galactan provides the possibility to trace the metabolism of certain polymers in certain cell types, while analyzing whole stem portions.

The nascent galactan of developing flax fibers was well characterized by methylation analysis (Gorshkova *et al.*, 1996), NMR technique, immunolabelling, MALDI-TOF MS of the fragments, obtained by chemical and enzymatic hydrolysis (Gurjanov *et al.*, 2006, 2007). The polysaccharide is built as a RGI type polymer with high galactose content and very low arabinose content. The rhamnogalacturonan I backbone has a high degree of substitution (>90%). Side chains, mainly composed of (1→4)-β-D-galactose, are divergent in structure and length and include: 1) short branches of only one or two Gal residue(s); 2) long (linear) branches of up to 26 Gal residues; 3) mixed branches of between 3 and 12 Gal residues, that are resistant to galactanase cleavage; 4) side chains of at least 17 Gal residues, decorated with single Ara moieties. Gal/Rha ratio is the same in various parts of the galactan peak (Gurjanov *et al.*, 2007).

Nascent galactan can be obtained in significant amounts due to accumulation of Golgi-derived vesicles in fiber cytoplasm and the peculiar way of their content secretion.

Appearance of these specific vesicles accompanies initiation of gelatinous secondary cell wall formation. On the micrographs they have a characteristic bicolor appearance ('dark' center and 'light' periphery) and are easily distinguishable from the vesicles made in different tissues and during the other stages of fiber development (Salnikov *et al.*, 1993; Gorshkova *et al.*, 2005). Many of the bicolor vesicles fuse each other, forming large bicolor vacuoles, the size of which can reach several micrometers. Both bicolor vesicles and vacuoles are labelled with LM5 antibody. Vesicles and vacuoles later fuse with the plasma membrane and spread their content in a special, 'syringe-like', manner, covering a significant area of periplasm and forming 'dark' strips on the inner cell wall surface (Salnikov *et al.*, 2008).

The metabolism of nascent galactan of flax fibers was traced in pulse-chase studies (Gorshkova *et al.*, 1997, 2005). The label, coming from 40-min photosynthesis of intact plants with $^{14}CO_2$ (pulse) was chased through the various periods of plant growth without labeled substrate. The approach was designed to study the undisturbed metabolism of the polymer *in vivo*. The label was present in the nascent galactan already after the pulse, then accumulated up to the maximal levels at 8 hr of chase-period and later declined; the position of the galactan peak on the elution profile didn't change throughout the experiment (Figure 2). The ratio of galactose and rhamnose radioactivity in the polymer was similar to that in mass, kept at the same level at least through two days of chase period and then declined (Chemikosova, Mikshina, Gorshkova, unpublished data).

The half-life time of the nascent galactan within Golgi derivatives was determined to be around 5 hours (Gorshkova *et al.*, 2005), far exceeding the time, necessary for the polysaccharides to get to the cell wall through 'ordinary' Golgi vesicles in other tissues – which is usually 20-25 min (Fry, 2004). What can be the reason for such unusual behavior? It was suggested that the fiber-specific galactan interacts with the newly formed cellulose microfibrils before they get fully crystallized (Gorshkova and Morvan, 2006). To do so the galactan has to be at the place of cellulose chain synthesis and interaction. However, the cellulose synthase complex can be spatially separated on the plasma membrane with the place of Golgi vesicle fusion. We suggest that the special type of the galactan secretion, which permits to cover the large area of periplasm, is designed to increase the chances of the galactan to meet the cellulose microbirils, while they are still in the process of construction. This may be necessary for the assembly of specific cell walls with axial orientation of cellulose microfibrils to provide the basis for its peculiar mechanical properties.

4. Variability of the galactan structure. Blockwise arrangement?

Tissue- and stage- specificity of the galactan suggests its peculiar role in cell wall formation, suggesting that its structure is rather conserved. We examined the composition of the galactan collected from homogenization buffer of over 50 samples from stems of flax plants of 23 different genotypes (Chernova *et al.*, 2007).

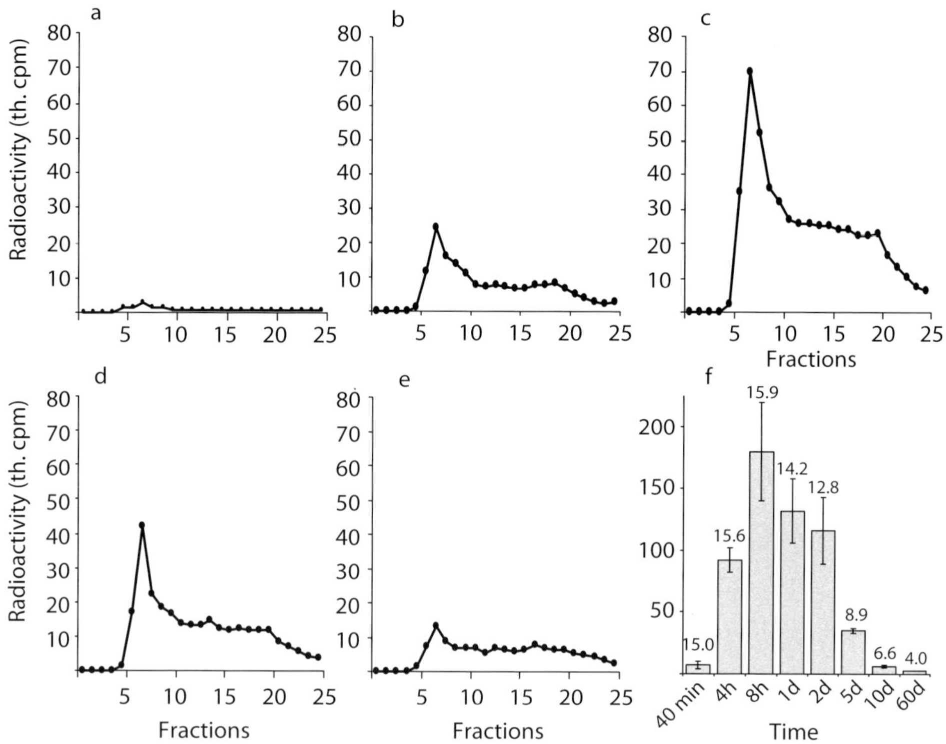

Figure 2. Radioactivity of the nascent galactan in pulse-chase experiments with intact plants. Radioactivity of fractions on elution profile (Sepharose 4B-CL) after a) 40-min photosynthesis of intact plant with $^{14}CO_2$ (pulse); after b) 4 hr, c) 8 hr, d) 1 d and e) 30 d of chase-period. f) Total radioactivity of the galactan (combined fractions 5-11 of the elution profile); number above the column-ratio of the galactan Gal and Rha radioactivity.

A comparison of the content and composition of tissue-specific galactan in fibers of different samples showed their significant variation. For the galactan content, it is not surprising: this parameter is determined by the ratio of the rates of the polymer synthesis and its deposition in the cell wall, and it might be considerably more dynamic than the content of polymer incorporated into the cell wall. However, not only content, but also the structure of this polymer considerably varied: the average length of side chains (Gal/Rha ratio) in the studied samples varied from 5 to 40 galactose residues (Figure 3). The values could be very different within one genotype (for instance, genotypes '*b*' and '*f*', Figure 3a). The Gal/Rha ratio of the galactan can be changed by environmental conditions, i.e. drought (Chemikosova *et al.*, 2006). Despite considerable variation in average length of side chains, the position of the polymer peak on the elution profile was always similar.

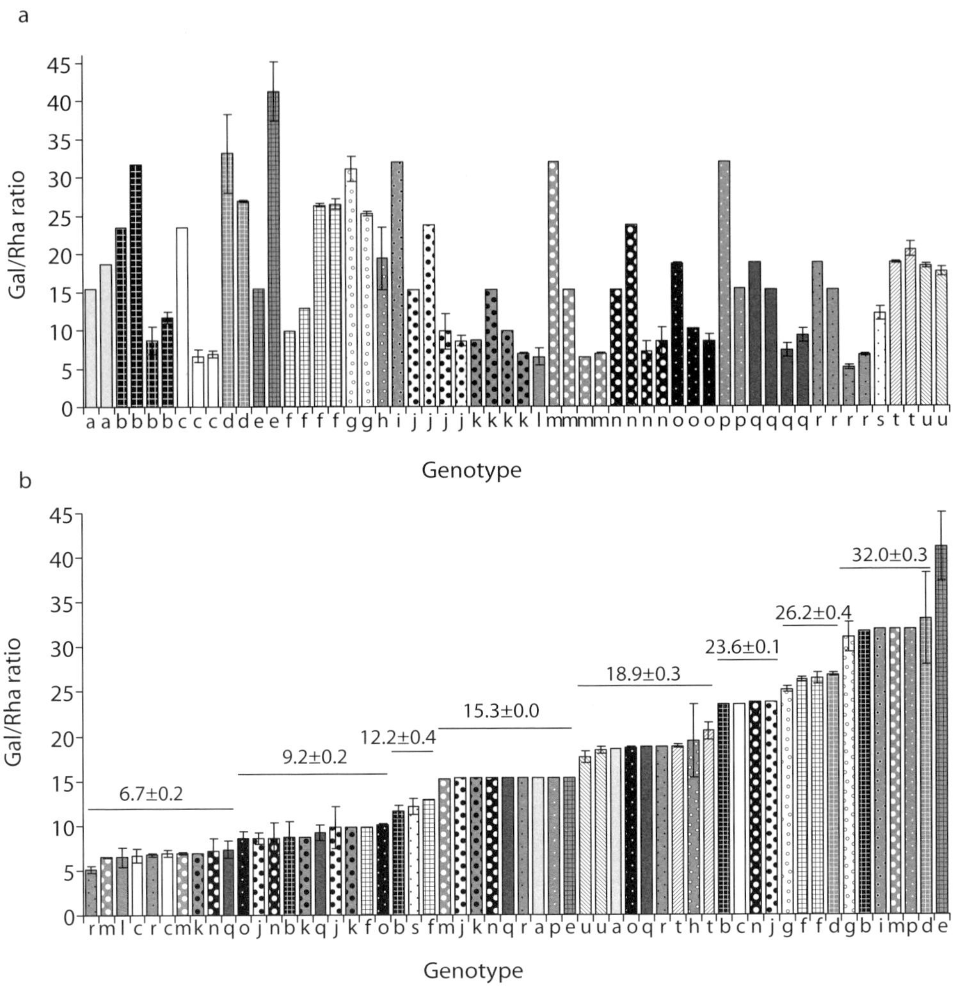

Figure 3. Average length of the nascent galactan side chains in fibers of various flax genotypes (for list of genotypes see Chernova et al.*, 2007). a) Ratio of galactose to rhamnose (mole%:mole%). Different genotypes are designated by different letter. Columns, representing samples of the same flax genotype, have the same filling. b) Ranked raw of the Gal/Rha ratio in various samples. Genotypes have same designation as in a). Numbers above the columns give average for the group values with standard deviations. The difference between the mean values for the groups is significant at* P*<0.001.*

The interesting outcome of this study was that the values of average length of the nascent galactan side chains were discrete: ranking of the Gal/Rha ratio of various samples revealed groups with very close values, which differed with high degree of significance from the neighboring ones (Figure 3b). A stepwise change in the average degree of side chain polymerization during assembly of a significant molecular mass polymer can be realized in the case, when elongation of side chain(s) goes not by addition of single monosaccharides one by one, but by incorporation of galactan block(s) of considerable length. Additional information is needed to prove whether this is the case.

Within flax samples of a single genotype, the level of paratypic variability between years of study and growth conditions was comparable to variability between different genotypes. At the same time the studied polysaccharide is a tissue-specific polymer of those cells, which form the cell wall of the gelatinous type. The question is which characteristics of the polymer determine its functional role?

5. Why the galactan?

The unusual properties of tissue-specific galactan from fibers, which form the cell wall of gelatinous type, can be revealed by treatment of this polymer with specific endogalactanase (Gurjanov *et al.*, 2007) or galactosidase (Mikshina *et al.*, 2008). Upon cleavage of considerable amounts of galactose residues the polymer still elutes at the same position of the profile (Figure 4a-b). The ability to keep the hydrodynamic volume of the molecule after significant shortening of side chains suggests a complex 'secondary' structure of the polymer.

What parts of the polymer could be involved in formation of the special secondary structure? Characteristic features of the galactan from gelatinous fibers could be the following: (a) high degree of polymerization; (b) high percentage of substituted rhamnose residues; (c) combination of long and short side chains in the polymer; (d) specific branching of side chains; (e) the presence of minor not yet characterized components (for instance, ferulic acid); and (f) small amounts of galactose residues with a linkage type, which is different from β-1,4 (for example, β-1,3 or β-1,6). Some of the galactose always remained even after prolong treatment with the endogalactanse or galactosidase (the Gal/Rha ratio didn't go below 3), suggesting that part of Gal residues are inaccessible to enzyme and are probably involved in keeping secondary structure. The phenomenon is more general, as similar properties were observed for the arabinan (Voragen, personal communication), and can be of fundamental importance.

6. Modifications of pectic galactan during cell wall assembly and maturation

Developing secondary cell wall of flax fiber contains two layers, the ratio of which changes with time. The inner layer of the fiber secondary cell wall appears on the micrographs as having a 'loosened' structure with 'dark' stripes (Salnikov *et al.*, 1993; Roland *et al.*, 1995). During fiber development this 'loosened' layer is transformed

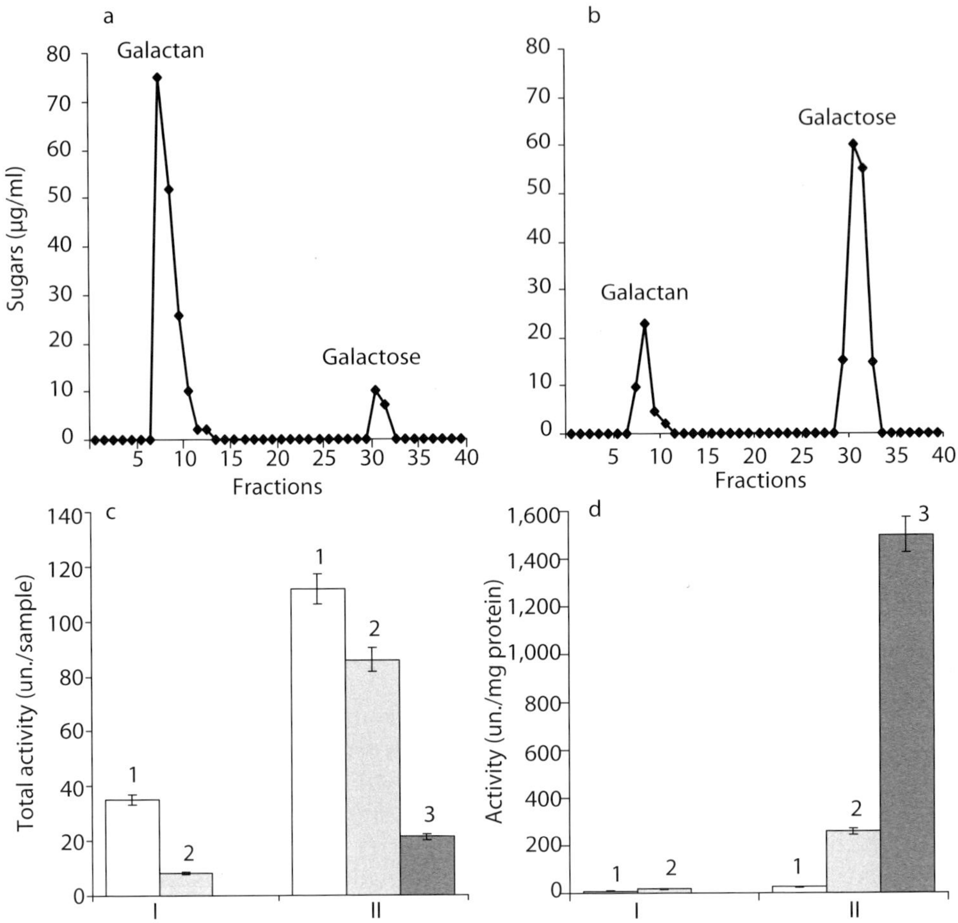

Figure 4. Activity of flax fiber β-galactosidase (Mikshina et al., 2008). a-b) Elution profile of the isolated nascent galactan on the Seharose 4B-CL column after a) 5 h and b) 48 h of incubation with β-galactosidase. c) Total and d) specific activity of β-galactosidase in flax stem parts, containing fibers at intrusive growth stage (I) or at secondary cell wall formation (II); 1: supernatant of tissue homogenate; 2: fraction of polymers, sedimented with 80% ethanol; 3: nascent galactan fraction. Activity of β-galactosidase was determined with o-nitrophenyl-β-D-galactopyranoside.

into a 'solid' one and finally the whole secondary wall becomes 'solid' at fiber maturity (Gorshkova *et al.*, 2004, 2005).

The important role in transformation of cell wall layer structure is ascribed to modification of tissue-specific galactan (Gorshkova and Morvan, 2006). The polymer has a molecular mass around 2000 kDa before incorporation into the cell wall. After incorporation it becomes significantly smaller (100-400 kDa) (Figure 1). Nascent and tightly bound galactans have similar monosacharide composition, types of linkages, ^{1}H and ^{13}C NMR spectrum and are both labelled with LM5, but the mass of tightly bound galactan is roughly 10 times lower, than of the nascent one (Gurjanov *et al.*, 2008). Having in mind the complicated structure of the polymer, which has a set of different side chains and the same Gal/Rha ratio in various parts of its peak on elution profile, the galactan, tightly bound to cellulose, has to be considered as the building block of high-molecular mass galactan from Golgi derivatives. The interaction of polymers within Golgi vesicles to form the high molecular mass complexes was described in several other cases (Crosthwaite *et al.*, 1994; Kerr and Fry, 2003), though the nature of the linkages is still not clear.

How can we imagine relations between 200 kDa and 2000 kDa galactans? Theoretically, there are several possibilities (Figure 5): (1) the larger molecule has longer side chains; (2) several 200 kDa blocks are combined together by covalent linkage of the backbone; (3) the 200 kDa blocks are associated together through interactions of the side chains; (4) the 200 kDa blocks are kept together by some 'third force', like proteins interacting with them, or some lipid envelope. The last option does not seem to be valid as neither the treatment with proteinase, nor extraction with chlorophorm-methanol mixture changed the position of high-molecular mass peak on elution profile. Separation of the polymers, performed in high-salt and low pH buffer, confirmed that the difference in position of the polymers on the elution profiles was not due to spontaneous association of smaller pectin molecules into larger one. It should also be noted that neither alkali, nor cellulase in the concentration and time of application, used to isolate the galactan, tightly bound to cellulose, had any noticeable effect on the isolated high-molecular mass galactan from homogenization buffer.

The probability of variation of the nascent galactan structure during plant development makes it difficult to directly compare its structure with the galactan, tightly bound to cellulose: the nascent galactan is sampled at certain time, while the cell wall galactan accumulated through the whole previous period of secondary wall formation. Still, comparison of the Gal/Rha ratio in these polymers makes it obvious that only trimming of Gal residues can not account for the change in molecular mass. The fragmentation most probably occurs in the RG backbone of the polysaccharide, since the monosaccharide compositions of 2000 kDa and 200 kDa polymers are rather similar. Another option is the breakage of non-covalent linkages, if they keep the building blocks together. The choice between the options is hampered by the ability of the polymer to keep hydrodynamic volume, which makes it difficult to evaluate its molecular mass and to calculate the amount of monosaccharide residues that differentiate the nascent galactan from the one, tightly associated with cellulose.

To check the reasons for such discrepancy in position of the nascent galactan and the one, tightly bound to cellulose, on elution profile, we performed the pulse-chase experiments with intact plants. Composition and label redistribution within several pools – the galactan form Golgi vesicles, the galactan from various fractions of cell wall and the galactose mono- and oligomers, which could be the products of

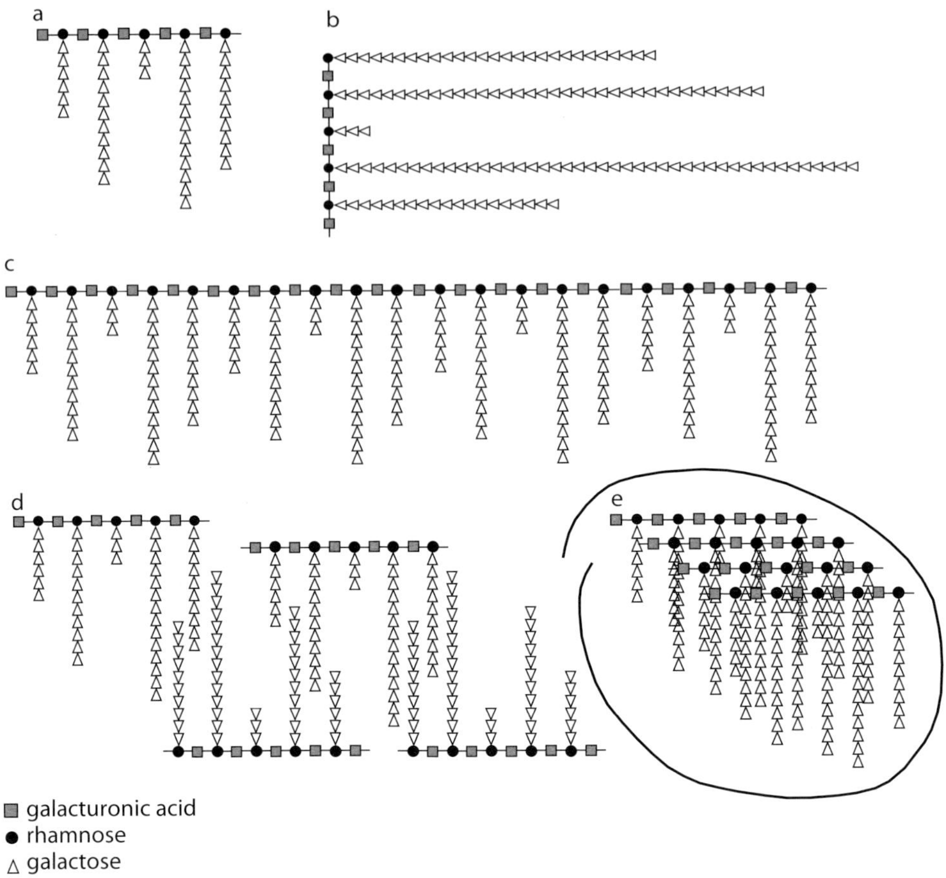

Figure 5. Possible relations between the galactan (200 kDa), tightly bound to cellulose, and the nascent galactan (2000 kDa) from homogenization buffer. a) the galactan (200 kDa), tightly bound to cellulose; b-d) the nascent galactan with b) longer side chains, c) covalent linkages of the backbones, c) interactions of the side chains and d) 'third force', keeping the molecules together.

partial galactan hydrolysis – were analyzed (Chemikosova, Mikshina, Gorshkova, unpublished data). It was found that right after the pulse the label was already detected in the galactan, tightly bound to cellulose. The labeled polymer eluted in the region of 100-400 kDa, meaning that the disassembly of the nascent galactan into smaller blocks occurred before or during interaction with cellulose.

7. Flax fibers contain a lot of free galactose at the stage of secondary cell wall formation

A significant amount of free galactose was found specifically in fibers and only at the stage of secondary cell wall formation (Figure 6) (Mikshina *et al.*, 2008). In flax plants free galactose was not present in the xylem and in those parts of the stem, which contained intrusively growing fibers. Concentration of free galactose, calculated per

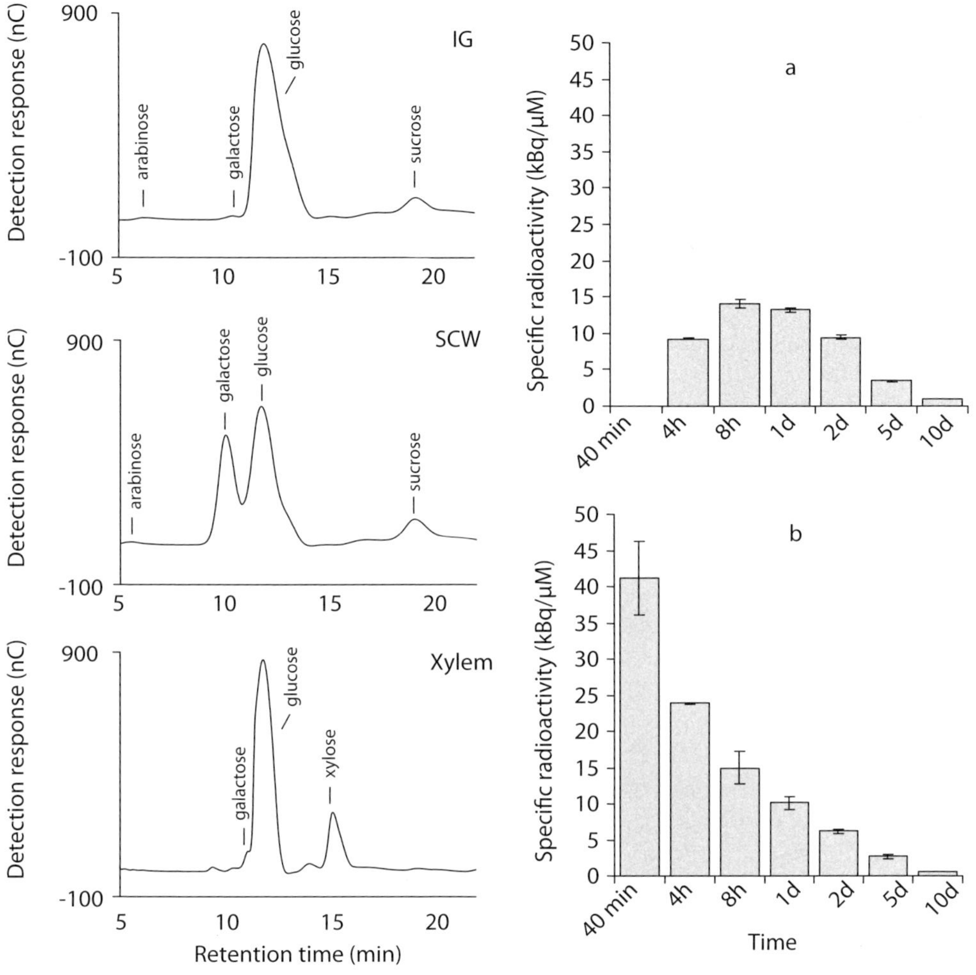

Figure 6. Free galactose in flax fibers at the stage of secondary cell wall formation (Mikshina et al., *2008). (Left) Low molecular sugars on chromatograms of homogenate supernatant in stem part with (top) intrusively growing fibers; (middle) fibers, developing secondary cell wall of gelatinous type; (bottom) xylem. Dynamics of specific radioactivity of a) free galactose and b) free glucose in peels, enriched in fibers, developing secondary cell wall, in pulse-chase experiments with intact flax plants.*

peels, enriched in fibers, developing secondary cell wall of gelatinous type, was 14 mM, so in fibers themselves we can expect around 50 mM. This is high, since the concentration of free galactose in plant tissues usually does not exceed 2 mM (Kim *et al.*, 2007), above 5 mM it is considered to be toxic due to the competitive interrelation of UDP-Gal with UDP-Glu in cellulose biosynthesis (Yamamoto *et al.*, 1988).

The dynamics of specific radioactivity of free galactose in pulse-chase experiments with intact plants was very different from that of primary products of photosynthesis, like glucose (Figure 6a-b), and indicated that this monosaccharide originated from a polymer hydrolysis. No galactose-containing oligomers were found at any stage of the experiment, suggesting that monomeric galactose was cleaved off by exoenzyme.

The activity of β-galactosidase, which corresponded in time and place with the content of free galactose, was measured (Mikshina *et al.*, 2008). The enzyme was able to effectively trim off the galactose residues from the tissue-specific β-1,4-D-galactan (Figure 4a-b). Moreover, the β-galactosidase interacts with this polymer as the enzyme, having molecular mass around 60 kDa, elutes together with the nascent galactan in the region of 2000 kDa. Activity of β-galactosidase in the stem part, which contained intrusively growing fibers, was significantly lower, than in the stem part with fibers, developing secondary cell wall (Figure 4c-d). Analysis of cDNA microarrays to compare the transcription profile in various parts of flax stem revealed the gene of β-galactosidase among the most up-regulated ones at the transition of phloem fibers from intrusive growth to cell wall thickening (Roach and Deyholos, 2007; Ebskamp and Gorshkova, unpublished data). Proteomic analysis of developing flax stem demonstrated that the abundance of β-galactosidase in fibres was notably higher than in the surrounding non-fibre cells of the cortex; the protein showed electrophoretic pattern consistent with post-translational modifications (Hotte and Deyholos, 2008). The prominent role of the β-galactosidase in formation of secondary cell wall of the gelatinous type can be inferred from the above data.

8. Concluding remarks

Formation of cell wall supramolecular structure is a complicated event. Partly it is based on the processes of polysaccharides' self assembly in specific surroundings. Besides, cell wall formation may involve the modification of the interacting polysaccharides. The data to back this assumption are very limited, since it is difficult to trace the individual polysaccharides and to compare it before and after incorporation into cell wall.

The special type of Golgi apparatus activity in flax fibers, while forming secondary cell wall of gelatinous type, permits to obtain the nascent pectic galactan before it is incorporated into cell wall. The fibers can be isolated from flax stem, giving a possibility to analyse cell wall of certain cell type. This model system provides a unique opportunity to trace the metabolism of individual cell wall polysaccharide in certain cell type.

We can assume that the pool of the galactan from Golgi derivatives is divided into two parts. One is entrapped by cellulose microfibrils during their formation or lateral interaction and gets tightly associated with cellulose. The remaining galactan is partly trimmed off by the galactosidase, present in the cell wall, giving rise to free galactose in the periplasmic space. The considerable amount of low molecular weight sugar provides the condition for gelation of the polymer or some other conformational changes. The process may be assisted by arabinogalactan proteins, which are always present in various tissues of the developing flax stem, and are set free in homogenization buffer together with nascent galactan; they are revealed in fibers with gelatinous cell wall both by immunochemistry (Meloche *et al.*, 2007) and biochemical isolation (Girault *et al.*, 2000).

Acknowledgements

The work was partially supported by grants of Russian Foundation for Basic Research (06-04-48853, 08-04-00845 and 09-04-97038).

References

Arend, M., 2008. Immunolocalization of (1,4)-β-galactan in tension wood fibers of poplar. Tree Physiology 28: 1263-1267.

Chemikosova, S.B., N.V. Pavlencheva, O.P. Gur'yanov and T.A. Gorshkova, 2006. The effect of soil drought on the phloem fiber development in long-fiber flax. Russian Journal of Plant Physiology 53: 656-662.

Chernova, T.E., O.P. Gur'yanov, N.B. Brach, A.V. Pavlov, E.A. Porokhovinova, S.N. Kutuzova, S.B. Chemikosova and T.A. Gorshkova, 2007. Variability in the Composition of Tissue-Specific Galactan from Flax Fibers. Russian Journal of Plant Physiology 54: 782-789.

Cosgrove, D.J., 1997. Assembly and enlargement of the primary cell wall in plants. Annual Review of Cell and Developmental Biology 13: 171-201.

Crosthwaite, S.K., F.M. MacDonald, E.A.H. Baydoun and C.T. Brett, 1994. Properties of a protein-linked glucuronoxylan formed in the plant Golgi apparatus. Journal of Experimental Botany 45: 471-475.

Davis, E.A., C. Derouet, H.C. Hervé du Penhoat and C. Morvan, 1990. Isolation and an NMR study of pectins from flax (*Linum usitatissimum* L.). Carbohydrate Research 197: 205-215.

Fry, S.C., 2004. Tansley review: primary cell wall metabolism: tracking the careers of wall polymers in living plant cells. New Phytologist 161: 641-675.

Girault, R., F. Bert, C. Rihouey, A. Jauneau, C. Morvan and M. Jarvis, 1997. Galactans and cellulose in Flax fibres: putative contributions to the tensile strength, International Journal of Boiligical Macromolecules 21: 179-188.

Girault, R., I. His, C. Andeme-Onzighi, A. Driouich and C. Morvan, 2000. Identification and partial characterization of proteins and proteoglycans encrusting the secondary cell wall of flax fibres. Planta 211: 256-264.

Gorshkova, T. and C. Morvan, 2006. Secondary cell-wall assembly in flax phloem fibers: role of galactans. Planta 223: 149-158.

Gorshkova, T.A., S.E. Wyatt, V.V. Salnikov, D.M. Gibeaut, M.R. Ibragimov, V.V. Lozovaya and N.C. Carpita, 1996. Cell-wall polysaccharides of developing flax plants. Plant Physiology 110: 721-729.

Gorshkova, T.A., S.B. Chemikosova, V.V. Lozovaya and N.C. Carpita, 1997. Turnover of galactans and other cell wall polysaccharides during development of Flax plants. Plant Physiology 114: 723-729.

Gorshkova, T.A., S.B. Chemikosova, V.V. Salnikov, N.V. Pavlencheva, T. Stolle-Smits and J.E.G. van Dam, 2004. Occurrence of cell-specific galactan is coinciding with bast fibre developmental transition in flax. Industrial Crops and Products 19: 217-224.

Gorshkova, T.A, M. Ageeva, S. Chemikosova and V. Salnikov, 2005. Tissue-specific processes during cell wall formation in flax fiber. Plant Biosystems 139: 88-92.

Gurjanov, O.P., T.A. Gorshkova, M. Kabel, H.A. Schols and J.E.G. van Dam, 2006. Structural characterization of tissue-specific galactan from flax fibers by 1H NMR and MALDI TOF mass spectrometry. Russian Journal of Bioorganic Chemistry 32: 558-567.

Gurjanov, O.P., T.A. Gorshkova, M. Kabel, H.A. Schols and J.E.G. van Dam, 2007. MALDI-TOF MS Evidence for the Linking of Flax Bast Fibre Galactan to Rhamnogalacturonan Backbone. Carbohydrate Polymers 67: 86-96.

Gurjanov, O.P., N.N. Ibragimova, O.I. Gnezdilov and T.A. Gorshkova, 2008. Polysaccharides, tightly bound to cellulose in cell wall of flax bast fibre: Isolation and identification. Carbohydrate Polymers 72: 719-729.

Hayashi, T., 1989. Xyloglucans in the primary cell wall. Annual Review of Plant Physiology and Plant Molecular Biology 40: 139-168.

His, I., Ch. Andème-Onzighi, C. Morvan and A. Driouich, 2001. Microscopic Studies on Mature Flax Fibers Embedded in LR White: Immunogold Localization of Cell Wall Matrix Polysaccharides. The Journal of Histochemistry and Cytochemistry 49: 1525-1535.

Hotte, N.S.C. and M.K. Deyholos, 2008. A flax fibre proteome: identification of proteins enriched in bast fibre cells. BMC Plant Biology 8: doi:10.1186/1471-2229-8-52.

Jones, L., G.B. Seymour and J.P. Knox, 1997. Localization of pectic galactan in tomato cell wall using a monoclonal antibody specific to (1→4)-β-D-galactan. Plant Physiology 113: 1405-1412.

Kerr, T.V. and S.C. Fry, 2003. Pre-formed xyloglucans and xylans increase in molecular weight in three distinct compartments of a maize cell-suspension culture. Planta 217: 327-339.

Kim H.-O., C. Hartnett, C.H. Scaman, 2007. Free galactose content in selected fresh fruits and vegetables and soy beverages. Journal of Agriculture and Food Chemistry 55: 8133-8137.

McDougall, G.J., I.M. Morrison, D. Stewart, J.D.B. Weyers and J.R. Hillman, 1993. Plant fibres: botany, chemistry and processing for industrial use. Journal of the Science of Food and Agriculture 62: 1-20.

Mellerowicz, E.J., M. Baucher, B. Sundberg and W. Boerjan, 2001. Unravelling cell wall formation in the woody dicot stem. Plant Molecular Biology 47: 329-274.

Meloche, C.G., J.P. Knox and K.C. Vaughn, 2007. A cortical band of gelatinous fibers causes the coiling of redvine tendrils: a model based upon cytochemical and immunocytochemical studies. Planta 225: 485-98.

Mikshina, P.V., S.B. Chemikosova, N.E. Mokshina, N.N. Ibragimova and T.A. Gorshkova, 2008. Free galactose and β-galactosidase activity on different stage of flax fiber formation. Russian Journal Plant Physiol 56: 58-67.

Mooney, C., T. Stolle-Smits, H. Schols and E. de Jong, 2001. Analysis of retted and non retted flax fibres by chemical and enzymatic means. Journal of Biotechnology 89: 205-216.

Morvan, C., C. Andème-Onzighi, R. Girault, D.S. Himmelsbach, A. Driouich and D.E. Akin, 2003. Building flax fibres: More than brick in the walls. Plant Physiology and Biochemistry 41: 935-944.

Müller, M., M. Burghammer and J. Sugiyama, 2006. Direct investigation of the structural properties of tension wood cellulose microfibrils using microbeam X-ray fibre diffraction. Holzforschung 60: 474-479.

Pauly, M., P. Albersheim, A. Darvill and W.S. York, 1999. Molecular domains of the cellulose/ xyloglucan network in the cell walls of higher plants. Plant Journal 20: 629-639.

Pilate, G., A. Déjardin, F. Laurans and J.-Ch. Leplé, 2004. Tension wood as a model for functional genomics of wood formation. New Phytologist 164: 63-72.

Prodhan, A.K.M.A., R. Funada, J. Ohtani, H. Abe and K. Fukazawa, 1995. Orientation of microfibrils and microtubules in developing tension-wood fibres of Japanese ash (Fraxinus mandshurica var. japonica). Planta 196: 577-585.

Roach, M. and M. Deyholos, 2007. Microarray analysis of flax (*Linum usitatissimum* L.) stems identifies transcripts enriched in fibre-bearing phloem tissues. Molecular Genetics and Genomics 278: 149-165.

Roland, J.-C., M. Mosiniak and D. Roland, 1995. Dynamique du positionnement de la cellulose dans les parois des fibres textiles du lin (*Linum usitatissimum*). Acta Botanica Gallica 142: 463-484.

Rose, J. K.C., K.A. Hadfield, J.M. Labavitch and A.B. Bennett, 1998. Temporal sequence of cell wall disassembly in rapidly ripening melon fruit. Plant Physiology 117: 345-361.

Salnikov, V.V., M.V. Ageeva, V.N. Yumashev and V.V. Lozovaya, 1993. The ultrastructure of bast fibers. Russian Journal Plant Physiology 40: 458-464.

Salnikov, V.V., M.V. Ageeva and T.A. Gorshkova, 2008. Homofusion of Golgi vesicles in flax phloem fibers during formation of gelatinous secondary cell wall. Protoplasma.

Šturcová, A., I. His, D.C. Apperley, J. Sugiyama and M.C. Jarvis, 2004. Structural details of crystalline cellulose from higher plants. Biomacromolecules 5: 1333-1339.

Van Hazendonk, J.M., E.J.M. Reinerink, P. de Waard and J.E.G. van Dam, 1996. Structural analysis of acetylated hemicellulose polysaccharides from fibre flax (*Linum usitatissimum* L.). Carbohydrate Research 291: 141-154.

Yamamoto R., M. Inouhe and Y. Masuda, 1988. Galactose inhibition of auxin-induced growth of mono-and dicotyledonous plants. Plant Physiology 86: 1223-1227.

In vitro synthesis of polygalacturonic acid

K. Yasui, J. Jinno, T. Ohashi and T. Ishimizu
Department of Chemistry, Graduate School of Science, Osaka University, Machikaneyamacho 1-1, Toyonaka, Osaka 560-0043, Japan

Abstract

One of the basic backbone units of pectin is polygalacturonic acid (PGA), a homopolymer of α-1,4-linked galacturonic acid (GalA). PGA synthase (PGA α-1,4-galacturonosyltransferase) catalyzes the successive transfer of GalA onto the non-reducing end of oligogalacturonic acid to synthesize PGA. PGA is thought to consist of consecutive GalA residues with the degree of polymerization (DP) around 100 in native molecules. However, it has not been experimentally proven that PGA synthase enables elongation to such a length of PGA chain. *In vitro* synthesis of PGA has been partially successful with production of pyridylaminated oligogalacturonic acids (PA-OGAs) with DP of up to 27. In this article, we describe experimental conditions for generating longer PGA *in vitro*. A microsomal fraction prepared from azuki bean epicotyls was solubilized with CHAPS (20 mM) in a buffer and the solubilized enzyme reacted with PA-OGA of DP 12 and UDP-GalA in a buffer containing 10 mM potassium oxalate. The enzyme was able to transfer more than 30 GalA residues and elongated the chains up to PA-OGA of DP 45.

Keywords: galacturonosyltransferase, glycosyltransferase, oligogalacturonic acid, polygalacturonic acid, chain elongation

1. Introduction

Polygalacturonic acid (PGA), which is a homopolymer of α1,4-linked D-galacturonic acid (GalA), is the main backbone of homogalacturonan and rhamnogalacturonan II in pectin. The molecular sizes of PGAs have been estimated by several methods such as size exclusion chromatography, end group analysis, methylation analysis and electron paramagnetic resonance. The degree of polymerization (DP) of PGA in pectins is reported to be 40-138 (Chambat and Joseleau, 1980; Konno *et al.*, 1986; Irwin *et al.*, 1988; Thibault *et al.*, 1993; Yapo *et al.*, 2007). The question to be answered is therefore how such a long polymer is synthesized in a plant cell.

PGA biosynthesis can be roughly divided into three steps, namely initiation, elongation and termination of polysaccharide synthesis. It is not clear how PGA synthesis is initiated or terminated. The PGA α1,4-galacturonosyltransferase (referred to as PGA synthase) has been shown to be involved in the elongation step of PGA (Ishimizu and Hase, 2004; Mohnen, 2008) and this enzyme has been partially characterized. This glycosyltransferase requires UDP-GalA as a donor substrate (Villemez *et al.*, 1965; Lin *et al.*, 1966) and oligogalacturonic acid (OGA) as an acceptor substrate (Akita *et al.*, 2002; Ishii *et al.*, 2002). The apparent K_m for UDP-GalA varies between plant sources (10^{-6} to 10^{-4} M; Villemez *et al.*, 1966; Bolwel *et al.*, 1985; Doong and Mohnen, 1998; Takeuchi and Tsumuraya, 2001; Akita *et al.*, 2002). This variation may be attributed to the use of crude enzyme solutions. The apparent K_m for OGAs is reported to be in the order of 10^{-5} M (Akita *et al.*, 2002). The shortest OGA that works as an acceptor substrate for this enzyme is pyridylaminated (PA-) OGA of DP 4 (Akita *et al.*, 2002). Given that the reducing end residue of this oligosaccharide is open, this enzyme requires at least trigalacturonic acid for elongation activity. This also means that PGA synthase is not involved in the initial step of PGA synthesis. The enzyme requires submillimolar Mn^{2+} ions and several hundred millimolar of sucrose for maximum activity *in vitro* (Villemez *et al.*, 1965). PGA synthesis occurs at the Golgi membrane and the catalytic site of PGA synthase is directed to its lumenal side (Northcote and Pickett-Heaps, 1966; Zhang and Staehelin, 1992; Sterling *et al.*, 2001). Although this membrane protein can be solubilized with detergents such as CHAPS, TritonX-100, and zwittergent (Doong and Mohnen, 1998; Akita *et al.*, 2002; Crombie *et al.*, 2003) all attempts to purify the enzyme have resulted in failure, although partial purification of the enzyme has been reported (Crombie *et al.*, 2003; Sterling *et al.*, 2006; Ohashi *et al.*, 2007). This is due to extreme instability of the PGA synthase *in vitro*, as has been pointed out since its initial identification (Villemez *et al.*, 1966). This instability has hampered the characterization of PGA synthase. Although we found conditions that increase the *in vitro* stability of PGA synthase (Ohashi *et al.*, 2007), they are still not enough to allow purification. Although genes belonging to glycosyltransferase family 8 have been pointed out as being involved in PGA synthase (Bouton *et al.*, 2002; Leboeuf *et al.*, 2005; Orfila *et al.*, 2005; Sterling *et al.*, 2006), characterization of PGA synthase has been limited because its purification has not been accomplished. It therefore still remains to be clarified how many GalA residues are successively transferred by PGA synthase. We observed that PGA synthase transferred 13 GalA residues successively to OGA *in vitro* (Akita *et al.*, 2002). If this enzyme is involved in PGA chain elongation *in vivo*, it potentially transfers about 100 GalA residues consecutively to produce PGA. We are designing experimental conditions for PGA chain elongation by PGA synthase and in this paper, we describe conditions that allow observation of successive transfer of more than 30 GalA residues to OGA *in vitro*.

2. Results and discussion

2.1 PGA synthase assay method using fluorogenic OGAs

The traditional assay method developed in the 1960s uses radiolabeled UDP-GalA with the enzyme activity being estimated by measuring radioactivity incorporation into trichloroacetic acid or ethanol insoluble materials (Villemez *et al.*, 1965; Lin *et al.*, 1966; Villemez *et al.*, 1966). Saponified enzyme products are digested by a purified clostridial polygalacturonate eliminase, which provides proof that the radioactivity is incorporated into presynthesized pectic substances contained in the crude enzyme solution. However, this method cannot be used to evaluate the enzyme activity accurately because acceptor substrates are not unambiguous. Quantitative evaluation requires exogeneous homogeneous acceptor substrates and separation methods for substrates and products. To this end, we prepared OGAs labeled with a fluorogenic reagent, 2-aminopyridine (PA-OGA) (Akita *et al.*, 2002) (Figure 1).

These oligosaccharides were detected with high sensitivity and separated from one another on a Q-Sepharose anion-exchange column. The fluorescence intensity of each PA-OGA was almost the same (Akita *et al.*, 2002), meaning that PA-OGAs are suitable for quantitative analysis of PGA synthase activity. With respect to donor substrate, UDP-GalA is now available from Carbosource Services, although it is relatively high in price. UDP-GalA can be prepared in the lab by enzymatic methods using UDP-sugar pyrophosphorylase (Kotake *et al.*, 2004) which converts GalA 1-phosphate and

(a)

(b)

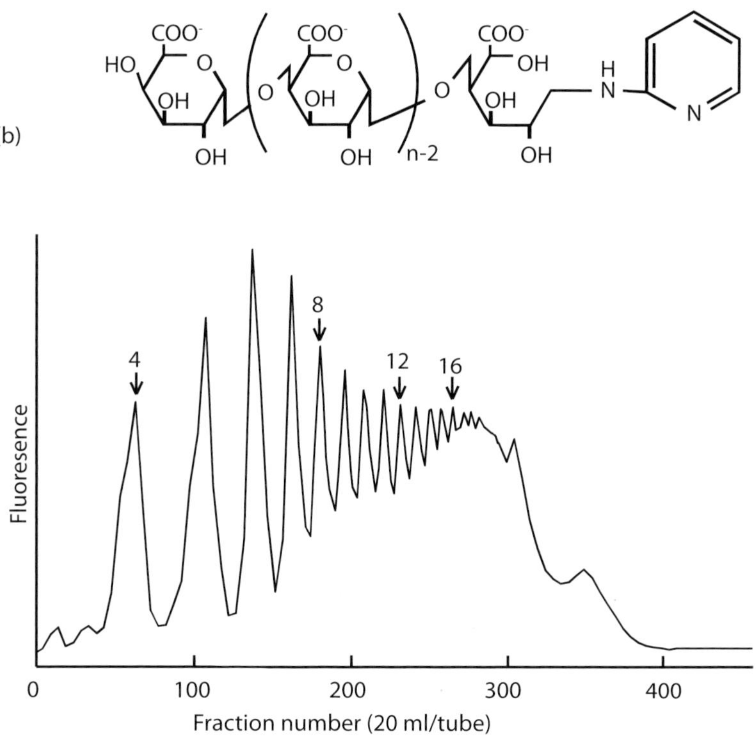

Figure 1. Preparation of PA-OGAs. a) Structure of PA-OGA of DP n. b) A mixture of PA-OGAs was separated on a Q-Sepharose column. Numbers and arrows indicate DP of PA-OGAs and their elution positions, respectively.

UTP to UDP-GalA and pyrophosphate (Ohashi *et al.*, 2006) (Figure 2). Although this conversion is an equilibrium reaction, UDP-GalA is generated quantitatively from GalA 1-phosphate and UTP in the presence of inorganic pyrophosphatase.

We chose the microsomal fraction of petunia pollen tube as an enzyme source, because its growth rate is very high and it is expected that PGA synthase is highly expressed in pollen tubes. When PA-OGA of DP 14 was used as an exogeneous acceptor substrate, the microsomal fraction solubilized with a detergent produced multiple products: PA-OGA of DP 15, 16, 17, and so on (Akita *et al.*, 2002) (Figure 3). With increasing incubation time, PA-OGA chains were elongated to DPs of up to 27. Thus, PGA

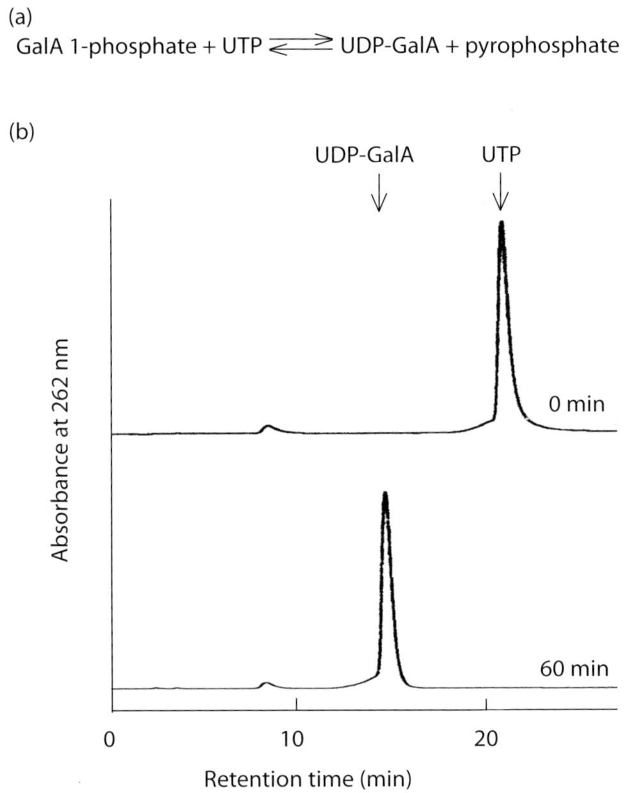

Figure 2. Preparation of UDP-GalA from UTP and GalA 1-phosphate using UDP-sugar pyrophosphorylase. a) Reaction scheme of UDP-GalA preparation using UDP-sugar pyrophosphorylase. In the presence of pyrophosphatase, the forward reaction preferentially occurs. b) UDP-GalA was quantitatively produced from UTP and GalA 1-phosphate. A mixture of 1 mM UTP and 1 mM GalA 1-phosphate was reacted with UDP-sugar pyrophosphorylase in the presence of inorganic pyrophosphatase for 60 min. The reaction was analyzed on a DEAE anion-exchange column by monitoring the absorbance at 262 nm. Arrows indicate the elution positions of UDP-GalA and UTP.

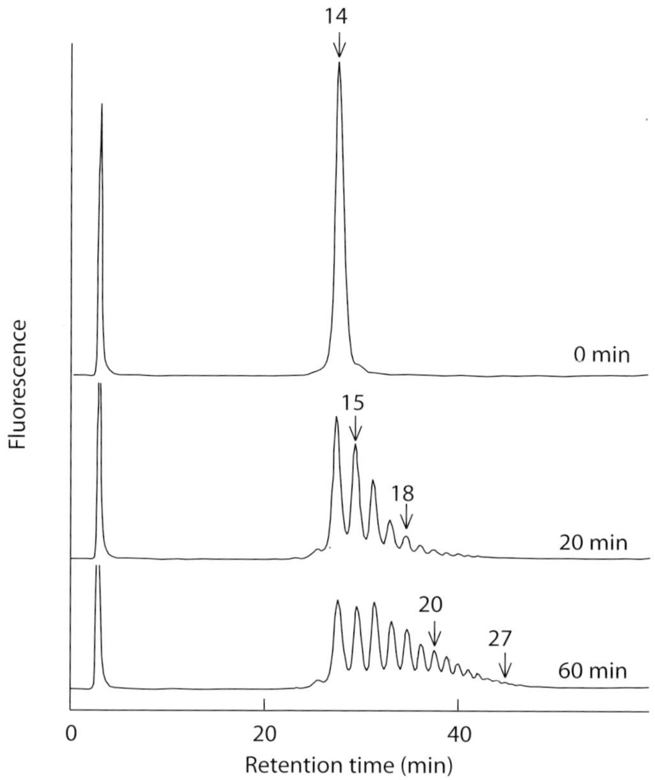

Figure 3. PGA synthase activity of the microsomal fraction from petunia pollen tubes. PA-OGA of DP 14 was used as an acceptor substrate. The enzyme reaction was performed for 0 to 60 min. Numbers and arrows indicate DP of PA-OGAs and their elution positions, respectively. Reprinted with permission from the American Society of Plant Biologists, Plant Physiology *(Akita* et al.*, 2002).*

synthase was shown to transfer GalA residues successively from UDP-GalA to the non-reducing end of OGA. In this system up to 13 GalA residues were transferred to PA-OGA in 1 h. This result is the first demonstration that PGA synthase successively transfers GalA residues resulting in the elongation of PGA chains. This pattern of sugar transfer shows that PGA synthase repeatedly transfers single GalA residues to the non-reducing end of PA-OGA in a non-processive fashion. This assay method enabled analysis of acceptor substrate specificity of the enzyme (Akita *et al.*, 2002). The best acceptor substrate of this enzyme is PA-OGA of DP 14 (out of substrates with DPs between 3 and 15).

This method has several advantages over the conventional method using radiolabelled UDP-GalA. PGA synthase gives multiple products even if homogeneous PA-OGA is used as an acceptor substrate. However, they are easily separated on an anion-exchange chromatography column and can be quantitatively evaluated. In addition, this method is highly sensitive with 1 pmol of fluorogenic OGAs being enough for detection and quantification. PA-OGAs stored in a freezer are stable and deterioration upon freezing and thawing has not been observed.

2.2. Chain elongation of PGA in vitro

Because the DP of PGA chains in nature has been reported to be around 100, PGA synthase is expected to transfer ~100 GalA residues consecutively to produce PGA. We developed an assay method for PGA synthase using fluorogenic OGAs, although it was only observed that up to 13 GalA residues were successively transferred (Akita *et al.*, 2002). There are several possible reasons to explain the inability to observe the expected activity of PGA synthase *in vitro*. First, the PGA synthase extracted from plant sources is not stable and its activity is gradually reduced even if the extract is kept at 4 °C. The microsomal fraction has to be used in the *in vitro* assay because purification of this enzyme has not been accomplished. Therefore it is important to obtain crude enzyme solution with higher specific activity for *in vitro* PGA chain elongation. Second, the microsomal fraction contains considerable polygalacturonase activity. The existence of this enzyme hampers characterization of PGA synthase in the microsomal fraction. Therefore the apparent PGA synthase activity in the microsomal fraction is lower than actual activity. If an inhibitor of this polygalacturonase is found, its use will be essential for obtaining a true measurement of PGA synthase activity. Another reason for not observing the expected activity of the enzyme is the solubility of longer OGAs. The HEPES buffer used in our assay method is not able to solubilize longer OGAs (DP > 30).

To obtain an enzyme solution with higher specific activity, we investigated the specific activity of PGA synthase from epicotyls of three legumes, namely pea, mung bean and azuki bean. The epicotyls of legumes rapidly grow at a rate of several cm per day and are expected to express PGA synthase at high levels. Of these plants, azuki bean epicotyls exhibited the highest activity of the enzyme.

We searched for inhibitors for the microsomal polygalacturonase. A relatively high concentration (5 mM) of trigalacturonic acid totally inhibited polygalacturonase activity against PA-OGA with DP 6. Fortunately, this compound showed little or no effect on PGA synthase activity.

In order to improve the solubility of longer OGAs, the buffering reagent was addressed. Instead of the HEPES buffer, oxalate buffer that was used for separation of OGAs with DP 50 on an anion-exchange column (Hotchkiss and Hicks, 1990) appears to be effective in solubilizing longer OGAs in the PGA synthase assay.

On the basis of these results, we refined the *in vitro* reaction conditions for PGA synthase to allow generation of longer PGA chains. Trigalacturonic acid (5 mM) as an inhibitor of the microsomal polygalacturonase and potassium oxalate (10 mM) to enhance the solubility of longer OGAs were added to the reaction buffer established previously (Akita *et al.*, 2002). As enzyme source, we used the microsomal fraction from azuki bean epicotyls. When PA-OGA of DP 12 was used as an acceptor substrate, this improved method generated PA-OGAs of DP up to 45 in a 2 h reaction (Figure 4), meaning that PGA synthase transfers more than 30 GalA residues consecutively *in vitro*.

Thus, we improved conditions for the PGA chain elongation reaction *in vitro* and successfully observed that PGA synthase can transfer more than 30 GalA residues successively and produced PA-OGA of DP up to 45. However, PGA synthase potentially produces PGA of DP around 100 and there is thus further refinement of *in vitro* conditions for PGA synthesis to be performed. This refinement will clarify the further characteristics of PGA synthase. With reference to these characteristics, it is expected to open the way for purification of this unstable enzyme and allow study of the structure of the enzyme which is considered be part of a protein complex (Ohashi *et al.*, 2007).

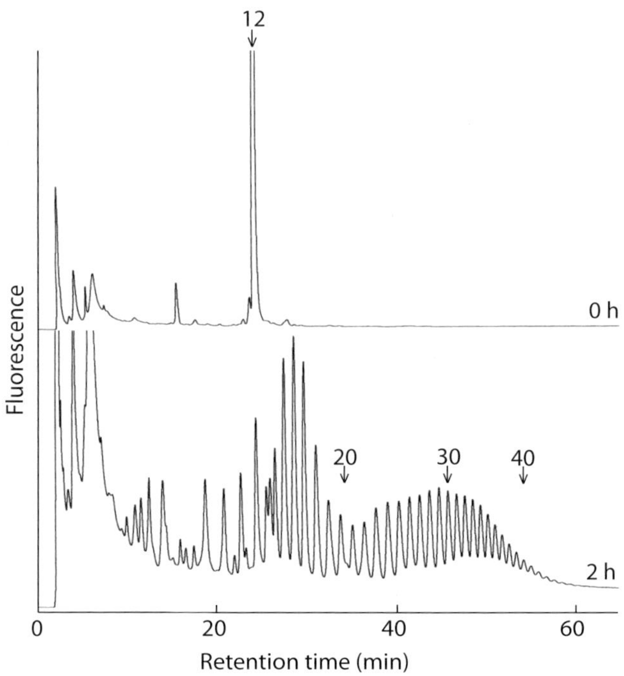

Figure 4. PGA synthase activity of the microsomal fraction from azuki bean epicotyls in the improved assay method. PA-OGA of DP 12 was used as an acceptor substrate. The enzyme reaction was performed for 2 h. Numbers and arrows indicate DP of PA-OGAs and their elution positions, respectively.

Acknowledgements

This work was supported by the Grants-in-Aid for Special Research on Priority Areas (no. 19043009) and the Grant-in-Aid for Scientific Research for Plant Graduate Students from the Nara Institute of Science and Technology from the Ministry of Education, Culture, Sports, Science, and Technology of Japan.

References

Akita, K., T. Ishimizu, T. Tsukamoto, T. Ando and S. Hase, 2002. Successive glycosyltransfer activity and enzymatic characterization of pectic polygalacturonate 4-α-galacturonosyltransferase solubilized from pollen tubes of *Petunia axillaris* using pyridylaminated oligogalacturonates as substrates. Plant Physiol. 130: 374-379.

Bolwell, G.P., G. Dalessandro and D.H. Northcote, 1985. Decrease of polygalacturonic acid synthase during xylem differentiation in sycamore. Phytochemistry 24: 699-702.

Bouton, S., E. Leboeuf, G. Mouille, M.-T. Leydecker, J. Talbotec, F. Granier, M. Lahaye, H. Höfte and H.-N. Truong, 2002. *QUASIMODO1* encodes a putative membrane-bound glycosyltransferase required for normal pectin synthesis and cell adhesion in Arabidopsis. Plant Cell 14: 2577-2590.

Chambat, G. and J.-P. Joseleau, 1980. Isolation and characterization of a homogalacturonan in the primary walls of *Rosa* cells cultures *in vitro*. Carbohydr. Res. 85: C10-C12.

Crombie, H.J., C. Scott and J.S. Reid, 2003. Detergent-solubilization of a homogalacturonan galacturonosyltransferase from mung bean. In: F. Voragen, H. Schols and R. Visser (eds.) Advances in pectin and pectinase research. Kluwer, Dordrecht, the Nederlands, pp.35-45.

Doong, R.L. and D. Mohnen, 1998. Solubilization and characterization of a galacturonosyltransferase that synthesizes the pectic polysaccharide homogalacturonan. Plant J. 13: 363-374.

Irwin, P.L., M.D. Sevilla and W. Chamulitrat, 1988. Homopolygalacturonan molecular size in plant cell wall matrices via paramagnetic ion and nitroxyl amide dipolar spin-spin interactions. Biophys. J. 54: 337-344.

Ishii, T., J. Ichita, H. Matsue, H. Ono and I. Maeda, 2002. Fluorescent labeling of pectic oligosaccharides with 2-aminobenzamide and enzyme assay for pectin. Carbohydr. Res. 337: 1023-1032.

Ishimizu, T. and S. Hase, 2004. Pectin biosynthesis. In: A. Hemantaranjan (ed.) Advances in Plant Physiology Vol. 7. Scientific Publishers, Jodhpur, India, pp. 175-193.

Hotchkiss Jr, A.T. and K.B. Hicks, 1990. Analysis of oligogalacturonic acids with 50 or fewer residues by high-performance anion-exchange chromatography and pulsed amperometric detection. Anal. Biochem. 184: 200-206.

Konno, H., Y. Yamasaki and K. Katoh, 1986. Enzymatic degradation of pectic substances and cell walls purified from carrot cell cultures. Phytochemistry 25: 623-627.

Kotake, T., D. Yamaguchi, H. Ohzono, S. Hojo, S. Kaneko, H.K. Ishida and Y. Tsumuraya, 2004. UDP-sugar pyrophosphorylase with broad substrate specificity toward various monosaccharide 1-phosphates from pea sprouts. J. Biol. Chem. 279: 45728-45736.

Leboeuf, E., F. Guillon, S. Thoiron and M. Lahaye, 2005. Biochemical and immunohistochemical analysis of pectic polysaccharides in the cell walls of *Arabidopsis* mutant *QUASIMODO1* suspension-cultured cells: implications for cell adhesion. J. Exp. Bot., 56: 3171-3182.

Lin, T.-Y., A.D. Elbein and J.-C. Su, 1966. Substrate specificity in pectin synthesis. Biochem. Biophys. Res. Commun. 22: 650-657.

Mohnen, D., 2008. Pectin structure and biosynthesis. Curr. Opin. Plant Biol. 11: 266-277.

Northcote, D.H. and J.D. Pickett-Heaps, 1966. A function of the Golgi apparatus in polysaccharide synthesis and transport in the root-cap cells of wheat. Biochem. J. 98: 159-167.

Ohashi, T., N. Cramer, T. Ishimizu and S. Hase, 2006. Preparation of UDP-galacturonic acid using UDP-sugar pyrophosphorylase. Anal. Biochem. 352: 182-188.

Ohashi, T., T. Ishimizu, K. Akita and S. Hase, 2007. Stabilization and minimum active component of polygalacturonic acid synthase involved in pectin biosynthesis. Biosci. Biotechnol. Biochem. 71: 2291-2299.

Orfila, C., S.O. Sorensen, J. Harholt, N. Geshi, H. Crombie, H.N. Truong, J.S.G. Reid, J.P. Knox and H.V. Scheller, 2005. *QUASIMODO1* is expressed in vascular tissue of *Arabidopsis thaliana* inflorescence stems, and affects homogalacturonan and xylan biosynthesis. Planta, 222: 613-622.

Sterling, J.D., H.F. Quigley, A. Orellana and D. Mohnen, 2001. The catalytic site of the pectin biosynthetic enzyme α-1,4-galacturonosyltransferase is localized in the lumen of the Golgi. Plant Physiol. 127: 360-371.

Sterling, J.D., M.A. Atmodjo, S.E. Inwood, V.S.K. Kolli, H.F. Quigley, M.G. Hahn and D. Mohnen, 2006. Functional identification of an Arabidopsis pectin biosynthetic homogalacturonan galacturonosyltransferase. Proc. Natl. Acad. Sci. USA. 103: 5236-5241.

Takeuchi, Y. and Y. Tsumuraya, 2001. *In vitro* biosynthesis of homogalacturonan by a membrane-bound galacturonosyltransferase from epicotyls of azuki bean. Biosci. Biotechnol. Biochem. 65: 1519-1527.

Thibault, J.-F., C.M.G.C. Renard, A.V. Axelos, P. Roger and M.-J. Crépeau, 1993. Studies of the length of homogalacturonic acid regions in pectins by acid hydrolysis. Carbohydr. Res. 238: 271-286.

Villemez, C.L., T.-Y. Lin and W.Z. Hassid, 1965. Biosynthesis of the polygalacturonic acid chain of pectin by a particulate enzyme preparation from *Phaseolus aureus* seedlings. Proc. Natl. Acad. Sci. USA. 54: 1626-1632.

Villemez, C.L., A.L. Swanson and W.Z. Hassid, 1966. Properties of a polygalacturonic acid-synthesizing enzyme system from *Phaseolus aureus* seedlings. Arch. Biochem. Biophys. 116: 446-452.

Yapo, B.M., P. Lerouge, J.-F. Thibault and M.-C. Ralet, 2007. Pectins from citrus peel cell walls contain homogalacturonans homogenous with respect to molar mass rhamnogalacturonan I and rhamnogalacturonan II. Carbohydr. Polym. 69: 426-435.

Zhang, G.F. and LA. Staehelin, 1992. Functional compartmentation of the Golgi apparatus of plant cells. Plant Physiol. 99: 1070-1083.

Which mango processing residues are suitable for pectin recovery in terms of yield, molecular and techno-functional properties of extractable pectins?

S. Neidhart[1], S. Sirisakulwat[1,2], A. Nagel[1], P. Sruamsiri[2] and R. Carle[1]
[1]Institute of Food Science and Biotechnology, Chair of Plant Foodstuff Technology, Hohenheim University, Garbenstr. 25, 70599 Stuttgart, Germany
[2]Department of Horticulture, Faculty of Agriculture, Chiang Mai University, Chiang Mai 50200, Thailand

Abstract

Mimicking industrial pectin recovery, two typical procedures of hot-acid extraction were applied to exemplarily selected mango by-products on a laboratory scale to explore the suitability of the latter in terms of crude and starch-corrected yield, composition as well as molecular and techno-functional properties of the pectins obtained. To exemplify residues from pulp production, fruits of monoembryonic cv. 'Tommy Atkins' at two representative ripeness stages were processed into purées on a pilot-plant scale with recovery of pectins from peels and pulp wastes. With regard to canning and drying waste, pectins from peels of polyembryonic cv. 'Kaew Khiew' were compared to those obtained from the mesocarp, also at two typical ripeness stages. The broad range of recoverable rapid-set to ultra-rapid-set mango pectins with respect to yield, molecular structure, and behaviour was illustrated and compared to commercial non-standardized apple pectin with a similar degree of methylation. Dried mango peels resulted in starch-adjusted yields of ~13% on the average, but the gelling and thickening capacities of peel pectins were limited by a characteristic, almost monodisperse fraction with a peak molecular weight of ~15000-22000 considerably reducing the average molecular weights. From mango pulp, pectins with very good gelling and thickening properties were partly produced, but the yields were insufficient (≤ 9.3%). For some mango by-products, non-starch neutral sugars limited the galacturonic acid contents to levels below the legal minimum of 65%. Since native starch did not always degrade completely until processing of the fruit, co-extraction of starch partly played a significant role.

Keywords: cultivar, fruit processing, *Mangifera indica* L., maturity

1. Introduction

Among tropical fruits and fruits at large, mango (*Mangifera indica* L.) belongs to the most important cash crops with a high export potential, both as fresh and processed fruit. The global production of mango fruits was ~28×10^6 Mt in year 2005 (FAOSTAT, 2006). After India (38.6%) and China (13.1%), Thailand was the third largest mango producing country with 6.4%, closely followed by Mexico (5.4%). The global production of mango juice and pulp in the year 2005 was 153×10^3 Mt and 931×10^3 Mt, respectively. Compared to fruits of temperate zones, considerably higher ratios of by-products arise from mango processing due to higher amounts of inedible waste material such as peels (15-18% of total fruit weight) and seeds (13-29%) (Vásquez-Caicedo *et al.*, 2002). Increasing mango processing in most producing countries has recently boosted the worldwide interest in value-adding utilization of industrial mango residues, also stimulating investigations into pectin recovery from mango (Koubala *et al.*, 2008; Sirisakulwat *et al.*, 2008). Mango has been deemed a promising, but not yet exploited alternative pectin source (Thibault and Ralet, 2003; Endress *et al.*, 2006). Unlike the major sources for pectin recovery, multifarious processing of mangoes at different scales implies residues from quite heterogeneous raw material in terms of cultivars and maturity, ranging from mature-green varieties, sliceable fruits for canning and drying to soft, full-ripe fruits for pulping (Vásquez-Caicedo *et al.*, 2004). After considerable accumulation of starch during fruit growth, the climacteric mango fruit is usually harvested at the mature-green stage and subjected to post-harvest ripening until appropriate ripeness for processing (Neidhart *et al.*, 2007), associated with solubilization of pectic polymers and degradation of hemicelluloses and starch to various extents, depending on the cultivar and the part of the fruit. Fruit maturity and co-extraction of other polymers during production of pectins for commercial use may affect yields and techno-functional properties of crude pectic polymers much more than in case of citrus peels and apple pomace.

Therefore, the suitability of different types of mango residues for pectin recovery by conventional hot-acid extraction was to be discussed. The study was based on the evaluation of 78 mango pectins, which were recovered from the dried peels and mesocarp of four cultivars at different fruit ripeness as well as from industrial peel waste and the by-products of pulp production. With respect to concepts for integrated processing into high-quality mango products, focus was put on raw material rich in pro-vitamin A (Vásquez-Caicedo *et al.*, 2005, 2007). The aim of this contribution was to explore possible resource-immanent obstacles to the exploitation of mango by-products. For this purpose, the variability of mango pectins in terms of yield and quality was to be illustrated on the basis of 16 exemplarily selected mango pectins, reflecting possible sources and typical extraction modes.

2. Material and methods

2.1 Sample material and experimental design

Mature-green mango fruits cv. 'Kaew Khiew' (KAk; ~400 kg) were obtained from an orchard in the San Sai district, Chiang Mai, Thailand, during main harvest season in May 2004 and ripened for 9 days at 25 ± 2 °C and 71-80% relative humidity (RH) at

Chiang Mai University, Chiang Mai, until full ripeness. Fruit samples of ~2 kg were randomly collected for daily monitoring of post-harvest ripening (Sirisakulwat *et al.*, 2008). At selected ripeness stages (Table 1), ~180 kg of fruit were manually peeled and de-stoned, while small fruit portions were processed consecutively. The peels were immediately quick-frozen in liquid nitrogen, vacuum-packed in aluminium pouches, and stored at -80 °C. The mesocarp was cut into cubes of ~1 cm^3 with stainless steel knives prior to quick-freezing. Portions of ~20 kg were pre-dried in a cabinet dryer at 80 °C as single layers on 56×99 cm trays until a moisture content of ~25%, then vacuum-packed and stored at -20 °C until final freeze-drying, because of limited lyophilization capacity. The lyophilized samples were sent deep-frozen by airfreight to Hohenheim University, Stuttgart, Germany. After further lyophilization upon arrival, the brittle samples were manually crushed, vacuum-packed into PE pouches, and stored with desiccant at ~22 °C until pectin extraction.

By analogy, almost mature-green mango fruits cv. 'Tommy Atkins' (TA; ~60 kg), obtained from a wholesaler in Stuttgart, Germany, were ripened for 11 days at 20 ± 3 °C and 62-84% RH at Hohenheim University, Stuttgart. To study the pectins of by-products from purée production, purées were produced on a pilot plant scale at two different ripeness stages (Table 1) from fruit batches of 17 and 20 kg according to the procedure detailed by Vásquez-Caicedo *et al.* (2007) with slight modifications. The peels were obtained by thermal-manual peeling, involving steaming of the fruit for 7 min. After subsequent manual de-stoning, the pulping residues were collected from a two-stage pulping and finishing process at 10 and 0.4 mm. The peels were immediately dried in a fluidized-bed laboratory dryer (80 °C; Rapid dryer TG 100, Retsch, Haan, Germany), the pulping residues concurrently in a drying cabinet (80 °C).

A commercial non-standardized apple pectin with a degree of esterification (DE) of ~72% (Herbstreith & Fox, Neuenbürg, Germany) was characterized for comparison (sample AP1).

Table 1. Ripeness of the fruits used for pectin extraction.

Cultivar	RT [days]	Ripeness stage	TSS [g/100 g]	TSS/TTA
Kaew Khiew	7	A_K[1]	21.6	55.2
(KAk)	9	B_K	22.1	94.7
Tommy Atkins	8	A_T	14.3	43.0
(TA)	11	B_T	14.2	107.2

RT, post-harvest ripening time; TSS, total soluble solids expressed as sucrose (°Brix); TSS/TTA, sugar/acid ratio of the mesocarp, as calculated from TSS and the titratable acids (TTA) quantified as citric acid (pH 8.1) in g/100 g of mesocarp.

[1] In the legends of Figures 1-5, the ripeness stages are referred to as A for A_K and A_T and as B for B_K and B_T irrespective of the differences between the respective fruit batches.

2.2 Pectin extraction

Pectin was recovered from the dried raw material (DRM) on a laboratory scale, using two standard protocols for extraction at pH 1.5 and 2.0, respectively (Sirisakulwat *et al.*, 2008). Per run, 40 g of DRM were filled up with distilled water to a total of 800 g and the protocols were thus applied on a double scale throughout. After boiling for 15-20 min and subsequent re-cooling, the pH was adjusted to 1.5 with H_2SO_4 (25% w/w) or pH 2.0 with 2 N H_2SO_4 prior to extraction (90 °C, 2.5 h) in an oil bath. Precipitation in propan-2-ol (≥99.9% v/v) and drying (60 °C, 1-2 h) of the pectins were performed as detailed by Sirisakulwat *et al.* (2008). Three runs per extraction method were carried out for each DRM sample from TA fruits, 25-27 runs per variant for samples from KAk fruits. For each run, the yield of dried crude pectin was calculated as alcohol insoluble substance (AIS, g/100 g of DRM). Per sample and extraction method, the AIS of all runs was pooled and ground in a centrifugal mill type ZM 1 (Retsch, Haan, Germany) to particle sizes <0.25 mm. The starch-adjusted pectin yield ($AIS_{S\text{-}corr}$, g/100 g of DRM) was calculated after enzymatic quantification of co-extracted starch (g/100 g of AIS) by means of the respective test combination (R-Biopharm, Darmstadt, Germany).

2.3 Pectin analyses

The methodology detailed by Sirisakulwat *et al.* (2008) was applied to characterize the ground AIS and the reference pectin (AP1) in terms of composition as well as molecular and techno-functional properties. In brief, the galacturonide content was titrimetrically quantified as galacturonic acid (M_r = 194.1 g/mol) on an ash- and moisture-free basis according to the legal pectin specification (JECFA, 2007). Based on the recovery rate after AIS purification in acidified propan-2-ol, it was expressed for the original AIS sample on a starch-free basis ($GalUA_{titr}$, g/100 g of $AIS_{S\text{-}corr}$) and additionally converted into the anhydrogalacturonic acid content (M_r = 176.13 g/mol) of the original sample (AUA_{titr}, g/100 g of AIS). The degree of esterification (DE) resulted as the percentage of the titrimetric saponification equivalents relative to the total of neutralization and saponification equivalents. Furthermore, the anhydrouronic acid content (M_r = 176.13 g/mol) was colorimetrically quantified with *m*-hydroxydiphenyl (MHDP) and expressed on a starch-free basis (AUA_{MHDP}, g/100 g of $AIS_{S\text{-}corr}$), while the respective neutral sugar contents were determined gaschromatographically as their alditol acetates. Based on the simultaneous HPLC analysis of methanol and acetic acid, the degrees of methylation (DMe) and acetylation (DAc) were calculated as the respective molar ratios to AUA_{MHDP}.

Intrinsic viscosities and molecular size distributions were explored in buffers that mimicked the gel systems used for the gelation studies in terms of pH and concentration. The intrinsic viscosity ([η], ml/g) was assessed as an average parameter influenced by all polymers of an AIS sample. Based on the Solomon-Ciuta equation, it was computed from the specific viscosities of different AIS solutions as a mean of 10 records and verified by the Huggins plot and the Kraemer plot. A 0.28 M buffer (pH 3.0 ± 0.05) of potassium acetate (KAC: 4.57 g/l) and lactic acid [LA (90%): 23.59 g/l] served as solvent.

Molecular size distributions were analyzed by high-performance size exclusion chromatography (HP-SEC) with refractive index (RI) detection relative to dextran after calibration with 7 DIN-accredited standards of the types 5000-670000 (Sigma-Aldrich Chemie, Munich, Germany). A series of TSK-GEL columns G6000PW$_{XL}$, G3000PW$_{XL}$, and G2500PW$_{XL}$ (300×7.8 mm) was used in combination with a TSK guard column PW$_{XL}$ (40×6.0 mm) (Tosoh Bioscience Division, Tokyo, Japan). The columns were eluted at 40 °C and a flow rate of 0.6 ml/min with a 0.294 M buffer (pH 3.0) of potassium acetate (KAC: 4.046 g/l) and lactic acid [LA (~90%): 25.25 g/l] that contained 0.6 mM potassium citrate as complexing agent.

To study the flow properties of standardized AIS solutions, the viscosities of aqueous AIS solutions (2% w/w) were measured at 20 °C over a shear rate range of 1-100 s^{-1} by using a Bohlin CVO 120 HR rheometer (Malvern Instruments, Herrenberg, Germany) with a modified double-gap (DG 40/50) device.

For pectin-sucrose gels (TSS = 65 ± 0.5 °Brix, pH 3.0 ± 0.04) with AIS doses (c_p) of 0.15-1 g/100 g in a KAC/LA buffer (0.25 mol/kg of gel), the setting temperatures were assayed by dynamic rheometric analyses in duplicate, using the Bohlin CVO 120 HR rheometer with a Searle cylinder device (C 25) during cooling from 90 to 20 °C at a constant rate (-1.028 ± 0.006 K/min).

The gel properties were characterized by the breaking strengths of pectin-sucrose gels (TSS = 65 °Brix, pH 3.0 ± 0.05) with AIS doses (c_p) of 0.2-0.35 g/100 g in a KAC/LA buffer (0.28 mol/kg of gel) on the basis of an empirical tension test established for the standardization of pectins and the comparative evaluation of pectin sources (Endress and Dilger, 1990). By using the Herbstreith-Pectinometer type Mark III (Herbstreith & Fox, Neuenbürg, Germany), the breaking strength (BS) was recorded in device-specific units (HPE). The AIS dose required for a standardized gel (BS_o = 530 HPE) was calculated by interpolation as the breaking capacity of the AIS (BC_{530HPE}, g/100 g of gel). TSS and pH of all gel samples were experimentally verified. For its normalization to constant AUA levels, BC_{530HPE} was related to the AUA_{titr} content of the AIS and expressed as AUA_{titr} at BC_{530HPE} in g of AUA/100 g of gel.

3. Results and discussion

3.1 Pectin yields obtainable from mango peels and pulp

Commercial pectins are typically recovered by hot-acid extraction (pH 1-3, 50-90 °C), aiming at maximized yield through partial hydrolysis of rhamnopyranosyl and arabinofuranosyl linkages, solubilization of ionically bound pectins, and hydrolysis of rhamnogalacturonan II (RGII) borate esters (MacDougall and Ring,

2004). Irrespective of cultivar and fruit ripeness, the peels of the mango fruits studied yielded considerably more pectin than the pulp, with the crude pectin yields (AIS) ranging at 12.0-15.7 and 5.8-9.5 g/100 g of dried raw material, respectively (Table 2). Under the conditions applied, the pH of the extracting agent hardly influenced the yield. But when applied to peels of mature-green fruits, the more acidic agent turned out to be more efficient in pectin solubilization than extraction at pH 2 (Sirisakulwat *et al.*, 2008).

Evaluation of co-extractable starch has been among the unique features of this extensive study on mango pectins. As shown by the 'Kaew Khiew' samples, starch may markedly contribute to the AIS mass, particularly to pulp AIS (Table 2). It has to be noted that, during post-harvest ripening until stages useable for eating or processing (Vásquez-Caicedo *et al.*, 2004; Pott *et al.*, 2005), starch may not degrade completely. Thus, pectin recovery from mangoes might partly require additional enzymatic treatments for the artificial removal of residual starch, similar to the exploitation of

Table 2. Yields of alcohol-insoluble substance (AIS) extracted from dried peels and pulp of mango fruit cvs. 'Kaew Khiew' and 'Tommy Atkins' at two different ripeness stages as well as starch-corrected yields ($AIS_{S\text{-}corr}$).

Cultivar	Extraction: pH	Ripeness stage	AIS (g/100 g DRM)	Starch (g/100 g AIS)	$AIS_{S\text{-}corr}$ (g/100 g DRM)
Dried peels:					
Kaew Khiew	1.5	A_K	15.7 ± 0.2[1]	18.4 ± 0.4	12.9 ± 0.2
		B_K	13.8 ± 0.2	4.8 ± 0.2	13.1 ± 0.2
	2.0	A_K	13.5 ± 0.2	16.0 ± 1.4	11.4 ± 0.2
		B_K	13.1 ± 0.2	4.5 ± 0.4	12.5 ± 0.2
Tommy Atkins	1.5	A_T	12.4 ± 0.04	1.1 ± 0.2	12.3 ± 0.04
		B_T	13.4 ± 0.3	0.6 ± 0.1	13.3 ± 0.3
	2.0	A_T	12.0 ± 0.2	0.9 ± 0.03	11.9 ± 0.2
		B_T	13.0 ± 0.2	0.4 ± 0.2	12.9 ± 0.2
Dried mesocarp:					
Kaew Khiew	1.5	A_K	7.8 ± 0.2	35.2 ± 0.7	5.1 ± 0.1
		B_K	5.8 ± 0.1	15.2 ± 0.3	4.9 ± 0.1
	2.0	A_K	8.1 ± 0.1	31.4 ± 0.7	5.6 ± 0.1
		B_K	6.6 ± 0.1	13.7 ± 0.1	5.7 ± 0.1
Tommy Atkins	1.5	A_T	8.4 ± 0.04	2.7	8.2
		B_T	8.7 ± 0.2	2.8	8.5
	2.0	A_T	8.4 ± 0.2	1.7	8.3
		B_T	9.5 ± 0.2	1.7	9.3

[1]Means of 3 (Tommy Atkins) and 25-27 (Kaew Khiew) extractions per sample. DRM, dry raw material.

unripe apples (Endress *et al.*, 2006). Without such treatments, comparison of pectin yields is only acceptable on a starch-free basis after yield correction ($AIS_{S\text{-corr}}$).

Despite very different starch contents, the starch-adjusted net pectin yields ($AIS_{S\text{-corr}}$) were rather uniform, though still highly specific for the part of the fruit. In view of net pectin yields of 5-9 g/100 g (Table 2) confirming earlier findings (Kratchanova *et al.*, 1991) despite other recovery conditions, the pulp of mango fruits can hardly be recommended from an economic point of view for conventional pectin recovery through hot-acid extraction. The polyuronides were assumed to be too tightly bound to the insoluble matrix (Kratchanova *et al.*, 1991). Unlike the pulp, the peels of mango fruits proved to be a reasonable source due to net pectin yields of ~13 g/100 g (Table 2), thus coming closer to apple pomace (10-15%) than to citrus peels (20-30%) (Endress *et al.*, 2006). Reported pectin yields from peel waste and experimental peel samples greatly varied (6.4-24.5%) as recently reviewed by Sirisakulwat *et al.* (2008), whereas the net pectin yields of mango peels shown in Table 2 are deemed representative irrespective of cultivar and provenance. Reported maximum yields of 21-24.5% (Kratchanova *et al.*, 1991; Berardini *et al.*, 2005; Sirisakulwat *et al.*, 2008) occurred rarely.

3.2 Gelling and gel properties of mango pectins

Although the mango peels were comparable to apple pomace as regards the net pectin yields (cf. 3.1), mango AIS mostly produced much softer gels than the commercial non-standardized apple pectin (AP1) used as a reference (Figure 1). Among the samples shown, the maximum gelling capacity of AP1 was indicated by the minimum pectin dose (c_P = 0.17 g/100 g of gel) needed for a standard gel with a breaking strength of 530 HPE (BS_o). AIS from TA fruits partly came notably close to AP1, especially when recovered by extraction at pH 1.5 (Figure 1a). But strikingly high doses of AIS from KAk fruits would have been necessary for a BS_o gel. Whereas the pectin yield was hardly influenced by the extracting agent, the gelling capacity was clearly enhanced by the recovery of mango AIS at pH 1.5 instead of pH 2, consistent with the findings of Sirisakulwat *et al.* (2008).

Among the TA samples, the AIS dose was minimum when pulp of the less ripe fruit (stage A_T; Table 1) had been extracted at pH 1.5 (Figure 1a). Despite the poor net pectin yields of the pulp (Table 2), TA pulp surpassed TA peels in the gelling capacity of the AIS. By contrast, it was the peel of fully ripe fruits (stage B_K; Table 1) that yielded the AIS, again by extraction at pH 1.5, with the highest gelling capacity among the KAk samples (Figure 1a). However, the starch contents greatly varied among the latter group, while TA AIS only contained residual starch amounts (Table 2). Therefore, the poor gelling capacity of KAk AIS might at least partly be ascribed to pectin dilution by starch as particularly suggested by Figure 1a.

A direct comparison of the gelling capacities is provided by Figure 2, showing the breaking capacity (BC_{530HPE}) expressed as the AIS doses needed for BS_o gels as

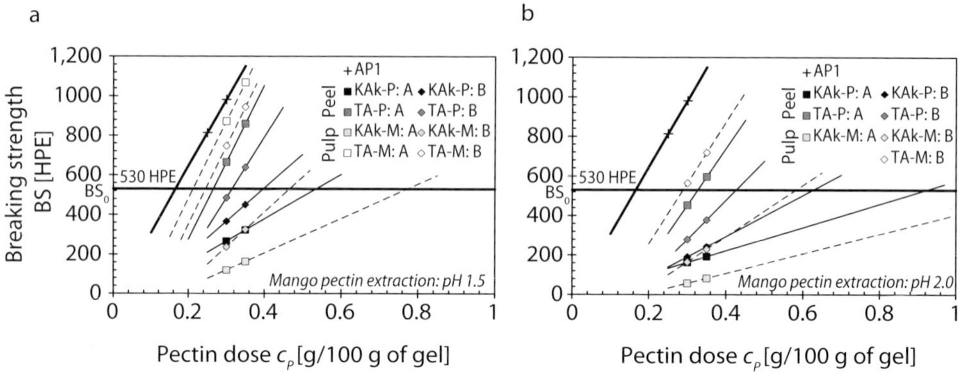

Figure 1. Gel properties of mango pectins: breaking strength (BS) of gels (65 °Brix, pH 3.0) at different doses (c_p) of a commercial non-standardized apple pectin (AP1) and alcohol-insoluble substance (AIS) samples extracted at a) pH 1.5 and b) pH 2.0 from dried peels (P) and pulp (mesocarp, M) of mango fruit cvs. 'Kaew Khiew' (KAk) and 'Tommy Atkins' (TA) at two different ripeness stages (A, B; cf. Table 1).

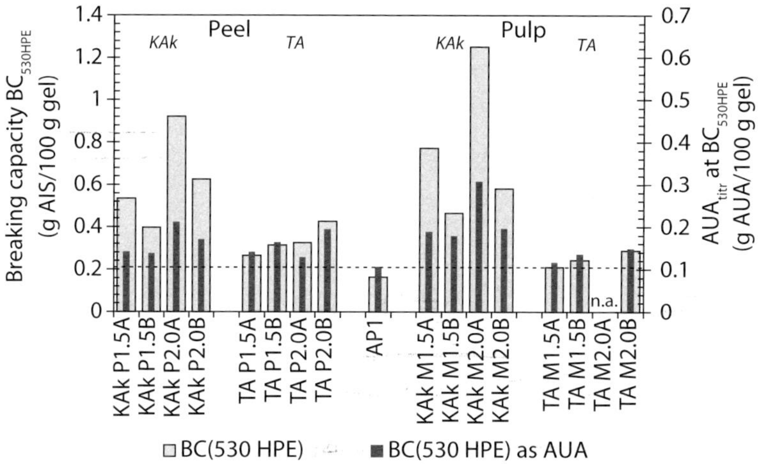

Figure 2. Gel properties of mango pectins: breaking capacity (BC_{530HPE}) of a commercial non-standardized apple pectin (AP1) and alcohol-insoluble substance (AIS) samples extracted at pH 1.5 or 2.0 from dried peels (P) and pulp (mesocarp, M) of mango fruit cvs. 'Kaew Khiew' (KAk) and 'Tommy Atkins' (TA) at different ripeness stages (A, B; cf. Table 1). BC_{530HPE} = AIS dose required for a standard gel (65 °Brix, pH 3.0) with a breaking strength (BS_0) of 530 HPE; AUA_{titr} at BC_{530HPE} = normalized BC_{530HPE} expressed as the corresponding AUA_{titr} dose according to the titrimetric anhydrogalacturonic acid content (AUA_{titr}) of the AIS or pectin used; n.a., not analyzed.

customary. To verify the dilution hypothesis mentioned above, the different starch contents of the AIS samples were considered through normalization of the breaking capacities according to the titrimetric anhydrogalacturonic acid contents (AUA_{titr}) of the AIS. When BC_{530HPE} was expressed as the respective AUA_{titr} dose required for a BS_o gel (AUA_{titr} at BC_{530HPE} in Figure 2), nearly all TA samples and KAk peel AIS recovered at pH 1.5 came rather close to AP1. Hence, the limited gelling capacities were indeed caused by pectin dilution through co-extracted starch to some extent. However, the mango pectins were overall less valuable than AP1 in terms of the gelling capacity for other reasons.

On the other hand, despite poor gel strengths, the gelling samples of mango AIS, especially those recovered at pH 2.0, displayed extremely high setting temperatures exceeding those of respective gelling samples prepared from AP1 (data not shown), as extensively discussed previously for KAk peel pectins (Sirisakulwat *et al.*, 2008). Hence, mango AIS had a great capacity for junction zone formation, but only limited capacity of solvent immobilization in the gel, thus overall producing rather soft gels.

3.3 Structural differences between mango pectins obtained from peel and pulp

As expected (Koubala *et al.*, 2008; Sirisakulwat *et al.*, 2008), hot-acid extraction of dried mango pulp and peel uniformly yielded high-methoxyl pectins (DE ~78%) that were hardly acetylated (DAc ~4%) (Table 3). In this respect, mango AIS resembled the apple pectin AP1 that was characterized by a DE of 72.0 ± 0.2% and a DAc of 6.1 ± 0.2% (Sirisakulwat *et al.*, 2008). Consistently, mango fruit has been deemed a good source for rapid-set to ultra-rapid-set pectins, similar to apple pomace.

However, even on a starch-free basis ($AIS_{S\text{-}corr}$), the galacturonide contents of mango AIS were limited, despite hot-acid extraction, to a range of ~30-60 g/100 g, as mutually confirmed by titrimetric analysis and colorimetric quantification ($GalUA_{titr}$ and AUA_{MHDP}; Table 3). By contrast, the respective results obtained for the apple pectin AP1 were 71.2 ± 0.1 g/100 g of $AIS_{S\text{-}corr}$ ($GalUA_{titr}$) and 85.1 ± 2.2 g/100 g of $AIS_{S\text{-}corr}$ (AUA_{MHDP}), in compliance with the minimum galacturonic acid content (65%) specified for commercial pectins on an ash- and moisture free basis (Endress *et al.*, 2006; JECFA, 2007).

Table 3 shows one of the present key problems limiting the broad exploitation of this source. Many of the mango AIS samples, such as KAk AIS, may not meet the legal specification, even after artificial starch removal. Among the mango samples studied, maximum $GalUA_{titr}$ contents (63 g/100 g of $AIS_{S\text{-}corr}$ corresponding to 80 g/100 g of AIS on an ash- and moisture-free basis after acidic purification) were observed for AIS obtained from TA pulp at pH 1.5, also revealing maximum gelling capacity (Figures 1a and 2). Extraction at pH 1.5 always resulted in higher galacturonide contents compared to the recovery at pH 2.0 (Table 3), again consistent with the higher gelling capacities of those samples (Figure 2). However, AIS obtained from KAk peel usually revealed slightly lower galacturonic acid levels than AIS from the

Table 3. Esterification of pectins and their galacturonic acid contents on a starch-free basis ($AIS_{S\text{-}corr}$) after recovery as alcohol-insoluble substance (AIS) from dried peels and pulp of mango fruit cvs. 'Kaew Khiew' and 'Tommy Atkins' at two different ripeness stages.

Cultivar	Extraction: pH	Ripeness stage	$GalUA_{titr}$ (g/100 g $AIS_{S\text{-}corr}$)	AUA_{MHDP} (g/100 g $AIS_{S\text{-}corr}$)	DE (%)	DAc (%)
Dried peels:						
Kaew Khiew	1.5	A_K	35.8 ± 0.7	37.9 ± 1.5	73.0 ± 0.4	3.9 ± 0.1
		B_K	40.2 ± 0.8	34.2 ± 0.9	71.8 ± 0.1	3.6 ± 0.1
	2.0	A_K	30.1 ± 1.1	32.6 ± 1.2	77.2 ± 0.9	3.9 ± 0.1
		B_K	31.4 ± 0.6	30.9 ± 1.6	75.7 ± 0.4	3.9 ± 0.2
Tommy Atkins	1.5	A_T	59.6 ± 1.3	59.6 ± 1.7	78.0 ± 0.7	2.6 ± 0.1
		B_T	57.8 ± 2.1	53.2 ± 2.2	77.8 ± 0.3	2.2 ± 0.1
	2.0	A_T	44.0 ± 8.0	49.8 ± 2.5	77.6 ± 4.0	3.9 ± 0.2
		B_T	50.6 ± 1.4	44.1 ± 1.9	80.4 ± 0.3	3.3 ± 0.1
Dried mesocarp:						
Kaew Khiew	1.5	A_K	41.6 ± 1.6	37.9 ± 1.8	72.4 ± 0.7	5.7 ± 0.2
		B_K	50.0 ± 1.1	46.6 ± 1.3	74.5 ± 0.4	3.9 ± 0.1
	2.0	A_K	39.6 ± 2.4	30.4 ± 1.3	79.4 ± 0.3	8.9 ± 0.3
		B_K	43.2 ± 1.5	40.5 ± 1.6	75.7 ± 2.3	5.7 ± 0.2
Tommy Atkins	1.5	A_T	62.7 ± 0.1[1]	60.2 ± 1.1[1]	78.5 ± 0.2	3.8 ± 0.1
		B_T	63.0 ± 0.6[1]	58.3 ± 1.8[1]	77.5 ± 0.1	3.1 ± 0.1
	2.0	A_T	55.0 ± 0.2[1]	52.8 ± 1.3[1]	82.5 ± 0.1	5.0 ± 0.1
		B_T	58.0 ± 0.4[1]	52.5 ± 1.3[1]	82.1 ± 0.2	4.0 ± 0.1

$GalUA_{titr}$, titrimetric galacturonic acid content (M_r = 194.1 g/mol); AUA_{MHDP}, colorimetric anhydrogalacturonic acid content (M_r = 176.13 g/mol); DE, degree of esterification (titrimetric); DAc, degree of acetylation (based on AUA_{MHDP} and acetate quantification by HPLC).
[1] Standard error estimated without error propagation (single starch value).

respective pulp, despite the higher gelling capacity of the former, thus indicating any further differences.

From Table 3, the question arises which other components of mango AIS, in addition to starch, contribute to the observed limitation as to the galacturonide contents. For AIS obtained from peels, this was mainly attributed to notably high levels of galactose residues (~10-25 g/100 g of $AIS_{S\text{-}corr}$; Table 4), which were almost equal to the galacturonic acid contents in case of AIS from KAk peels (Figure 3b). Galactose occurred at remarkably constant levels within each group, depending on the mango cultivar and the part of the fruit, but irrespective of ripeness and extraction modes (Table 4). The samples presented in Table 4 reflect the great variability of mango AIS.

Table 4. Major neutral sugars on a starch-free basis ($AIS_{S\text{-}corr}$) in the alcohol-insoluble substance (AIS) recovered from dried peels and pulp of mango fruit cvs. 'Kaew Khiew' and 'Tommy Atkins' at two different ripeness stages.

Cultivar	Extraction: pH	Ripeness stage	Ara (g/100 g $AIS_{S\text{-}corr}$)	Gal (g/100 g $AIS_{S\text{-}corr}$)	$Glc_{S\text{-}corr}$ (g/100 g $AIS_{S\text{-}corr}$)
Dried peels:					
Kaew Khiew	1.5	A_K	2.1 ± 0.04	25.1 ± 0.5	10.6 ± 0.5
		B_K	2.5 ± 0.1	23.7 ± 0.5	4.5 ± 0.2
	2.0	A_K	8.0 ± 0.2	24.9 ± 0.6	11.5 ± 1.7
		B_K	8.1 ± 0.1	24.1 ± 0.4	4.3 ± 0.4
Tommy Atkins	1.5	A_T	2.4 ± 0.01	11.3 ± 0.1	3.0 ± 0.2
		B_T	2.5 ± 0.1	10.4 ± 0.4	2.1 ± 0.1
	2.0	A_T	9.5 ± 0.2	11.5 ± 0.3	2.6 ± 0.1
		B_T	10.0 ± 0.2	11.4 ± 0.3	2.5 ± 0.2
Dried mesocarp:					
Kaew Khiew	1.5	A_K	2.0 ± 0.1	4.6 ± 0.2	22.4 ± 1.4
		B_K	1.5 ± 0.04	4.7 ± 0.1	10.9 ± 0.4
	2.0	A_K	6.8 ± 0.2	4.1 ± 0.1	15.6 ± 1.1
		B_K	6.9 ± 0.2	4.3 ± 0.1	8.0 ± 0.2
Tommy Atkins	1.5	A_T	2.2[1,2]	3.4[1,2]	4.3[1,2]
		B_T	2.1[1,2]	3.6[1,2]	4.1[1,2]
	2.0	A_T	8.8[1,2]	3.2[1,2]	4.6[1,2]
		B_T	8.4[1,2]	3.3[1,2]	4.6[1,2]

Ara, arabinose; Gal, galactose; $Glc_{S\text{-}corr}$, glucose (without glucose constituting starch).
[1]Standard error (SE) estimated without error propagation (single starch value).
[2]SE ≤ 0.01 g/100 g.

For peel AIS obtained from other Thai mango cultivars, galactose contents of 17-30 g/100 g of $AIS_{S\text{-}corr}$ were found (Sirisakulwat *et al.*, 2008).

As a major component of pulp AIS, glucose residues, not included in starch, reached levels up to 22 g/100 g of $AIS_{S\text{-}corr}$ in samples of less ripe KAk fruits (ripeness stage A_K). Though occurring at lower levels in peel AIS, such non-starch glucose residues also greatly contributed to AIS of the less ripe KAk peels (~11 g/100 g of $AIS_{S\text{-}corr}$). Their occurrence might be attributed to matrix glycans (Sirisakulwat *et al.*, 2008), since xyloglucan fragments (Yashoda *et al.*, 2005), which were obviously extractable by hot acid from mature-green fruits due to the early onset of xyloglucan degradation during maturation, gradually degraded during post-harvest ripening.

As expected (McDougall and Ring, 2004), pectin isolation was facilitated by extraction at pH 1.5 compared to the less acidic procedure (Table 3) due to improved hydrolysis

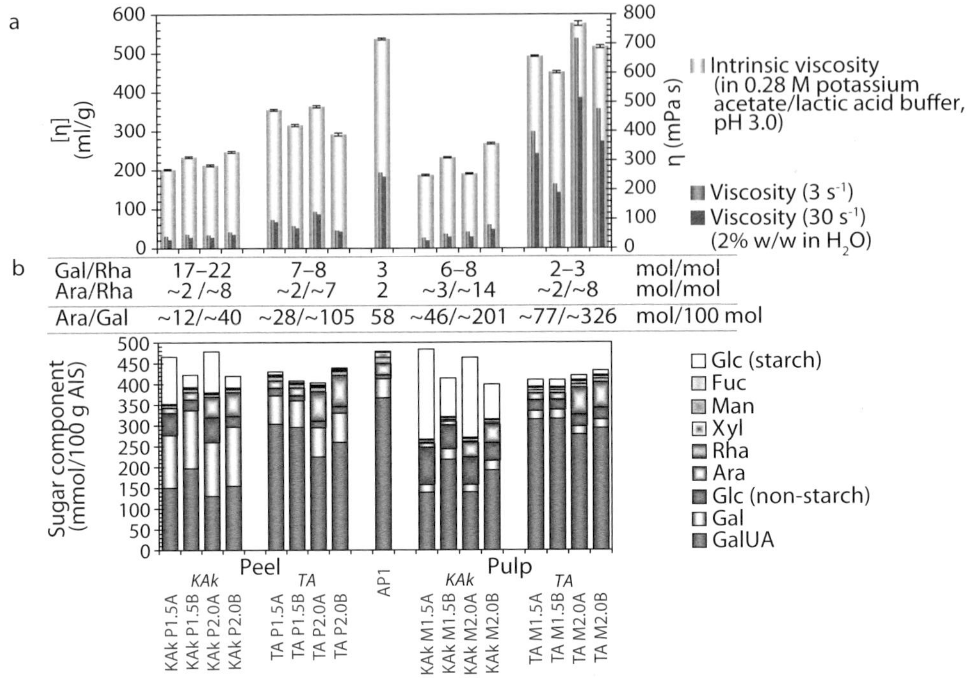

Figure 3. Structural indices of a commercial non-standardized apple pectin (AP1) and alcohol-insoluble substance (AIS) samples extracted from dried peels and pulp of mango fruit cvs. 'Kaew Khiew' (KAk) and 'Tommy Atkins' (TA) at two different ripeness stages (A, B; Table 1): a) Intrinsic viscosities and shear rate-specific viscosities of aqueous solutions (2% w/w) as well as b) the molar sugar composition. Ara, arabinose; Fuc, fucose; Gal, galactose; GalUA, galacturonic acid (titrimetric); Glc, glucose; Man, mannose; Rha, rhamnose; Xyl, xylose.

of the highly acid-labile arabinofuranosyl linkages. The retention of arabinose residues at levels of 7-10 g/100 g of $AIS_{S\text{-}corr}$ in AIS obtained at pH 2.0 (Table 4) gave a rough impression of the abundance of this neutral sugar in mango cell wall material. The contents of the other neutral sugars were hardly affected by the extracting agent (Figure 3b).

Depending on the cultivar, the part of the fruit, ripeness, and the extraction mode, galactose was the most abundant neutral sugar, followed by varying levels of non-starch glucose and arabinose (Figure 3b). They amounted to a total of up to 44.4 g/100 g of $AIS_{S\text{-}corr}$ (sample obtained at pH 2.0 from KAk peel at ripeness stage A_K). For comparison, the apple pectin AP1 contained galactose, arabinose, and non-starch glucose only at levels of 7.5 ± 0.1, 3.6 ± 0.1, and 1.5 ± 0.02 g/100 g of $AIS_{S\text{-}corr}$, respectively (Sirisakulwat *et al.*, 2008). Due to hot-acid extraction, commercial pectins are rich in galacturonic acid (75-90%; Thibault and Ralet, 2003; MacDougall and

Ring, 2004), but the molecules differ in their sizes, the methylation and acetylation degrees, intramolecular distribution of methyl-esterified units, and side chain lengths (Neidhart *et al.*, 2003; Guillotin *et al.*, 2005). The neutral sugars detected (Figure 3b) may imply a simplifying backbone model that consists of homogalacturonans (HG) and rhamnogalacturonans I (RGI), ignoring any occurrence of RGII, though minor RGII fractions might even be found in acid-extracted pectins (Yapo *et al.*, 2007). The molar ratios of AUA_{MHDP} to rhamnose (Rha) increased from 23-29 (KAk peel AIS) over 26-37 (TA peel AIS) and 35-45 (TA pulp AIS) to 50-63 (KAk pulp AIS). Whereas the AUA_{MHDP}/Rha ratios found for mango peel AIS were thus similar to those reported for apple and citrus pectins (Neidhart *et al.*, 2003; Coenen *et al.*, 2007), those of mango pulp AIS were slightly higher.

If all galactose residues detected were ascribed to galactans with one chain branching off from each rhamnose unit of RGI, long and/or highly ramified galactan side chains of 17-22 residues would occur in AIS from KAk peels according to the molar Gal/Rha ratios (Figure 3b). But the respective ratios observed for AIS from TA peel (7-8), KAk pulp (6-8), and TA pulp (2-3) indicated a broad variability among mango pectins. Likewise, the major fraction of cold-water-soluble cell wall polysaccharides of mango pulp (cv. 'Alphonso'), an acidic polysaccharide of 1818-1000 kDa, comprised AUA, galactose, arabinose, and rhamnose at recomputed ratios of 5.8/7.7/6.8/1 (mature-green fruit) and 3.2/4.5/4.5/1 (ripe fruit) (Yashoda *et al.*, 2005). Commercial apple and citrus pectins, including AP1 (3.0), only revealed Gal/Rha ratios < 3.3 (Neidhart *et al.*, 2003), similar to the RGI isolated from the acid-extracted fraction of citrus peel cell wall material (Yapo *et al.*, 2007).

On the other hand, a neutral arabinogalactan-type polymer ranked, beside two rhamnogalacturonans, among the major fractions of EDTA-soluble mango pulp pectins (Prasanna *et al.*, 2004). Consistently, hot-water extraction of alcohol-insoluble mango peel residues yielded pectins notably rich in galactose and arabinose (AUA/Gal/Ara/Rha ~ 15/19/7/1 and 23/11/12/ 1), when compared to hot-acid extraction (Koubala *et al.*, 2008).

The great variability of mango AIS as regards the sugar composition entailed a similar variability as to the intrinsic viscosity (Figure 3a). Despite the analysis in different solvents, the intrinsic viscosity of the apple pectin AP1 (537 ml/g) was within the range reported for commercial pectins of similar DE (Neidhart *et al.*, 2003) and a less methylated lime pectin (Koubala *et al.*, 2008). Figure 3 points out to a weak correlation (Pearson correlation coefficient $r = 0.85$) between the intrinsic viscosity and the galacturonic acid content ($GalUA_{titr}$ in mmol/100 g of AIS) of the 17 samples. Only the mango AIS samples obtained from TA pulp, which came most close to AP1 regarding the composition, revealed intrinsic viscosities in the same range (452-576 ml/g). The almost starch-free AIS from TA peels, characterized by medium galactose contents, had intermediate intrinsic viscosities (292-364 ml/g). KAk AIS, rich in neutral sugars throughout, showed minimum levels (187-268 ml/g). The notably low intrinsic viscosities of the latter two groups might thus rather be attributed to the presence of branched neutral polysaccharide elements than to a low average molar mass of pectin molecules.

Consistent with their high intrinsic viscosities, only the mango AIS samples obtained from TA pulp displayed high viscosities of respective standard solutions, partly even exceeding those of the AP1 solution (Figure 3a). Especially the thickening properties of TA pulp AIS recovered at pH 2.0 was remarkable. Beside comparatively high $GalUA_{titr}$ contents (55-58 g/100 g of $AIS_{S\text{-}corr}$; i.e. 80-83% on an ash- and moisture-free basis according to the legal specification), both samples were rather rich in arabinose residues (Figure 3b). Branched arabinan side chains of RGI and arabinogalactans may contribute to thickening through macromolecular entanglements, but might sterically disturb the formation of extended junction zones between homogalacturonans in sugar-acid gels. Consequently, the corresponding pulp AIS samples extracted at pH 1.5 were superior as regards the gelling capacity (Figure 2), despite slightly lower average molecular weights (793×10^3 - 906×10^3 vs. 1107×10^3 - 1254×10^3; Table 5).

Again, it was only TA pulp AIS recovered at pH 2.0 that showed average molecular weights similar to that of AP1. Table 5 reflects the broad range that was observed among the mango AIS samples (422×10^3 - 1254×10^3). Pulp AIS always revealed clearly higher average molecular weights than respective peel AIS. Whereas AIS extracted at pH 1.5 was always richer in galacturonic acid, higher average molecular weights were achieved by extraction at pH 2.0.

For mango AIS, characteristic molecular weight profiles were observed by analytical SEC when the AIS had been obtained by hot-acid extraction from the dried peel (Figure 4a). Unlike mango pulp AIS (Figure 4b), the former were governed by an

Table 5. Average molecular weights of a commercial non-standardized apple pectin (AP1) and alcohol-insoluble substance (AIS) samples extracted at pH 1.5 and 2.0 from dried peels and pulp of mango fruit cvs. 'Kaew Khiew' and 'Tommy Atkins' at two different ripeness stages (A_i, B_i; cf. Table 1).

	Weight-average molecular weight[1] (M_w) [$\times10^3$]								
	cv. Kaew Khiew				AP1	cv. Tommy Atkins			
	pH 1.5		pH 2.0			pH 1.5		pH 2.0	
	A_K	B_K	A_K	B_K		A_T	B_T	A_T	B_T
Peel	422	528	665	598		579	571	844	660
Pulp	422	566	790	807		906	793	1,254	1,107
Peel/pulp					1,192				

[1]relative to dextran standards.

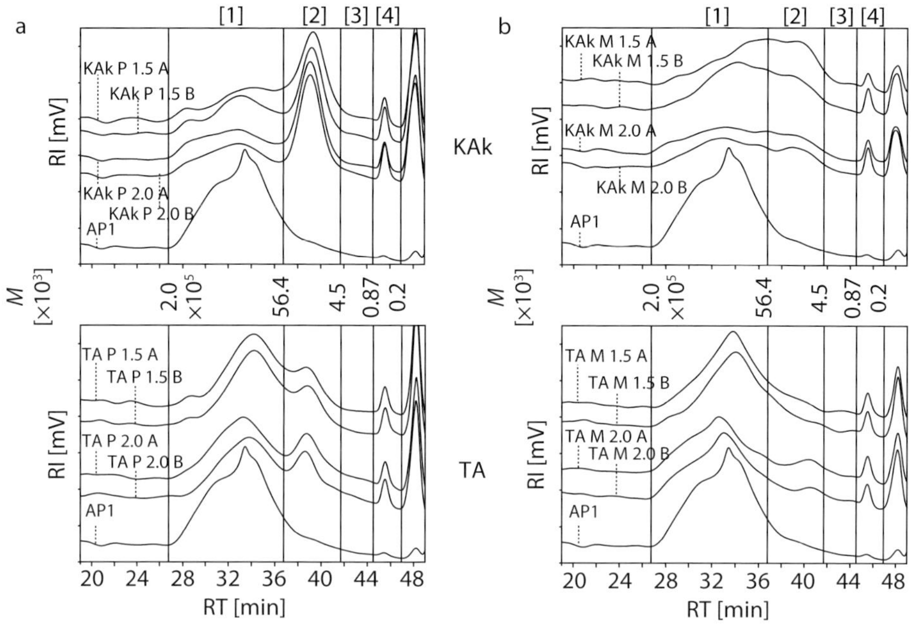

Figure 4. Molecular weight distribution of a commercial non-standardized apple pectin (AP1) and alcohol-insoluble substance (AIS) samples extracted at pH 1.5 and 2.0 from (a) dried peels (P) and (b) dried pulp (mesocarp, M) of mango fruit cvs. 'Kaew Khiew' (KAk) and 'Tommy Atkins' (TA) at two different ripeness stages (A, B; cf. Table 1). Molecular weights relative to dextran standards. The full calibration range comprised four zones ([1], [2], [3], [4]) that were empirically defined according to the characteristic peaks observed in zones [2] and [4] for all profiles of mango peel pectins studied.

almost monodisperse fraction with a peak molecular weight (M_p) of ~15000-22000 (zone [2] in Figure 4a). This fraction was predominant in KAK peel AIS, consistent with peel AIS of other Thai cultivars (Sirisakulwat *et al.*, 2008). In TA peel AIS, it was found to a much lesser, but important extent.

As shown by the comparison with AP1, where this fraction was absent, mango peel AIS also contained a corresponding high-molecular weight fraction (zone [1] in Figure 4a) in addition to the fraction in zone [2]. In terms of this high-molecular weight fraction, the profiles of TA peel AIS even came rather close to that of AP1. It was thus the fraction with a M_p of ~15000-22000 that caused the comparatively lower average molecular weights of mango peel AIS (Table 5) and increased polydispersity significantly.

As exemplarily shown by TA pulp AIS, especially after recovery at pH 2.0 (Figure 4b), the molecular weight distribution of pectins obtainable from mango pulp might be very similar to that of AP1. Minor amounts eluting in zone [2] were detected, but the characteristic monodisperse fraction, significantly lowering the average molecular weight of peel AIS, was absent. Compared to TA pulp AIS, the samples obtained from KAk pulp, which had a more heterogeneous composition (Figure 3b), were highly polydisperse. Hence, their average molecular weights were overall low (Table 5), but the fraction with a M_p of ~15000-22000 found in AIS of the respective peels was either absent or masked as a minor component. Similar differences in polydispersity of pulp pectins like those between AIS samples from TA and KAk pulp were reported for four other cultivars, leading to the assumption that the molecular weights of pulp pectins from polyembryonic cultivars exceed those of monoembryonic ones (Ollé *et al.*, 1996). However, among the samples shown in Figure 4, it was the pulp of the monoembryonic variety (TA) that yielded the purer pectins having higher molecular weights. Unlike the polyembryonic KAk fruits that had been subjected to ripening directly after harvest, the imported TA fruits chosen for purée production revealed notably low TSS (Table 1) even after ripening, possibly due to limited starch accumulation until harvest. Measures restricting the climacteric rise to extend the shelf life required for export marketing might have affected post-harvest ripening, thus possibly interfering with a mere cultivar effect.

The overall available data suggest that the impact of the cultivar on the variability of mango pectin quality may be rather specific, i.e. independent of the classification into mono- and polyembryonic varieties. Moreover, the cultivar seems to be less important than the part of the fruit used for extraction as well as fruit maturation and ripening. During post-harvest ripening, polygalacturonase (PG), β-galactosidase, and (1→4)-β-glucanase significantly increased in the pulp, while pectin methylesterase (PME) declined (Ali *et al.*, 2004). In the peels, however, starch and matrix glycan fragments almost degraded completely during ripening, but depolymerization of pectins and galactans was insignificant (Sirisakulwat *et al.*, 2008), complying with fairly low and rather constant activities of PG and PME in mango peel compared to the pulp (Yanru *et al.*, 1995).

4. Conclusions

Hot-acid extraction of dried mango by-products was shown to yield very different pectins. Only some of them were directly suitable for gelling or thickening agents, despite generally high to very high methylation degrees and minor acetylation. The broad variability of mango AIS in terms of yield and quality was documented by the presented examples. In view of starch-adjusted yields of ~13% on the average, a recovery potential might be expected for the peels only. However, the gelling and thickening capacities of peel pectins were limited by a characteristic, almost monodisperse fraction with a M_p of ~15000-22000 considerably reducing the average molecular weights. On the other hand, the TA samples exemplarily indicated that mango pectins with very good gelling and thickening properties might be produced from the pulp, but with insufficient yields (≤ 9.3%). On the whole, major obstacles to pectin recovery from mango by-products by conventional hot-acid extraction include the great variability of pectin quality, poor yields from dried pulp, and partly occurring galacturonic acid contents below the legal minimum due to high non-starch neutral sugar contents. Despite the described peculiarities, dried mango by-products might be comparable to apple pomace in some respect. Co-extracted starch can be significant as reflected by the KAk samples studied.

Acknowledgements

The authors thank Herbstreith & Fox KG, Neuenbürg, Germany, in particular Hans-Ulrich Endress and Christine Rentschler, for providing laboratory facilities for pectin extraction. Berit Jöns, Klaus Mix, and Martin Leitenberger are acknowledged for their technical assistance with the experiments at Hohenheim University, Dagmar Hettrich and Andrea Reinartz for their assistance in monitoring of TA fruit ripening and purée production on the pilot plant scale, and the staff members of the Uplands Program at Chiang Mai University for their assistance with the harvest and preprocessing of KAk fruits.

This research was funded by Deutsche Forschungsgemeinschaft (DFG), Bonn, Germany: project SFB 564-E2.2. It is part of the Special Research Program 'Research for Sustainable Land Use and Rural Development in Mountainous Regions of Southeast Asia' (The Uplands Program).

References

Ali, Z.M., L.-H. Chin and H. Lazan, 2004. A comparative study on wall degrading enzymes, pectin modifications and softening during ripening of selected tropical fruits. Plant Science 167: 317-327.

Berardini, N., R. Fezer, J. Conrad, U. Beifuss, R. Carle and A. Schieber, 2005. Screening of mango (*Mangifera indica* L.) cultivars for their contents of flavonol *O*- and xanthone *C*-glycosides, anthocyanins, and pectin. Journal of Agricultural and Food Chemistry 53: 1563-1570.

Coenen, G.J., E.J. Bakx, R.P. Verhoef, H.A. Schols and A.G.J. Voragen, 2007. Identification of the connecting linkage between homo- or xylogalacturonan and rhamnogalactuonan type I. Carbohydrate Polymers 70: 224-235.

Endress, H.-U. and S. Dilger, 1990. Checking pectin jelly strength with the Pektinometer. Food Technology International 1: 279-282.

Endress, H.-U., F. Mattes and K. Norz, 2006. Pectins. In: Y.H. Hui (ed.) Handbook of Food Science, Technology, and Engineering. CRC Press, Taylor & Francis Group, Boca Raton, FL, USA, Vol. 3, pp. 140/1-140/35.

FAOSTAT, 2006. FAOSTAT Agricultural data, last updated December 2005 (Agriculture & Food Trade) and January 2006 (Agricultural Production). In: Economic and Social Department, Statistics Division (ed.) Statistical Databases of the Food and Agriculture Organization of the United Nations. Food and Agriculture Organization of the United Nations, Rome, Italy. http://www.fao.org.

Guillotin, S.E., E.J. Bakx, P. Boulenguer, J. Mazoyer, H.A. Schols and A.G.J. Voragen, 2005. Populations having different GalA blocks characteristics are present in commercial pectins which are chemically similar but have different functionalities. Carbohydrate Polymers 60: 391-398.

JECFA (Joint FAO/WHO Expert Committee on Food Additives), 2007. Compendium of Food Additive Specifications. FAO JECFA Monographs 4. Food and Agriculture Organization of the United Nations, Rome, Italy. http://www.fao.org/docrep/010/a1447e/ a1447e00.htm.

Kratchanova, M., C. Bénémou and C. Kratchanov, 1991. On the pectic substances of mango fruits. Carbohydrate Polymers 15: 271-282.

Koubala, B.B., G. Kansci, L.I. Mbome, M.-J. Crépeau, J.-F. Thibault and M.-C. Ralet, 2008. Effect of extraction conditions on some physicochemical characteristics of pectins from 'Améliorée' and 'Mango' mango peels. Food Hydrocolloids 22 (7): 1345-1351.

MacDougall, A.J. and S.G. Ring, 2004. Pectic Polysaccharides. In: P. Tomasik (ed.) Chemical and functional properties of food saccharides. CRC Press, Boca Raton, FL, USA, pp. 181-195.

Neidhart, S., C. Hannak and K. Gierschner, 2003. Sol-gel transition of high-esterified pectins and their molecular structure. In: A.G.J. Voragen, H.A. Schols and R. Visser (eds.) Advances in pectin and pectinase research. Kluwer Academic Publishers, Dordrecht, the Netherlands, pp. 431-448.

Neidhart, S., A.L. Vásquez-Caicedo, B. Mahayothee, I. Pott, W. Mühlbauer, P. Sruamsiri and R. Carle, 2007. The control of postharvest ripening processes and its implications for the productivity of mango processing. In: F. Heidhues, L. Herrmann, A. Neef, S. Neidhart, J. Pape, P. Sruamsiri, D.C. Thu and A. Valle Zárate (eds.) Sustainable land use in mountainous regions of Southeast Asia: Meeting the challenges of ecological, socio-economic and cultural diversity. Springer-Verlag, Berlin, Heidelberg, New York, pp. 134-146, 163-171.

Ollé, D., Y.F. Lozano and J.-M. Brillouet, 1996. Isolation and characterization of soluble polysaccharides and insoluble cell wall material of the pulp from four mango (*Mangifera indica* L.) cultivars. Journal of Agricultural and Food Chemistry 44: 2658-2662.

Pott, I., S. Neidhart, W. Mühlbauer and R. Carle, 2005. Quality improvement of non-sulphited mango slices by drying at high temperatures. Innovative Food Science and Emerging Technologies 6 (4): 412-419.

Prasanna, V., T.N. Prabha and R.N. Tharanathan, 2004. Pectic polysaccharides of mango (*Mangifera indica* L): structural studies. Journal of the Science of Food and Agriculture 84: 1731-1735.

Sirisakulwat, S., A. Nagel, P. Sruamsiri, R. Carle and S. Neidhart, 2008. Yield and quality of pectins extractable from the peels of Thai mango cultivars depending on fruit ripeness. Journal of Agricultural and Food Chemistry 56: 10727-10738.

Thibault, J.F. and M.-C. Ralet, 2003. Physico-chemical properties of pectins in the cell walls and after extraction. In: A.G.J. Voragen, H.A. Schols and R. Visser (eds.) Advances in pectin and pectinase research. Kluwer Academic Publishers, Dordrecht, the Netherlands, pp. 91-105.

Vásquez-Caicedo, A.L., S. Neidhart, P. Pathomrungsiyounggul, P. Wiriyacharee, A. Chattrakul, P. Sruamsiri, P. Manochai, F. Bangerth, R. Carle, 2002. Physical, chemical and sensory properties of 9 Thai mango cultivars and evaluation of their technological and nutritional potential. International Symposium 'Sustaining Food Security and Managing Natural Resources in Southeast Asia: Challenges for the 21st Century', 8-11 January 2002, Chiang Mai, Thailand. https://www.uni-hohenheim.de/sfb564/project_area_e/e2/e2_publications. php.

Vásquez-Caicedo, A.L., S. Neidhart and R. Carle, 2004. Postharvest ripening behavior of nine Thai mango cultivars and their suitability for industrial applications. Acta Horticulturae (ISHS) 645: 617-625.

Vásquez-Caicedo, A.L., P. Sruamsiri, R. Carle and S. Neidhart, 2005. Accumulation of all-*trans*-β-carotene and its 9-*cis* and 13-*cis* stereoisomers during postharvest ripening of nine Thai mango cultivars. Journal of Agricultural and Food Chemistry 53: 4827-4835.

Vásquez-Caicedo, A.L., S. Schilling, R. Carle and S. Neidhart, 2007. Effects of thermal processing and fruit matrix on β-carotene stability and enzyme inactivation during transformation of mangoes into purée and nectar. Food Chemistry 102 (4): 1172-1186.

Yanru, Z., M. Pandey, N.K. Prasad and G.C. Srivastava, 1995. Ripening associated changes in enzymes and respiratory activities in three varieties of mango (*Mangifera indica* L.). Indian Journal of Plant Physiology 38 (1): 73-76.

Yapo, B.M., P. Lerouge, J.-F. Thibault and M.-C. Ralet, 2007. Pectins from citrus peel cell walls contain homogalacturonans homogenous with respect to molar mass, rhamnogalacturonan I and rhamnogalacturonan II. Carbohydrate Polymers 69: 426-435.

Yashoda, H.M., T.N. Prabha and R.N. Tharanathan, 2005. Mango ripening - chemical and structural characterization of pectic and hemicellulosic polysaccharides. Carbohydrate Research 340: 1335-1342.

Extraction and functional properties of 'green labelled' pectin from plant by-products

A. Zykwinska, M. Panouillé, J.-F. Thibault and E. Bonnin
INRA, UR 1268 Biopolymères, Interactions, Assemblages, 44000 Nantes, France

Abstract

Pectin is a family of complex and heterogeneous polysaccharides, which can be extracted from plant cell walls and used in food industry as additives due to their gelling, stabilising and thickening properties. At the industrial scale, pectins are extracted in acidic conditions most often from apple pomace and citrus peels. However, acidic extraction generates large amounts of effluents that require to be treated. Moreover, the consumer demand for 'green' products stimulates the search for alternative means of extraction. Therefore, enzymes could represent an alternative and environmentally friendly way to extract pectins. Following this idea, the use of proteases and cellulases preparations was investigated in order to extract 'green labelled' pectins from plant by-products. Different enzymatic preparations were characterized by their activities towards proteins, cellulose and pectins. These preparations were then tested regarding pectin extraction from different plant by-products (chicory roots, citrus peel, …). The yield of enzymatic extraction of pectins was lower (12%) compared to acidic one (28%). However, the enzyme-extracted high-methoxy (HM) pectins presented better gelling properties in the presence of sucrose at low pH (3) than the acid-extracted pectins. These HM pectins can be demethylated using Pectin Methyl Esterases (PMEs). Demethylation could be done after protease and cellulase extraction (2-steps treatment) or PME can directly be added to the protease and cellulase mixture (1-step treatment). In both cases, low-methoxy (LM) pectins present good gelling properties in the presence of calcium. It is therefore possible to obtain 'green labelled' HM and LM pectins with real functional properties.

Keywords: enzymatic extraction, by-product upgrading, cellulase, protease, cell wall polysaccharides

1. Introduction

Plant cell walls are mainly constituted of proteins and polysaccharides, among which cellulose, hemicelluloses and pectins. Pectin is a family of complex and heterogeneous polysaccharides consisting of distinctive covalently linked domains, namely homogalacturonans (HG) and rhamnogalacturonans (RG) (Voragen *et al.*, 1995, Vincken *et al.*, 2003). Industrial pectins are extracted from fruit juice industry by-products, mainly apple pomace and citrus peels. The most common industrial procedure for pectin extraction requires mild acidic conditions. Further chemical de-esterification leads to production of whole range of pectins, which vary essentially by their degree of methylation (DM). Thus, High Methoxy (HM) and Low Methoxy (LM) pectins may be obtained. In the food industry, HM pectins are widely used as gelling agents in the production of jams, marmalades, jellies, confectionery, while LM pectins are often used as stabilisers in acidified milk drinks including sour milk products and mixtures of fruit juice and milk (May, 1990; Rolin *et al.*, 1998). The industrial method used for the pectin manufacture has however some limitations. It generates large amounts of acidic effluents that need further treatment before being removed. It may also provoke undesired degradation of pectins during extraction. Following the growing demand for so-called 'green-chemistry', encouraging processes that reduce or eliminate the use or generation of hazardous reagents, a new eco-friendly method was developed for the pectin extraction. Assuming that pectins are embedded in the cell walls together with a cellulose-hemicellulose network and a protein network, the use of enzymes able to degrade each network respectively, was investigated. The enzymatic extraction of pectins was extended to plant by-products other than apple pomace and citrus peels, such as chicory root, cauliflower florets and leaves, endive pulp and sugar beet pulp. These by-products are also available in large quantities and are particularly rich in pectins.

2. Material and methods

2.1 Raw materials and cell wall material

Chicory roots were obtained from Lemaître SA, France. Dried citrus peels were obtained from Cargill Texturizing Solutions (France), cauliflower sample (florets and leaves) from Bretagne Biotechnologie Végétale (France), dried endive pulp from Orafti (Belgium), and fresh sugar beet pulp from Cagny sugar factory (France).

Cell wall material (CWM) was prepared from each by-product by boiling in five volumes of 96% (v/v) ethanol for 30 min. The slurry was filtered on a G3 sintered glass and the insoluble material was washed with 70% (v/v) ethanol. The solid was then dried by solvent exchange (ethanol, acetone) and left overnight at 40 °C.

2.2 Enzymes

Proteases. Neutrase (N) preparation produced by *Bacillus subtilis* was supplied by Novozymes (Copenhagen, Denmark). Promod 298L, 648L, 24L preparations produced by *Bacillus sp.* were provided by Biocatalyst (Cardiff, Wales, UK).

Cellulases. Cellulyve TR400 (CL), a solid preparation produced by solid-state fermentation by *Trichoderma reesei*, was provided by Lyven (Colombelle, France). A mixture of purified *Trichoderma reesei* cellulases (PC) containing cellobiohydrolases

(CBH I, CBH II) and endoglucanases (EG I, EG II) was prepared as described previously (Suurnäki *et al.*, 2000).

Pectin Methyl Esterases (PMEs). The orange peel PME (PME-O) was purchased from Sigma (France), as a solid preparation. A fungal PME from *Aspergillus aculeatus* (PME-A) was expressed in *Aspergillus oryzae*, and provided by Novozymes as a solid preparation. Solid preparations were dissolved at 10 mg/ml in sodium acetate buffer (50 mM, pH 5.5), with the exception for PME-O dissolved at 0.5 mg/ml

2.3 Pectin extractions

Acidic extraction of pectins. 5 grams of chicory root CWM were extracted with 500 ml of HCl (pH 1.3) for 1 hour at 85 °C. The extraction was performed three times. After centrifugation, the extract was adjusted to pH 5.5 with 1 M NaOH and extensively dialyzed against water before freeze-drying.

Enzymatic extraction of HM pectins. 5 grams of CWM were suspended in 500 ml of sodium acetate buffer (50 mM, pH 5.5) for 30 min at room temperature. 250 µl of protease and 250 µl of cellulase were then added. Two incubation conditions were applied: (*1*) 16 h at 40 °C and (*2*) 4 h at 50 °C. Enzymes were inactivated by boiling for 10 min. The extract separated from the residue was extensively dialyzed against water before freeze-drying.

Enzymatic demethylation of HM pectins with PMEs. HM pectins (150 mg) were solubilized in 30 ml of sodium acetate buffer (50 mM, pH 5.5) for the treatment with the PME-A (240 nkat) or 30 ml of sodium phosphate buffer (100 mM, pH 7) for the treatment with the PME-O (180 nkat). Pectin solutions were incubated at 40 °C with PME-A and at 30 °C with PME-O under continuous stirring. A new enzyme portion was added after 2 h. After 4 h of incubation, solutions were dialyzed against water and freeze-dried.

Enzymatic extraction of LM pectins. 5 grams of CWM were suspended in 250 ml of sodium acetate buffer (50 mM, pH 5.5) for 30 min at room temperature. 125 µl of protease and 125 µl of cellulase were then added. 125 µl of the PME-A was added in two times, at the beginning and after 2 h of incubation. After 4h at 50 °C, enzymes were inactivated by boiling for 10 min. The extract separated from the residue was extensively dialyzed against water before freeze-drying

2.4 Analytical methods

The individual neutral sugars were analyzed as their alditol acetate derivatives by gas-liquid chromatography (Blakeney *et al.*, 1983) using inositol as the internal standard. Soluble samples were hydrolyzed by 2 M H_2SO_4 at 100 °C for 2 h. Insoluble samples were first pre-hydrolysed by 72% (w/v) H_2SO_4 at 25 °C for 1h30. Uronic acid content (as GalA) was determined colorimetrically by the automated *m*-hydroxybiphenyl method (Thibault, 1979).

Methanol and acetic acid were released by alkaline deesterification in the presence of $CuSO_4$ and quantified by HPLC on a C18 column according to Levigne *et al.* (2002a), using isopropanol as the internal standard. Degree of methylation (DM) and degree of acetylation (DAc) were calculated as the molar ratio of methanol and acetic acid to GalA, respectively.

Molar mass (M_W, weight-average molar mass) was determined by light scattering detection after high performance size exclusion chromatography (HPSEC) on a Shodex SB-G pre-column followed by two Shodex OH-pak SB HQ 805 and 804 columns in series with a multiangle laser light scattering detector (MALLS; mini Dawn, Wyatt, Santa Barbara, CA, USA), and a differential refractometer (ERC 7517 A). Elution was performed with 50 mM $NaNO_3$ containing 0.02% NaN_3 at a flow rate of 0.7 ml/min at room temperature. The system was calibrated using pullulan standards. The weight-average molar mass was calculated with ASTRA™ 1.4 software (Wyatt Technology) using a refractive index increment d*n*/dc = 0.146 g/ml.

2.5 Preparation of gels

HM pectins. 25 mg of chicory pectin were dissolved in 2 ml of a citrate buffer (50 mM, pH 3). After stirring overnight at 4 °C, 3 g of sucrose was added. The mixtures were then heated under continuous stirring at 90 °C for 30 min in an oil bath prior to be loaded onto a controlled-strain rheometer (AR2000, TA Instruments) equipped with a Peltier temperature controller and with a cone-plane device (40 mm diameter, 4° angle).

LM pectins. 50 mg of chicory pectin were dissolved in 4.8 ml of distilled water by stirring overnight at room temperature. After solubilization, the pH was adjusted to 6 with NaOH and the final volume was adjusted to 5 ml. 2 ml of the pectin solution were then heated under continuous stirring at 70 °C for 15 min in an oil bath. 2 ml of a preheated $CaCl_2$ solution (2 mM or 6 mM, pH 6) was added to the samples dropwise under continuous stirring prior to be loaded onto the AR2000 rheometer.

2.6 Dynamic oscillatory measurements

Time sweep oscillatory measurements were performed at a frequency of 1 rad/s for a strain amplitude of 1%. Frequency sweep oscillatory measurements were performed at the end of 24 h kinetic. The frequency was swept from 100 rad/s to 0.01 rad/s. Different pectin systems prepared were loaded onto the preheated rheometer maintained at the temperature of solution. The temperature was then decreased and was maintained at 20 °C during the measurements. At the end of experiments, a strain sweep test was performed in order to check that the measurements were done within the linearity limits of the visco-elastic behaviour.

3. Results and discussion

3.1 Is it possible to extract pectin by the use of cellulase and protease?

Chicory root CWM was used to compare acid and enzymatic extractions (Table 1). It is very rich in galacturonic acid (27.9%), rhamnose (1.4%), arabinose (7.2%) and galactose (3.9%), suggesting that pectins are the major polysaccharides in chicory roots. CWM also contains a large quantity of glucose (23.2%), resulting mainly from cellulose and some minor xyloglucans (xylose 3.3%). These results are in accordance with the data from Femenia *et al.* (1998).

Acidic extraction leads to a yield of 27.8% (dry weight / dry weight CWM). Pectins are rich in GalA (53.1%), which represents more than a half of GalA present in CWM. They contain 1.7% rhamnose and are also rich in neutral sugars, especially in arabinose (12.8%). The DM is around 50, which classifies the acid pectin between HM and LM pectins. A non-negligible part of proteins is co-extracted with the pectin (7.5%).

After testing different combinations of cellulases and proteases (Panouillé *et al.*, 2006), the enzymatic extraction was performed with Neutrase and Cellulyve for 16 h at 40 °C. This treatment leads to a higher yield than the acidic extraction (34.6% *vs* 27.8%) and the enzymatic extract is richer in GalA 58.6% *vs* 53.1%. The DM is slightly higher (53 *vs* 49), showing that the pectin structure is more preserved by enzymatic than by acidic means. The enzymatic extraction allows the extraction of more than 80% of the pectic sugars.

HPSEC patterns for pectins extracted either by acid or by enzymes are shown in Figure 1. The average molar mass (M_w) of pectins, calculated for the pectin fraction eluted between 11 and 16 ml, was determined to be 500 kDa for the acid pectin and 200 kDa for the enzyme pectin. Taken together, the chemical composition and the macromolecular features of the extracts show high pectin heterogeneity, suggesting

Table 1. Extraction yields and chemical composition of chicory root CWM and chicory pectins extracted by hydrochloric acid and by enzymatic combination Neutrase + Cellulyve for 16 h at 40 °C.

	Yield[1] (%)	Composition (%)							DE		M_W (kDa)
		Rha	Ara	Xyl	Gal	Glc	GalA	Protein	DM	DAc	
CWM	5.4	1.4	7.2	3.3	3.9	23.2	27.9	7.4	68	36	-
Acid	27.8	1.7	12.8	0.3	5.9	0.9	53.1	7.5	49	12	500
Enzyme	34.6	2.1	9.1	0.8	4.0	2.9	58.6	2.2	53	17	200

[1]Yield is expressed as a percentage of fresh material for the CWM, and as a percentage of dry CWM for the pectins.

the presence in the cell wall of different pectin populations, as already reported for sugar beet (Levigne *et al.*, 2002b) or potato (Zykwinska *et al.*, 2006).

The ability to gel was investigated for the two pectins by rheological measurements. Due to the intermediate value of DM, gelling ability was studied at 1% pectin concentration in the presence of 60% sucrose at pH 3. The acid pectin forms a weak gel and shows a storage modulus G' around 10 Pa, whereas the enzymatic one does not show any gelling ability (data not shown). This is likely due to both the DM, which seems to low to permit the efficient hydrophobic interactions to be set up as described for HM native pectin (Oakenfull and Scott, 1984), and to the molar mass, which seems too low compared to the acid pectin.

Consequently, this new process, allowing the extraction of pectins by enzymatic means does not lead to extraction of pectins with functional properties. Therefore, some new conditions were searched in order to prepare functional pectins.

3.2 New conditions for the extraction of functional pectins

The enzymatic activities were measured on the commercial preparations that were selected for the extraction (Table 2). The use of various cell wall polysaccharides as

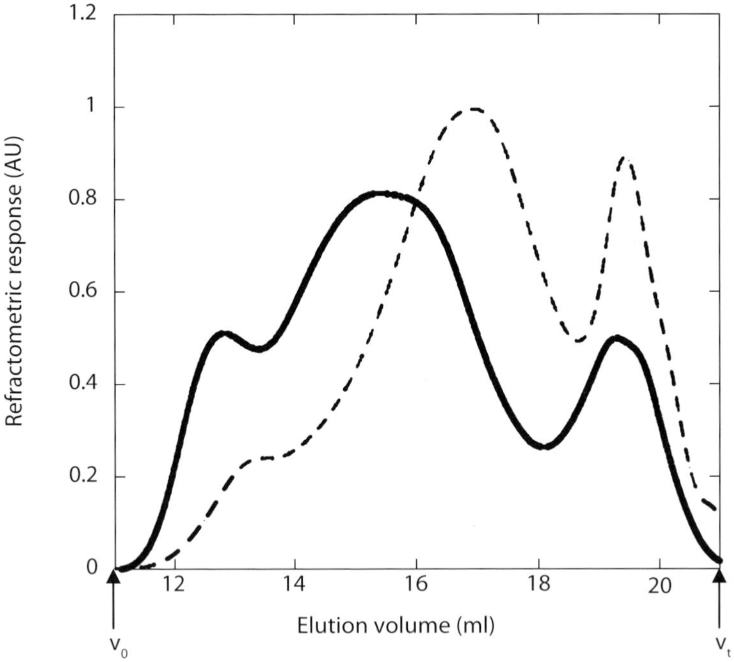

Figure 1. HPSEC profiles of pectins extracted with acid (full line) or with a combination of Neutrase + Cellulyve (dotted line).

Table 2. Enzymatic activities in the commercial preparations (protease activity expressed in mg protein/ml, the others in nkat/ml).

	Protease	CMC	XG	PG	RG	Ara	Gal
Neutrase	135	1	<1	1	9	14	9
Promod 24L	1,375	0	0	5	1	2	8
Cellulyve	<1	670	380	11	9	<1	<1
Cellulase PC	0	1,500	605	0	<1	<1	0

The solid preparations (Neutrase and Cellulyve) were solubilised at 10 mg/ml.

substrates shows that Neutrase and Cellulyve contain measurable levels of pectin degrading enzymes. By testing separately the activities towards the different pectic domains, it can be shown that Neutrase can degrade the hairy regions, as it contains rhamnogalacturonanases, arabinanases and galactanases. Moreover, Cellulyve contains polygalacturonase. Thus, the association of these two preparations are able to degrade pectin progressively. To avoid this undesired phenomenon, some other preparations were selected for extraction, namely the protease Promod 24L and a cellulase mixture designed by the VTT (Cellulase PC).

The protease Promod 24L was tested in combination with either Cellulyve or Cellulase PC, and the results were compared to the extraction with Neutrase and Cellulyve (Table 3). Moreover, the temperature and time of incubation were modified to limit, as far as possible, the contaminant activities to act. The incubation was run for 4h at 50 °C, allowing simultaneously to meet the industrial requirements for microbiological safety of extracted products.

Table 3. Extraction yields, chemical composition, and macromolecular features of pectins extracted from chicory root CWM with enzymes for 4h at 50 °C.

	Yield (%)	Composition (%)							DE (%)		Mw (kDa)	G' (Pa)
		Rha	Ara	Xyl	Gal	Glc	GalA	Protein	DM	DAc		
Neutrase +Cellulyve	11.4	0.6	6.1	0.4	3.6	3.0	52.8	1.9	45	9	230	40
Promod 24L +Cellulyve	13.1	0.6	3.2	0.4	3.9	3.2	55.1	2.3	52	11	260	65
Promod 24L +Cellulase PC	12.0	0.6	11.7	0.3	3.3	3.0	45.9	2.3	59	11	250	70

Similar extraction yields (11.4, 12.0 and 13.1% of dry CWM) are obtained whatever the enzyme association used. Chemical composition analysis reveals that extracted pectins are mainly composed of GalA, which represents from 20% to 26% of the initial GalA present in chicory root CWM. The extraction using Promod 24L and Cellulyve leads to pectin with the richest GalA and the poorest neutral side chains (arabinan and galactan) contents. The DMs of pectins extracted by Promod 24L are much higher than the DM obtained when Neutrase was used. Similarly, the average molar mass is decreased by the use of Neutrase. The use of Cellulase PC instead of Cellulyve induces a higher content in arabinose, showing longer arabinan side chains.

Once again, the gel properties of these pectins were investigated at 1% dry weight in the presence of 60% sucrose at pH 3. Figure 2 shows a typical example of storage modulus G' and loss modulus G" as a function of applied frequencies. It shows that G' is higher than G" in the whole range of frequency, meaning that the system gels. A slight frequency dependence observed for G' suggests that the system forms a weak gel.

The comparison of G' modulus of the three pectins shows that the G' increases when the DM increases (Table 3). These findings allow us to conclude that pectin extracted with Promod 24L and Cellulyve for 4h at 50 °C may gel in the conditions required for HM pectins.

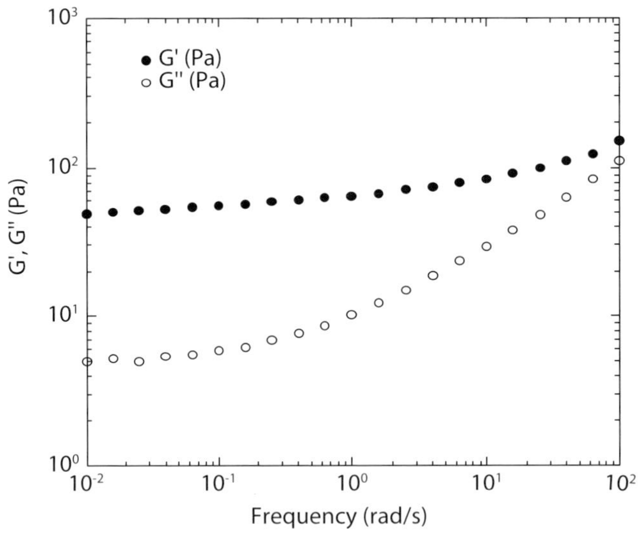

Figure 2. Rheological behaviour of the chicory pectin obtained with Promod 24L and Cellulyve (4h at 50 °C).

3.3 Is it possible to produce LM pectins from chicory root?

The controlled modification of HM pectins produced from chicory root was investigated. For this purpose, the pectin extracted with Promod 24L and Cellulyve was further treated by equal amount of two different PMEs, one from orange (PME-O) and one from *Aspergillus aculeatus* (PME-A). In the first case, the DM decreased from 52 to 22 and in the second case, it decreased from 52 to 11. The sugar composition of the demethylated pectins was similar to that of the corresponding native pectins (data not shown).

Due to the very low DM of the demethylated pectins, their gelling ability was tested at 1% pectin concentration in the presence of 2 mM $CaCl_2$ at pH 6 (Figure 3).

No frequency dependence was observed for both demethylated pectins as G' was constant in the whole range of frequency (Figure 3). Such behaviour is common for rigid gels. Moreover, the highest G' (~120 Pa) was observed for the pectin demethylated with the PME-A, with the lowest DM. Thus, PME-A seems the most efficient PME to decrease the DM of pectins, leading to formation of a strong gel.

Thus it was demonstrated that it is possible to produce LM chicory pectins that are able to gel. Two successive steps, the enzymatic extraction of native HM pectins followed by their enzymatic demethylation were however necessary. These encouraging results led us to test the one step extraction of LM pectins from chicory root CWM by adding the PME-A directly to the Promod 24L and Cellulyve mixture.

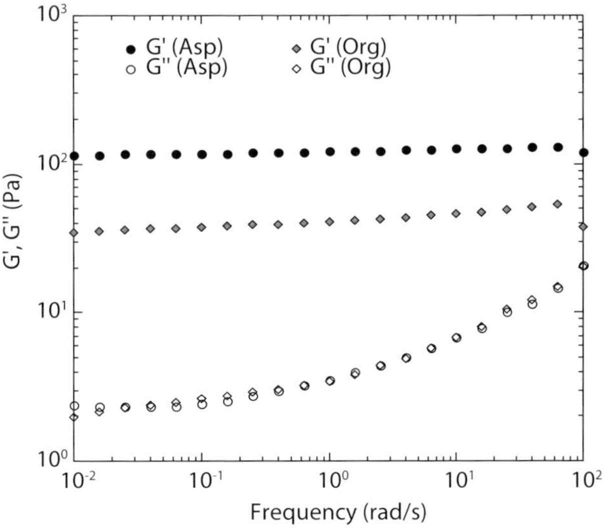

Figure 3. Rheological behaviour of LM chicory pectin extracted with Promod 24L and Cellulyve demethylated by PME-O (circles) or PME-A (diamonds).

In this case, the extraction yield is around 10%, the extracted fraction has a DM of 23 but it contains less GalA compared to native pectin subsequently demethylated with the PME-A (50% *vs* 54.7%). The amount of other sugars is similar (data not shown). The HPSEC profile of LM pectin extracted with Promod 24L, Cellulyve and PME-A differs only slightly from that of HM pectin extracted with Promod 24L and Cellulyve (Figure 4). The weight-average molar mass of LM pectin is around 330 kDa.

LM chicory pectin extracted with Promod 24L, Cellulyve and PME-A forms a weaker gel at 1 wt% pectin concentration (G' ~30 Pa) compared to LM chicory pectin extracted with Promod 24L and Cellulyve, and further demethylated (G' ~120 Pa). Higher amount of calcium is also required (6 mM). These differences in gelling ability can be explained by higher DM of LM pectin extracted in one step (23 *vs.* 11). It is likely that high pectin entanglement with other cell wall components during extraction and demethylation, may lead to lower accessibility of the PME-A to the pectin backbone.

Taken all together, these results lead to the conclusion that the enzymatic way using Promod 24L, Cellulyve and PME-A seems interesting for the extraction in only one step of LM chicory pectins which are able to gel.

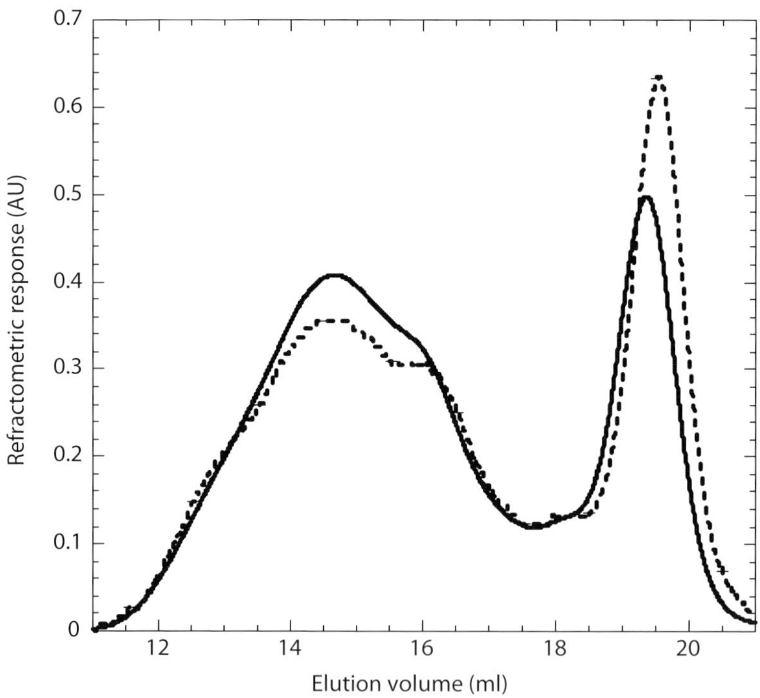

Figure 4. HPSEC profiles of chicory pectins extracted with Promod 24L+Cellulyve (full line) and Promod 24L+Cellulyve+PME-A (dotted line).

3.4 Is it possible to apply this new extraction procedure to other by-products?

According to the results obtained on chicory root CWM, the enzyme association Promod 24L and Cellulyve was tested to extract pectin from citrus peel, cauliflower florets and leaves, endive and sugar beet pulps. Yields, chemical compositions and macromolecular characteristics of extracted pectins are presented in Table 4.

Extraction yields vary from 4% for sugar beet to 12.6% for citrus peel. The extracted pectins have different compositions according to their origin. The highest GalA content is found in citrus pectin (75.6%), which corresponds to 29% of the initial GalA. Cauliflower, endive and sugar beet extracts contained less GalA. All extracts are rather rich in arabinan, especially sugar beet extract, and possess some galactan. Citrus and sugar beet pectins have DM of 68, while cauliflower and endive pectins have lower DM of 56 and 54, respectively. Cauliflower extract has a noticeably high DAc and the lowest molar mass (140 kDa). Citrus pectin has intermediate molar mass (180 kDa), whereas sugar beet pectin has the highest one (310 kDa). The weight-average molar mass of endive pectin was difficult to assess due to its high heterogeneity.

From these results, it appears that the enzyme combination selected for chicory root is also very efficient to extract pectin from citrus peel. It can therefore be concluded that citrus peel and chicory root seem promising sources for the recovery of high amounts of enzyme-extracted pectins. The enzymatically-extracted citrus pectin has similar characteristics as commercially available citrus pectins extracted by acid (Rollin, 1998, Morris *et al.*, 2000). However, the enzyme combination is rather inefficient on sugar beet or cauliflower by-products. Therefore, it is unlikely to set up one enzymatic method suitable for any plant matrix.

Table 4. Extraction yields, chemical composition and macromolecular features of enzyme-extracted pectins from different by-products.

	Yield (%)	Composition (%)							DE (%)		Mw (kDa)
		Rha	Ara	Xyl	Gal	Glc	GalA	Protein	DM	DAc	
Citrus	12.6	1.0	7.1	0.7	3.4	1.2	75.6	0.4	68	2	180
Cauliflower	6.3	0.4	5.3	0.8	6.3	1.5	15.6	3.3	56	50	140
Endive pulp	7.8	0.8	8.0	0.5	4.1	4.5	23.3	0.9	54	19	nd
Beet pulp	4.0	1.1	10.3	0.4	4.0	5.2	40.3	1.7	68	44	310

4. Conclusion

In this work, various plant by-products were tested as pectin sources. Different combinations of cellulases and proteases were tested and the gelling properties of the extracted pectins were investigated. It can be concluded that proteases and cellulases can be efficiently used to extract pectins having gelling properties. However, the best combination of enzymes has to be determined for each by-product, since the most suitable combination demonstrated on chicory root was rather inefficient on some other by-products. Depending on the concerned activities, it is possible to extract native HM pectin as well as LM pectin. It is also possible to use accessory enzymes to tailor the native pectin and thus control their functional properties. In any case, these enzymatic-extracted pectins will deserve a 'green label' as they are produced following an eco-friendly process avoiding the use or generation of hazardous substances.

References

Blakeney, A.B., P.J. Harris, R.J. Henry and B.A. Stone, 1983. A simple and rapid preparation of alditol acetates for monosaccharides analysis. Carbohydrate Research 113: 291-299.

Femenia, A., J.A. Robertson, K.W. Waldron and R.R. Selvendran, 1998. Cauliflower (*Brassica oleracea L*), globe artichoke (*Cynara scolymus*) and chicory witloof (*Chicorium intybus*) processing by-products as sources of dietary fibre. Journal of the Science of Food and Agriculture 77: 511-518.

Levigne, S., M. Thomas, M.-C. Ralet, B. Quéméner and J.-F. Thibault, 2002a. Determination of the degrees of methylation and acetylation of pectins using a C18 column and internal standards. Food Hydrocolloids 16: 547-550.

Levigne, S., M.-C. Ralet and J.-F. Thibault, 2002b. Characterisation of pectins extracted from fresh sugar beet under different conditions using an experimental design. Carbohydrate Polymers 49: 14153.

May, C.D., 1990. Industrial Pectins: Sources, Production and Applications. Carbohydrate Polymers 12: 79-99.

Morris, G.A., T.J. Foster and S.E. Harding, 2000. The effect of the degree of esterification on the hydrodynamic properties of citrus pectin. Food Hydrocolloids 14: 227-235.

Oakenfull, D and A. Scott, 1984. Hydrolphobic interactions in the gelation of high methoxyl pectins. Journal of Food Science 49: 1093-1098.

Panouillé, M., J.-F. Thibault and E. Bonnin, 2006. Cellulase and protease preparations can extract pectins from various plant byproducts. Journal of Agricultural and Food Chemistry 54: 8926-8935.

Rolin, C., B.U. Nielsen, P.-E. Glahn, 1998. Pectin. In: S. Dimitrin (ed.) Polysaccharides, structural diversity and functional versatility. Marcel Dekker, London, pp 377-431.

Suurnäkki, A., M. Tenkanen, M. Siika-Aho, M.L. Niku-Paavola, L. Viikari and J. Buchert, 2000. *Trichoderma reesei* cellulases and their core domains in the hydrolysis and modification of chemical pulp. Cellulose 7: 189-209.

Thibault, J.-F., 1979. Automatisation du dosage des substances pectiques par la méthode au méta-hydroxydiphényl. Lebensmittel Wissenschaft und Technologie 12: 247-251.

Vincken, J.-P., H.A. Schols, R.J.F.J. Oomen, M.C. McCann, P. Ulvskov, A.G.J. Voragen and R.G.F. Visser, 2003. If homogalacturonan were a side chain of rhamnogalacturonan I. Implications for cell wall architecture. Plant Physiology 132: 1781-1789.

Voragen, A.G.J., W. Pilnik, J.-F. Thibault, M.A.V. Axelos, and C.M.G.C. Renard, 1995. Pectins. In: A.M. Stephen (ed.) Food polysaccharides and their applications. Marcel Dekker, New-York, pp 287-340.

Zykwinska, A., C. Rondeau, C. Garnier, J.-F. Thibault and M.-C. Ralet, 2006. Alkaline extractability of pectic arabinan and galactan and thier mobility in sugar beet and potat cell walls. Carbohydrate Polymers, 65, 510-520.

Part 4

Functionality and applications of pectins

Pectin – β-lactoglobulin complex formation: influence of pectin overall charge and local charge density

B.L.H.M. Sperber[1,2], M.A. Cohen Stuart[2], H.A. Schols[1], W. Norde[2,3] and A.G.J. Voragen[1]
[1]Laboratory of Food Chemistry, Department of Agrotechnology and Food Science, Wageningen University, P.O. Box 8129, 6700 EV Wageningen, the Netherlands
[2]Laboratory of Physical Chemistry and Colloid Science, Department of Agrotechnology and Food Sciences, Wageningen University, P.O. Box 8038, 6700 EK Wageningen, the Netherlands
[3]Department of Biomedical Engineering, University Medical Centre Groningen, Antonius Deusinglaan 1, 9713 AV Groningen, the Netherlands

Abstract

The influence of the overall charge (degree of methyl esterification) and local charge density (degree of blockiness) of pectin on the complex formation between β-lactoglobulin and pectin is investigated by creating state diagrams to identify under what conditions soluble complexes are formed. Binding isotherms show the composition of these complexes. Pectins with a high local charge density are most tolerant to ionic strength in the formation of soluble complexes as opposed to pectin with a low local charge density. With this knowledge the adsorption of β-lactoglobulin and pectin, both from mixtures and sequential, to a hydrophobic surface is investigated at pH 4.25, where soluble complexes are formed over a large range of mixing ratio's. The adsorption to a hydrophobic surface from a mixed β-lactoglobulin – pectin solution leads to a reduction in the rate of adsorption as compared to a solution of β-lactoglobulin alone. This can be related to a lower free β-lactoglobulin concentration by means of the binding isotherms. When low weight ratio's of β-lactoglobulin / pectin are used, the ζ-potential is lower than for higher weight ratio's, due to charge compensation of β-lactoglobulin that is bound to pectin. The very low rate of adsorption makes the use of mixtures of β-lactoglobulin and pectin not interesting for emulsion stabilization as the newly created surface cannot be covered before coalescence of the droplets occurs. Sequential adsorption of β-lactoglobulin and pectin results in the build up of multi layers of β-lactoglobulin and pectin. When the multilayer is finished with a pectin with a high local charge density, it is most stable to the wash out of β-lactoglobulin from the multilayer. When the last layer is of a pectin with a low overall charge, the ζ-potential is lowest. This indicates that high methyl-esterified pectin protrude more into the solution than low methyl-esterified pectin. The above suggest that an high methyl-esterified pectin with a block wise distribution would work best to stabilize an emulsion consisting of a primary emulsion of β-lactoglobulin and a secondary emulsion of pectin.

Keywords: pectin, β-lactoglobulin, polysaccharide-protein complexes, electrostatic interaction, binding isotherm, adsorption, hydrophobic surface

1. Introduction

Pectin is widely used in the food industry to for its gelling properties in the production of jams and jellies, fruit juice, confectionary products and bakery fillings (Voragen *et al.*, 1995; May, 1997), but pectin is also added to acidified dairy drinks to stabilize the casein micelles from precipitation (Tholstrup Sejersen *et al.*, 2007). Pectin stabilizes the casein micelles by electrosorption to the micelle surface (Pereyra *et al.*, 1997; Maroziene and De Kruif, 2000; Tuinier *et al.*, 2002). It was first postulated that the negative charge of the pectin stabilised the casein micelle by electrostatic repulsion (Glahn, 1982), but ζ-potential measurements showed that the charge of the casein micelle-pectin complexes was not sufficient for stabilization (Kravtchenko *et al.*, 1995). For the stabilisation of acidified milk drinks, the choice of the type of pectin is very important, as different types have different stabilizing properties. The difference in stabilising properties of the pectins is related to the degree of methyl esterification (DM) and distribution of the methyl esters (DB) of galacturonic acid in the pectin backbone. High methyl-esterified (HM) pectin is found to be a better stabiliser than low methyl-esterified (LM) pectin (Pereyra *et al.*, 1997; Liu *et al.*, 2006). This better stabilization by HM-pectin is explained by the amount of the pectin molecule that is bound to the casein micelle. A LM-pectin has mainly non-methyl esterified GalA residues, resulting in the binding of the majority of the pectin molecule to the casein micelle. HM-pectin has much less binding places, resulting in a pectin molecule that protrudes more into the solution and provides a better stabilization of the casein micelle by steric repulsion (Pereyra *et al.*, 1997; Rolin, 2002; Boulenguer and Laurent, 2003). For a long term stability of the acidified milk drink it is important that a weak gel network is formed (Boulenguer and Laurent, 2003). Of the HM-pectins, calcium sensitive pectin is preferred over a non-calcium sensitive pectin as the blocks of non-methyl esterified GalA bind strongly to the casein micelle, while the random distributed non-calcium sensitive pectin hardly binds to the casein micelle (Glahn and Rolin, 1996; Laurent and Boulenguer, 2003).

Emulsions are often stabilised by proteins. Proteins adsorb on the oil-water interface, thereby protecting the emulsion from coalescence (Dickinson, 1999). Protein-stabilised emulsions are often not stable to environmental stresses like pH, salt, heating, freezing or shear forces. One way of improving the stability of the emulsion is the addition of a charged polysaccharide to the system (Dickinson, 1994; Güzey *et al.*, 2004). There are two ways of making the emulsion: a sequential adsorption of protein and polysaccharide, or the preparation of an emulsion from a mixed solution of protein and polysaccharide. Sequential adsorption, or layer-by-layer deposition, is achieved by making a primary emulsion with protein, to which a solution of polysaccharide is added. This obtains stable emulsions, although an ultrasound treatment seems to be a prerequisite to prevent flocculated droplets (Güzey *et al.*, 2004). Emulsions prepared from a mixture of protein and polysaccharide also give stable emulsions, but bridging flocculation seems to be less of a problem in these emulsions at low polysaccharide concentrations (Jourdain *et al.*, 2008). The characteristics of the adsorbed complexes and the structures of the adsorbed layers from sequential adsorption on the interface between oil and water are however still poorly understood (Dickinson, 2008).

To investigate the influence of overall charge and local charge density of the pectin molecule on the adsorption to hydrophobic surfaces of β-lg and β-lg – pectin complexes, state diagrams were created to identify the region, as function of pH and ionic strength, in which complexes between β-lg and pectin are formed. The functionality of these complexes was investigated by means of adsorption to a hydrophobic surface. Also the possibility of a layer-by-layer deposition of β-lg and pectin from their individual solutions by alternating solutions of β-lg and pectin was investigated.

2. Materials and methods

2.1 Pectin

Pectins were a kind gift of CP Kelco (Lille Skensved, Danmark) and obtained by acid extraction. Table 1 lists the characteristics of the pectins as can be found in (Daas *et al.*, 2001). The pectins were found to contain residual calcium from the isolation procedure (0.02-0.05 mol of calcium/mol of GalA). This calcium was removed by washing the pectins five times with 60% ethanol containing 5% of hydrochloric acid. Next, the pectins were washed with 60% ethanol until free of chloride-ions (tested with 0.2% silver nitrate). Pectins were subsequently washed with acetone and air dried. This treatment was sufficient for LMP and HMP_B, but HMP_R required further treatment. Therefore HMP_R was dissolved at 10g/l and mixed with AG 50W-X4 200-400 mesh ion exchange resin from Bio-Rad Laboratories (Hercules, CA). The ion-exchange resin was removed from the solution by centrifugation and HMP_B was lyophilized. This resulted in pectins that all three contained less than 0.001 mol of calcium/mol GalA.

Table 1. GalA content, DM and DB of pectin.

Pectin	Code Daas (Daas *et al.*, 2001)	GalA content (%)	DM	DB
LMP	C30	78.5	30.4	16.5
HMP_B	C70	84.5	69.8	10.9
HMP_R	C74	85.3	73.5	1.7

2.2 β-Lactoglobulin

β-Lg was isolated from bovine milk using a non-denaturing technique reported by (De Jongh *et al.*, 2001)). This resulted in a β-lg preparation that had a purity of over 98%, and a 60-40 ratio of the A and B isomers. β-Lg solutions were free of aggregated material as verified by dynamic light scattering. The iso-electric point (pI) of β-lg is 5.2, and the molecular weight 18.3 kDa. β-lg is predominately present in solution as a dimer.

2.3 State diagrams

State diagrams are constructed by means of a pH titration of a pectin – β-lg mixture, monitored by dynamic light scattering (DLS). The pH is lowered from approximately pH 7 to pH 3 with steps of about 0.1 pH unit. The concentration of pectin is kept constant at 0.2 g/l; the weight ratio (WR) of β-lg/pectin is set at 1, 2, 4 or 8. Ionic strength (I) is set with sodium chloride. From these DLS-titrations, where pH_{ϕ}, the state boundary between soluble complexes and separated phases, and pH_c, the state boundary between co-soluble polymers and soluble complexes the pH where a co-soluble are determined and presented in state diagrams. The procedure is described in detail by Sperber *et al.* (2009)

2.4 Fluorescence anisotropy

β-Lg was labelled with Alexa Fluor 430 carboxylic acid, succinimidyl ester (A10169, Invitrogen, Breda, the Netherlands), according to the 'Molecular Probes' protocol (www.probes.com). The degree of labelling was approximately 0.2 mol of dye per mol of β-lg.

Fluorescence anisotropy was measured on a TECAN Infinite F500 micro plate reader (TECAN, Austria), equipped with an excitation filter of 420 nm with a bandwidth of 10 nm and an emission filter of 540 nm with a bandwidth of 10 nm, corresponding to the excitation and emission maximum of Alexa Fluor 430. The fluorescence anisotropy is measured as follows:

$$A = \frac{I_{||} - I_{\perp}}{I_{||} + 2I_{\perp}} \qquad (1)$$

With A the fluorescence anisotropy and $I_{||}$ and $I_{\perp}$ the intensity of the fluorescent light in the parallel and perpendicular direction, respectively (Lakowicz, 2006). The measured fluorescence anisotropy is the average of the fraction β-lg bound (f_b) and the fraction of free β-lg (f_f). This allows f_b and f_f to be calculated as follows:

$$\Delta A = (f_b A_b + f_f A_f) - A_f \qquad (2)$$

with A_f the fluorescence anisotropy of free β-lg and A_b for bound β-lg. A_b is determined at low I and low weight ratio (WR), where virtually all β-lg is bound. The value found for A_b corresponds well with the value found for β-lg in the high viscosity medium glycerol. A_f is determined in solution with pectin at high I where complex formation is absent. β-Lg is assumed to bind as a dimer as there is no indication that the monomer-dimer equilibrium is disturbed upon binding to pectin.

2.5 Hydrophobization of solid surfaces

Silicium wafers are oxidized by placing them in an oven at 1000 °C for approximately 90 minutes. The wafers were cut to size and sonicated for 15 minutes in ethanol. Wafers were subsequently rinsed with ethanol and milli-Q water and dried in a stream of nitrogen. Next, the wafers were cleaned by immersion in a solution of concentrated sulphuric acid and 20% hydrogen peroxide (3:1). Wafers were washed with Milli-Q water and dried in a stream of air. The cleaned, dry wafers were hydrophobized by immersion in a 1% dichlorodimethyl-silane (DCDM-silane) solution in toluene. The strips were removed from the DCDM-silane solution and immersed in toluene. Next the wafers were washed, first with acetone and second with Milli-Q water. The wafers were stored in Milli-Q water until further use (always within 24 hours).

Microscope glass slide were hydrophobized using the same protocol outlined above, except for heating at 1000 °C and sonication.

2.6 Reflectometry

The adsorbed amount of material on the hydrophobized silica wafers is measured by reflectometry. Reflectometry is a simplified form of ellipsometry, where the changes in intensity of a reflected polarized laser beam are measured. For a detailed description of this technique, we refer to (Dijt *et al.*, 1990). Briefly, the technique can be described as follows: the output signal (S) is defined as the ratio of the parallel ($I_{||}$) and perpendicular intensities ($I_{\perp}$) of the reflected light:

$$S = \frac{I_{||}}{I_{\perp}} \tag{3}$$

Due to the adsorption of material S changes from its original value S_0 to a value $S_0 + \Delta S$. As S is set to zero for S_0 only ΔS is reported. ΔS can be converted to an amount of mass per unit area. For this, the composition of the adsorbed layer needs to be known, as well as the refractive index of the different layers: 3.85 for Si, 1.46 for SiO_2, 1.44 for protein and 1.33 for water. The refractive index increment (dn/dc) is 0.181 ml/g for β-lg (De Feiter *et al.*, 1978), 0.123 ml/g for LMP, 0.121 ml/g for HMP_B and 0.124 ml/g for HMP_R (as measured with a Shodex RI-72).

The experiments are started by a buffer rinse of the surface. When a stable baseline is obtained, the buffer flow is replaced by a sample flow.

2.7 Streaming potential

Streaming potentials are measured to calculate ζ-potential of the surface. Two hydrophobized glass slides are mounted in a cell separated by 0.1 mm. The sample solution is forced to flow through the cell by a mechanical pressure difference, Δp. The streaming potential (V_s) is measured, from which the ζ-potential can be calculated as follows:

$$\frac{V_s}{\Delta p} = \frac{\varepsilon_0 \varepsilon \zeta}{\eta K} \quad (4)$$

Where ε_0 is the dielectric permittivity of a vacuum, ε the dielectric constant of the solvent, η the dynamic viscosity of the solvent and K the conductivity of the cell filled with solution (Elgersma *et al.*, 1992). The ζ-potential of the surfaces is determined when the adsorption of material from the solution is complete.

3. Results

3.1 Complex formation between pectin and β-lg in solution

To study the formation of complexes between pectin and β-lg in solution two types of experiments were performed: state diagrams were constructed by means of DLS-titrations and binding isotherms were obtained with fluorescence anisotropy measurements. The state diagrams define which type of complexes is present as a function of the solution parameters pH and I. From this data, a pH is selected so that soluble complexes are formed over a large I range.

3.1.1 State diagrams

Figure 1 shows the state diagram of the complex formation between the three pectins and β-lg. pH_c is confirmed to be independent of the mixing ratio of β-lg and pectin and is found to be above the pI of β-lg at low I. This is a common observation for protein – polyelectrolyte systems. This is sometimes attributed to 'local charged patches' (Park *et al.*, 1992; Weinbreck *et al.*, 2003), but is shown to also be affected by the adjustment of the pK_a values of the protein ionic groups to favour binding to pectin (Sperber *et al.*, 2009).

For all three pectins pH_ϕ increases with increasing WR. As pH_ϕ is associated with the net charge neutralization of the complex, a higher WR will lead to more binding of β-lg to pectin and hence a charge compensation of the negative charge of pectin by the positive charge of β-lg at a higher pH. LMP shows a maximum in pH_ϕ for increasing I, while both HMP_B and HMP_R show a decrease with increasing I. LMP has only a DM of 30, compared to about 70 for the two HM-pectins. This means that the β-lg molecules that bind to the pectin need to be much closer together on LMP than for the two HM-pectins. By increasing I, both the attractive forces between pectin and β-lg

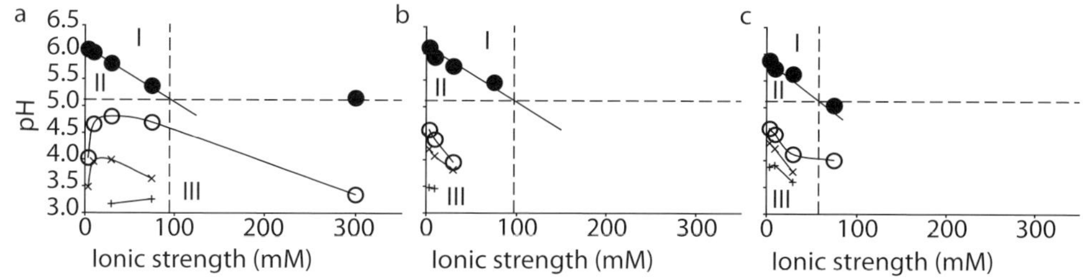

Figure 1. State diagrams of pectin and β-lg mixtures. a) LMP, b) HMP_B and c) HMP_R. ● pH_c, ○ pH_ϕ WR =8, × pH_ϕ WR =4 and + pH_ϕ WR =2. I is set with NaCl. Region I consists of co-soluble polymers, II of soluble complexes and III of insoluble complexes. $pI_{\beta\text{-}lg}$ is indicated with a horizontal dotted line, the critical salt concentrations are estimated to be 95mM for LMP, 98mM for HMP_B and 56mM for HMP_R

and the repulsive forces between pectin and β-lg, and also between two β-lg molecules are reduced (Seyrek *et al.*, 2003). This leads to an optimum in I for binding of β-lg to LMP. The increase in I prevents the formation of insoluble complexes within the measured pH range for the three pectins.

From the state diagrams, a critical salt concentration ($n_{s,c}$) can be determined by extrapolating the graph of pH_c against ionic strength until pH_c equals the pI of the protein. $n_{s,c}$ is the salt concentration where complexes are no longer formed above the pI of the protein. $n_{s,c}$ is one of the input parameters of a theoretical approach to estimate the local charge density on a polymer according to De Vries *et al.* (2003). In these estimates the protein is seen as a randomly charged surface with an average surface charge density, σ, as a number of elementary charges *e* per unit area. Δσ is the local variation around this average value. The polymer is assumed to be an ideal chain consisting of segments with a length l_K, carrying ν elementary charges per segment. This gives:

$$n_{s,c} \approx 0.4 \frac{\Delta\sigma\nu}{l_k} \tag{5}$$

Where l_K is estimated to be 2 nm for pectin (Axelos *et al.*, 1987). Computer simulation confirmed the linear dependence on the polyelectrolyte charge density (De Vries, 2004). An estimate for the charge fluctuations on the protein surface, at length scales in the order of the polymer segment length is:

$$\Delta\sigma^2 \approx -\frac{\mu}{\pi l_K^2} \frac{\delta\sigma}{\delta pH}\bigg|_{pH=pI} \tag{6}$$

where μ is a protein dependent numerical pre-factor, estimated to be 0.35, and $\partial\sigma/\partial pH$ at the pI of β-lg is estimated to be -0.25 nm^{-2} (De Vries *et al.*, 2003). With $n_{s,c}$ determined in Figure 1, 95 mM for LMP, 98 mM for HMP_B and 56 mM for HMP_R, this leads to a local charge density of 1.9 *e*/nm for LMP and HMP_B and 1.1 *e*/nm for

HMP_R. These numbers are rough estimates, but can be verified to be in the right range. The length of a galacturonic acid residue is 0.43 nm (Rees and Wight, 1971). This gives about 2.5 monomers per nm. With a DM of 30 for LMP, it can carry about 1.8 *e*/nm. For HMP_B and HMP_R this is about 0.75 *e*/nm. This clearly shows that the difference between the $n_{s,c}$ of HMP_R and HMP_B, respectively, can be interpreted as being due to a difference in the local charge density of a factor of almost 2, with HMP_B resembling LMP in its local charge density.

3.1.2 Binding isotherms of β-lg to pectin

Binding isotherms are useful for obtaining information about the amount of β-lg that can bind to the three pectins; also, information of the sensitivity of the binding of β-lg to pectin to ionic strength can be obtained. In Figure 2 the binding isotherms of β-lg to the three pectins are shown. For the binding isotherms pH 4.25 is selected, because it is a pH where soluble complexes of β-lg and pectin are formed over a large WR (Figure 1).

At low ionic strength, there is a very high affinity of β-lg to pectin as almost all added β-lg binds to the pectin. When saturation of the pectin is almost reached, the curve slopes of to a plateau (Γ_m). LMP can bind the most β-lg per gram of pectin, where HMP_B and HMP_R bind about equal amounts. Γ_m is determined by the overall charge of the pectin. As the ionic strength is increased, Γ_m decreases, until finally all binding of β-lg to pectin is suppressed. Binding of β-lg to LMP is suppressed at 200 mM, 100 mM for HMP_B and 75 mM for HMP_R. The local charge density of the pectin influences the tolerance to ionic strength for the binding of β-lg to pectin. A high local charge density is the most tolerant to ionic strength. These results were confirmed by isothermal titration calorimetry (results not shown).

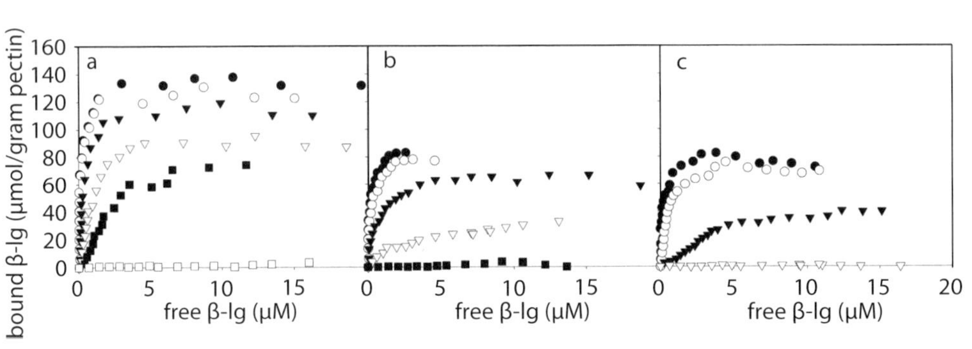

Figure 2. Binding isotherms of β-lg to a) LMP, b) HMP_B and c) HMP_R. 20mM NaAcetate buffer pH 4.25, I set with NaCl, ● 4 mM, ○ 10 mM, ▼ 30 mM, ▽ 75 mM, ■ 100 mM, □ 200 mM.

3.2 Adsorption of β-lg and pectin to a hydrophobic surface

The adsorption of β-lg, pectin and mixtures thereof, both from mixtures and sequentially, to hydrophobized silica surface is a model system for adsorption phenomena at hydrophobic surfaces in foods like oil-water or air-water interfaces. Fast adsorption is required to be able to stabilize new surface that is being formed when emulsions or foams are produced (Dickinson, 2003). With reflectometry, information about the rate of adsorption and the adsorbed amount can be determined. To prevent coalescence of emulsions the charge of the stabilized interface is important and can be determined by streaming potential measurements.

3.2.1 Adsorption of β-lg and pectin from homo-molecular solutions

Figure 3 shows the adsorption of β-lg and LMP to hydrophobized silica at pH 4.25. β-lg adsorbs very fast, to a plateau value that corresponds to approximately 1.3 mg/m^2. β-Lg is known to adsorb quickly to hydrophobic interfaces (Paulsson and Dejmek, 1992). The adsorbed amount of β-lg can be increased by increasing the ionic strength, which reduces electrostatic repulsion in the adsorbed layer, allowing the β-lg molecules to be packed closer together. Increasing the concentration of the β-lg solution from which the β-lg adsorbs will also increase the adsorbed amount, as the β-lg has less time to rearrange itself on the surface, allowing for more efficient use of the surface (Elofsson *et al.*, 1997). The ζ-potential of the hydrophobized silica surface was determined to be -10.0 mV. Not all of the silanol groups have reacted with DCDM-silane, leaving a negative charge on the hydrophobic surface. After adsorption of β-lg the ζ-potential of the surface is found to be +11.7 mV.

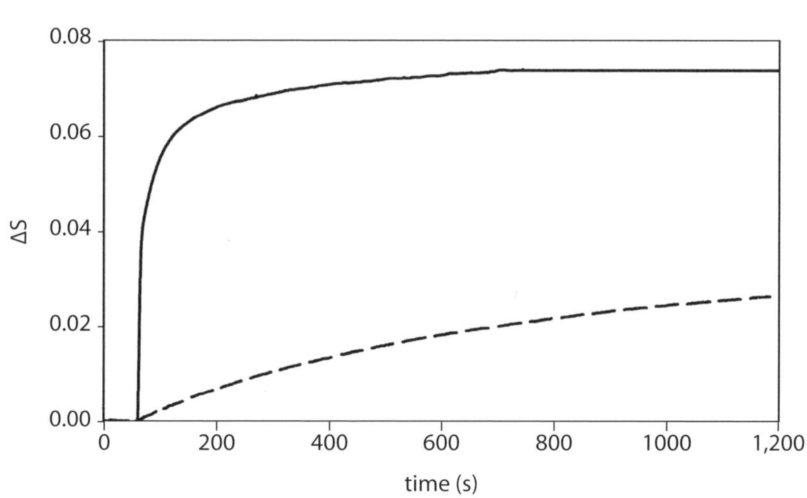

Figure 3. Adsorption of β-lg (solid) and LMP (long dash) to hydrophobic surface, pH 4.25, 20 mM NaAcetate, I is set with NaCl to 10 mM. [β-lg] = 10 mg/l [LMP] = 100 mg/l.

From a solution of LMP only very slow adsorption takes place. The negative charge of both the hydrophobic surface and the pectin slows the adsorption kinetics. Pectin is known to adsorb very poorly to hydrophobic surfaces and to have no emulsifying properties (Dickinson, 2003). As only very slow adsorption takes place from a pectin solution, the surface charge was not determined.

3.2.2 Adsorption of β-lg – pectin from mixed solutions

The adsorption on a hydrophobic surface from a mixed solution of β-lg and pectin is shown in Figure 4. For comparison, also the curve of β-lg from Figure 3 is shown. It is difficult to obtain an adsorbed amount from these graphs, as this depends on the refractive index increment (dn/dc) of the polymers. For adsorption from homo-solutions this is easy to do, since the composition of the adsorbed layer is known. With the adsorption from mixed solution both β-lg and pectin adsorb, but it is unknown in what ratio. In literature values for the dn/dc for pectin can be found ranging from 0.092 - 0.183 ml/g (Corredig *et al.*, 2000; Hunter and Wicker, 2005; Gulfi *et al.*, 2007). LMP, HMP_B and HMP_R were found to have nearly identical dn/dc of 0.123, 0.121 and 0.124 ml/g respectively. Possibly this is the cause of the extensive washing of the pectins to remove salt ions complexed with the carboxyl groups of the galacturonic acid. Since the pectins all have the same dn/dc, it is possible to qualitatively compare the results to each other.

The adsorption from mixed solutions of β-lg and LMP shows a considerable lowering of the adsorption speed for both a WR of 2 and 6. The WR 2 sample is located in the

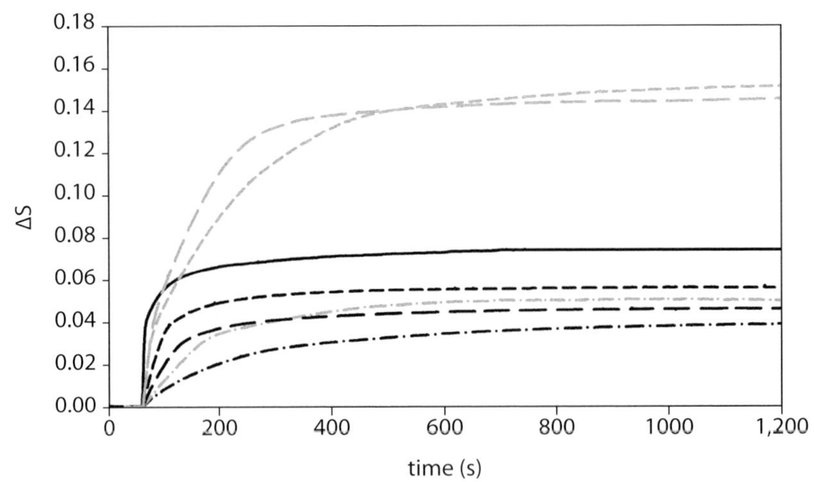

Figure 4. Adsorption of β-lg – pectin mixtures. β-lg – LMP WR 2 (– • –), WR 6 (– • –), β-lg – HMP_B WR 2 (— —), WR 6 (— —), β-lg – HMP_R WR 2 (– –), WR 6 (– –) and β-lg (——) to hydrophobic surface, pH 4.25, 20 mM NaAcetate, I is set with NaCl to 10 mM. [β-lg] = 10 mg/l.

beginning of the binding isotherm in Figure 2, virtually all added β-lg is bound to LMP here, with only 0.15% as free β-lg. WR 6 is the first point located in the plateau and 27% of the β-lg is present as free β-lg. The result of this considerable lowering of the $[\beta\text{-lg}]_{free}$ is a reduced rate of adsorption, as the adsorption rate is controlled by $[\beta\text{-lg}]_{free}$. This was also found for the adsorption to air-water interfaces from β-lg – LMP mixtures by (Ganzevles *et al.*, 2006a), where the change in surface pressure due to the adsorption of β-lg from a β-lg – LMP mixture could be mimicked by using a 10 times diluted β-lg solution compared to the β-lg – LMP mixture. The adsorbed amount from the mixture is lower than for adsorption from a β-lg solution alone. Since there is only very slow adsorption of β-lg on the surface, there is a lot of time for the adsorbed β-lg molecules to rearrange on the surface, leading to a much more efficient covering of the surface.

The ζ-potential of the surface is -19.9mV for WR 2 and -9.7mV for WR 6. As the surface is in equilibrium with the solution covering it, the surface layer is a representation of the β-lg – LMP solution. For WR 2, the binding sites on LMP are far from saturated, leaving a lot of negative charge on the pectin uncompensated by positive charge of the β-lg, thus giving a negative ζ-potential. At WR 6 LMP is close to saturation with β-lg, but the overall charge of the complex is still negative (see also (Ganzevles *et al.*, 2006b)).

For HMP_B the WR 2 sample also shows a considerable reduction in the adsorption rate. From the binding isotherm it follows that 6.5% is present as $\beta\text{-lg}_{free}$. For WR 6 this is 52%. Although the WR 6 sample shows some reduction in adsorption rate, it is a relatively small reduction, as can be expected from a reduction of $[\beta\text{-lg}]_{free}$ by only a factor 2. More pronounced is the difference in the amount of adsorbed mass in the WR 6 sample. The signal is approximately two times as high as for adsorption of β-lg alone. This implies that besides a surface coverage of β-lg that is comparable to a surface coverage of β-lg alone, a considerable amount of β-lg and HMP_B is adsorbed on top of the β-lg layer on the surface. The ζ-potential of the surface for WR 2 and 6 are -21.6mV and -13.9 respectively, following the same reasoning as for LMP.

The adsorption from mixtures of β-lg – HMP_R follows the reasoning as for HMP_B. For WR 2 13% of β-lg is not bound to HMP_R, 55% for WR 6. The higher $[\beta\text{-lg}]_{free}$ compared to HMP_B for WR 2 leads to an increased adsorption rate, compared to the adsorption from a solution of β-lg – HMP_B. The ζ-potential for WR 2 is -22.4mV, for WR 6 it is -10.8mV.

Figure 5 gives a model representation of the adsorption from a β-lg solution and from β-lg – pectin mixtures. The adsorption from a solution of β-lg is represented in Figure 5a. The β-lg adsorbs quickly leaving little time for rearrangement of the protein on the surface to reach the optimal conformation on the surface. Figure 5b shows the adsorption from mixtures of β-lg and pectin at a WR of 2. Since the $[\beta\text{-lg}]_{free}$ is low, adsorption of β-lg is slow allowing the protein to spread on the surface. The pectin adsorbing onto this β-lg layer is far from being saturated with

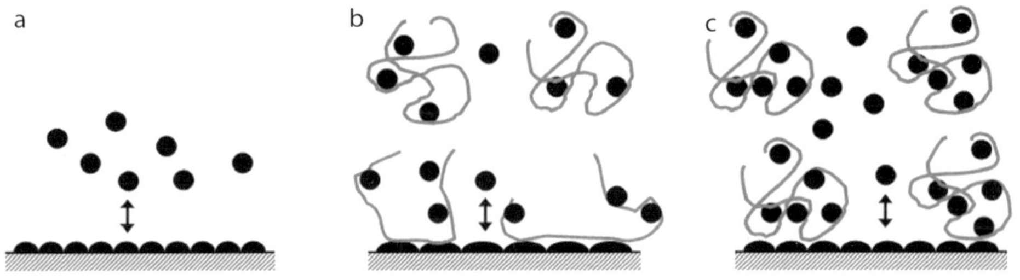

Figure 5. Model representation of adsorption on a hydrophobic surface from a solution of a) β-lg; b) β-lg – pectin mixture at WR 2 for all pectin and WR 6 for LMP; c) β-lg – pectin mixture at WR 6 for HMP_B and HMP_R.

β-lg, allowing the pectin to cover a larger surface area. Figure 5b also represent the adsorption from a solution of β-lg – LMP at WR 6, since $[\beta\text{-lg}]_{free}$ is also low under those conditions. The resulting pectin layer for LMP at WR 6 will contain more β-lg. Finally, Figure 5c shows the adsorption from mixtures of β-lg and HMP_B or HMP_R at WR of 6. As the concentration of $[\beta\text{-lg}]_{free}$ is considerably higher, the adsorption of β-lg is much faster, allowing for limited relaxation on the surface. The pectin is fully saturated with β-lg and the majority on the pectin protrudes into the solution, allowing for a considerable increase in the adsorbed mass compared to adsorption from a solution of β-lg.

3.2.3 Sequential adsorption of β-lg and pectin

Sequential adsorption of β-lg and pectin onto a hydrophobic surface can be used in the formation of secondary emulsions, which have an improved emulsion stability (Güzey and McClements, 2006), or for encapsulation purposes (De Kruif *et al.*, 2004). As pectin does not adsorb to a hydrophobic surface the first layer is made of β-lg. In total 4 layers of both β-lg and pectin are deposited, but more are possible.

Figure 6 shows the sequential adsorption of β-lg and pectin to a hydrophobic surface. As expected, the first layer of β-lg is identical for all three pectins, as no pectin has been in contact with the surface. A short buffer rinse removes the β-lg solution before pectin is offered to the surface. The ζ-potential of the β-lg layer is +12mV, as found before. After adsorption of pectin on the ζ-potential of the adsorbed layer has become negative: -28mV for HMP_B and HMP_R, but only -17mV for LMP. The difference in ζ-potential of the LMP layer compared to the HMP_B and HMP_R layer is attributed to the orientation of the pectin layer. As LMP contains much more carboxyl groups, it lays 'flatter' on the surface than the two HMP's. This results in a reduced amount of charge that is not compensated by β-lg. The adsorption of pectin on the β-lg layer has only a small effect on ΔS, mainly because of the low dn/dc compared to β-lg, but pectin is adsorbing on the β-lg surface, as indicated by the change of the ζ-potential.

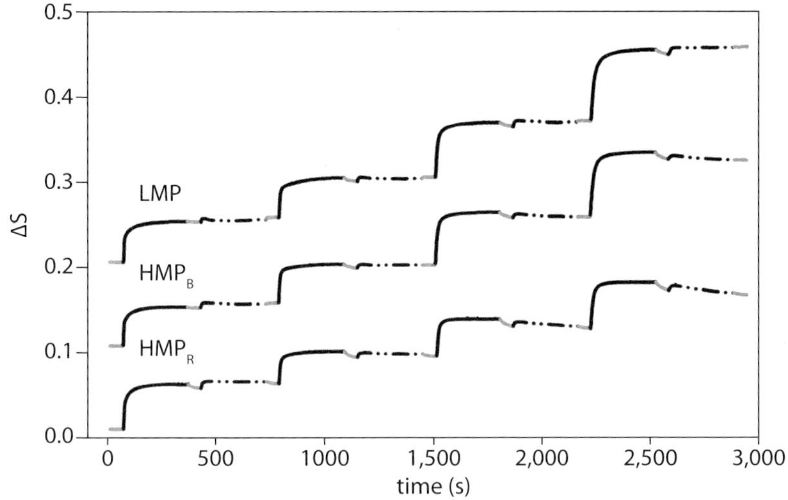

Figure 6. Sequential deposition on a hydrophobic surface of β-lg and LMP, HMP_B and HMP_R. [β-lg] = 10 mg/l, [pectin] = 100 mg/l, 20mM NaAcetate buffer pH 4.25, I is set with NaCl to 10 mM. Buffer (—), β-lg (—), pectin (— • • —)

The following layers for LMP all show the same trend, β-lg adsorbs on the LMP layer, LMP on the β-lg layer. The ζ-potential of these β-lg layers was determined to be -2mV. This shows that the no overcompensation of the LMP charge takes place. The LMP layers have a ζ-potential of about -21mV, so somewhat lower than the first layer of LMP. As adsorbing LMP molecules need to share β-lg molecules in order to attach to the existing layers, the charge compensation of β-lg is probably a bit more efficient for the subsequent layers of LMP. From the increase in ΔS, it seems that more β-lg adsorbs in subsequent layers. At the same time, the buffer rinse and LMP adsorption, after the initial plateau, seem to lower ΔS a bit. Although the binding isotherm shows a high affinity of β-lg for pectin at an ionic strength of 10mM, there will still be some desorption when the surface is in a solution that is void of either β-lg, LMP or both. Therefore, when β-lg is subjected for adsorption, it may also replenish some of the β-lg that was washed out during the buffer rinse and LMP adsorption. When comparing the buffer rinse after β-lg and LMP adsorption, the loss of mass is much greater after adsorption of β-lg than after adsorption of LMP, which implies that a saturated pectin looses the β-lg much easier than an unsaturated one.

HMP_B shows a similar pattern to LMP. The increase in ΔS is comparable to LMP, although the desorption of β-lg seems to be somewhat higher. Since HMP_B has a higher DM than LMP, this hints again to a different orientation of the pectin on the surface, where LMP protrudes less into the solution. The desorption of mass after the initial adsorption of HMP_B is caused by the β-lg not being bound as tightly to HMP_B than to LMP, allowing it to move through the HMP_B layer easier than for LMP.

The ζ-potential of the layers is -28mV for a HMP_B-terminal layer and -3mV after adsorption of β-lg, so no overcompensation of charge either.

For HMP_R the desorption of β-lg is even more pronounced, resulting in a decrease in the increment of ΔS. As HMP_R contains much less high affinity binding sites than LMP and HMP_B, the desorption of β-lg in a buffer solution or a solution of HMP_R is much quicker as compared to LMP and HMP_B. For the ζ-potential, a similar trend is observed: the HMP_R layers have a ζ-potential of -28mV, like HMP_B, but the ζ-potential after β-lg adsorption is -8mV. Where the charge of pectin was fully compensated for LMP and HMP_B, HMP_R cannot bind sufficient β-lg to compensate all negative charge on the pectin.

4. Conclusions

The stabilization of oil-water or air-water interfaces depends on the quick rate of adsorption of the surface-active component. In β-lg – pectin systems the surface-active component is β-lg, as pectin has no surface-active characteristics. Pectin and β-lg are shown to form soluble complexes in the pH region roughly between pH 5.5 and 3.5, depending on pectin type and ionic strength. Within this pH range, it is possible to use the formation of complexes between β-lg and pectin to enhance the surface stabilizing capacity of β-lg. To this end, it is important to choose the right type of pectin. LMP, a pectin with a high local charge density and high overall charge is shown to form complexes at the highest ionic strength of the tested pectins. HMP_B, a pectin with a low overall charge, but high local charge density due to a block wise distribution of the methyl esters, also has a good tolerance to ionic strength. HMP_R, with a low overall charge and a low local charge density, has the poorest tolerance to ionic strength. This shows that the ability to form complexes between β-lg and pectin is mainly determined by the local charge density.

Adsorption of β-lg to a hydrophobic surface is quick, but when pectin is added and adsorption has to take place from mixtures of β-lg and pectin, the adsorption is slower. This is caused by a lowering of the concentration of free β-lg, which is confirmed by the binding isotherms. This lowering in adsorption rate was also found by (Ganzevles *et al.*, 2006b) for adsorption to the air-water interface from mixed β-lg – pectin solutions. A negative value of the ζ-potential of the surface probes the presence of pectin, as a β-lg coated surface at pH 4.25 is positive. At WR of 6 (β-lg to pectin), $[\beta\text{-lg}]_{free}$ is relatively high and the adsorption is reasonably quick. The ζ-potential of these surfaces is only half compared to when a WR of 2 is used (-11mV vs. -20mV).

Sequential adsorption of β-lg and pectin leads to thick films in which adsorbed layers of β-lg and pectin are stacked in alternate fashion. When a pectin with a high local charge density is used (LMP or HMP_B), the adsorbed layer is much more stable against desorption of β-lg and pectin, than when a pectin with a low local charge density (HMP_R) is used. LMP and the two HM-pectins have a different conformation on the surface: the HM-pectin are thought to protrude more into the solution, due to less binding sites on the pectin, where LMP is more 'flat' on the surface.

The speed of adsorption and the surface characteristics make sequential adsorption the preferred technique over the adsorption from mixtures of β-lg and pectin to stabilize interfaces. Pectins with a high local charge density, or block wise distribution of the methyl esters, have the best tolerance for ionic strength, while pectins with a low overall charge, or high DM, protrude most into solution, giving the largest ζ-potential. This shows that HM, block wise pectin would be most suitable to enhance emulsion stability by means of a primary emulsion with β-lg and a secondary emulsion with a HMP_B pectin.

References

Axelos, M.A.V., J. Lefebvre and J.F. Thibault, 1987. Conformation of a low methoxyl citrus pectin in aqueous solution. Food Hydrocolloids, 1: 569-570.

Boulenguer, P. and M.A. Laurent, 2003. Comparison of the stabilisation mechanism of acid dairy drinks (ADD) induced by pectin and soluble soybean polysaccharide (SSP). In: A.G.J. Voragen, H.A. Schols and R. Visser (ed.), Advances in Pectin and Pectinase Research. Kluwer Academic Publishers, Dordrecht, the Netherlands, pp. 467-480.

Corredig, M., W. Kerr and L. Wicker, 2000. Molecular characterization of commercial pectins by separation with linear mix gel permeation columns in-line with multi-angle light scattering detection. Food Hydrocolloids, 14: 41-47.

Daas, P.J.H., B. Boxma, A.M.C.P. Hopman, A.G.J. Voragen and H. A. Schols, 2001. Nonesterified galacturonic acid sequence homology of pectins. Biopolymers, 58: 1-8.

De Feiter, J.A., J. Benjamins and F.A. Veer, 1978. Ellipsometry as a tool to study adsorption behavior of synthetic and biopolymers at the air-water interface. Biopolymers, 17: 1759-1772.

De Jongh, H.H.J., T. Gröneveld and J. De Groot, 2001. Mild isolation procedure discloses new protein structural properties of β-lactoglobulin. Journal of Dairy Science, 84: 562-571.

De Kruif, C.G., F. Weinbreck and R. De Vries, 2004. Complex coacervation of proteins and anionic polysaccharides. Current Opinion in Colloid and Interface Science, 9: 340-349.

De Vries, R., 2004. Monte Carlo simulations of flexible polyanions complexing with whey proteins at their isoelectric point. Journal of Chemical Physics, 120: 3475-3481.

De Vries, R., F. Weinbreck and C.G. De Kruif, 2003. Theory of polyelectrolyte adsorption on heterogeneously charged surfaces applied to soluble protein-polyelectrolyte complexes. Journal of Chemical Physics, 118: 4549-4659.

Dickinson, E., 1994. Protein-stabilized emulsions. Journal of Food Engineering, 22: 59-74.

Dickinson, E., 1999. Adsorbed protein layers at fluid interfaces: interactions, structure and surface rheology. Colloids and Surfaces B: Biointerfaces, 15: 161-176.

Dickinson, E., 2003. Hydrocolloids at interfaces and the influence on the properties of dispersed systems. Food Hydrocolloids, 17: 25-39.

Dickinson, E., 2008. Interfacial structure and stability of food emulsions as affected by protein–polysaccharide interactions. Soft Matter, 4: 932-942.

Dijt, J.C., M.A. Cohen Stuart, J.E. Hofman and G.J. Fleer, 1990. Kinetics of polymer adsorption in stagnation point flow. Colloids and Surfaces, 51: 141-158.

Elgersma, A.V., R.L.J. Zsom, J. Lyklema and W. Norde, 1992. Kinetics of single and competitive protein adsorption studied by reflectometry and streaming potential measurements. Colloids and Surfaces, 65: 17-28.

Elofsson, U.M., M.A. Paulsson and T. Arnebrant, 1997. Adsorption of β-lactoglobulin A and B in relation to self-association: Effect of concentration and pH. Langmuir, 13: 1695-1700.

Ganzevles, R.A., M.A. Cohen Stuart, T. Van Vliet and H.H.J. De Jongh, 2006a. Use of polysaccharides to control protein adsorption to the air–water interface. Food Hydrocolloids, 20: 872-878.

Ganzevles, R.A., K. Zinoviadou, T. Van Vliet, M.A. Cohen Stuart and H.H.J. De Jongh, 2006b. Modulating surface rheology by electrostatic protein/polysaccharide interactions. Langmuir, 22: 10089-10096.

Glahn, P.-E., 1982. Hydrocolloid stabilization of protein suspensions at low pH. In: G.O. Phillips, D.J. Wedlock and P.A. Williams (eds.) Progress in food and nutrition science. Pergamon Press, Oxford, UK, (Vol. 6), pp. 171-177.

Glahn, P.-E. and C. Rolin, 1996. Properties and food uses of pectin fractions. In: G.O. Phillips, P.A. Williams and D.J. Wedlock (eds.) Gums and Stabilisers for the Food Industry. Oxford University Press, Oxford, (Vol. 8), pp. 393-402.

Gulfi, M., E. Arrigoni and R. Amadò, 2007. *In vitro* fermentability of a pectin fraction rich in hairy regions. Carbohydrate Polymers, 67: 410-416.

Güzey, D. and D.J. McClements, 2006. IInfluence of environmental stresses on OAV emulsions stabilized by β-lactoglobulin-pectin and β-lactoglobulin-pectin-chitosan membranes produced by the electrostatic layer-by-layer deposition technique. Food Biophysics, 1: 30-40.

Güzey, D., H.-J. Kim and D.J. McClements, 2004. Factors influencing the production of o/w emulsions stabilized by β-lactoglobulin-pectin membranes. Food Hydrocolloids, 18: 967-975.

Hunter, J.L. and L. Wicker, 2005. De-esterification of pectin by alkali, plant and fungal pectinmethylesterases and effect on molecular weight. Journal of the Science of Food and Agriculture, 85: 2243-2248.

Jourdain, L., M.E. Leser, C. Schmitt, M. Michel and E. Dickinson, 2008. Stability of emulsions containing sodium caseinate and dextran sulfate: Relationship to complexation in solution. Food Hydrocolloids, 22: 647-659.

Kravtchenko, T.P., A. Parker and A. Trespoey, 1995. Colloidal stability and sedimentation of pectin-stabilized acid milk drinks. In: E. Dickinson and D. Lorient (eds.) Food macromolecules and colloids. Royal Society of Chemistry, Cambridge.

Lakowicz, J.R., 2006. Fluorescence anisotropy. In: J.R. Lakowicz (ed.), Principles of fluorescence spectroscopy. Springer, New York, pp. 353-382.

Laurent, M.A. and P. Boulenguer, 2003. Stabilization mechanism of acid dairy drinks (ADD) induced by pectin. Food Hydrocolloids, 17: 445-454.

Liu, J., A. Nakamura and M. Corredig, 2006. Addition of pectin and soy soluble polysaccharide affects the particle size distribution of casein suspensions prepared from acidified skim milk. Journal of Agricultural and Food Chemistry, 54: 6241-6246.

Maroziene, A. and C.G. De Kruif, 2000. Interaction of pectin and casein micelles. Food Hydrocolloids, 14: 391-394.

May, C.D., 1997. Pectins. In: A.P. Imeson (ed.), Thickening and gelling agents for food. Blackie Academic and Professional, London, pp. 230-260.

Park, J.M., B.B. Muhoberac, P.L. Dubin and J. Xia, 1992. Effects of protein charge heterogeneity in protein-polyelectrolyte complexation. Macromolecules, 25: 290-295.

Paulsson, M. and P. Dejmek, 1992. Surface-film pressure of beta-lactoglobulin, alpha-lactalbumin and bovine serum-albumin at the air–water-interface studied by Wilhelmy plate and drop volume. Journal of Colloid and Interface Science, 150: 394-403.

Pereyra, R., K.A. Schmidt and L. Wicker, 1997. Interaction and stabilization of acidified casein dispersions with low and high methoxyl pectins. Journal of Agricultural and Food Chemistry, 45: 3448-3451.

Rees, D.A. and A.W. Wight, 1971. Polysaccharide conformation. Part VII. Model building computations for -1,4 galacturonan and the kinking function of L-rhamnose residues in pectic substances. Journal of the Chemical Society B, 7: 1366-1372.

Rolin, C., 2002. Commercial pectin preparations. In: G.B. Seymour and J.P. Knox (eds.), Pectins and their Manipulation. Blackwell Publishing, Oxford, UK, pp. 222-241.

Seyrek, E., P.L. Dubin, C. Tribet and E.A. Gamble, 2003. Ionic strength dependence of protein-polyelectrolyte interactions. Biomacromolecules, 4: 273-282.

Sperber, B.L.H.M., H.A. Schols, M.A. Cohen Stuart, W. Norde and A.G.J. Voragen, 2009. Influence of the Overall Charge and Local Charge Density of Pectin on the Complex Formation between Pectin and β-Lactoglobulin. Food Hydrocolloids, 23: 765-772.

Tholstrup Sejersen, M., T. Salomonson, R. Ipsen, R. Clark, C. Rolin and S. Balling Engelsen, 2007. Zeta potential of pectin-stabilised casein aggregates in acidified milk drinks. International Dairy Journal, 17: 302-307.

Tuinier, R., C. Rolin and C.G. De Kruif, 2002. Electrosorption of pectin onto casein micelles. Biomacromolecules, 3: 632-638.

Voragen, A.G.J., W. Pilnik, J.F. Thibault, M.A.V. Axelos and C.M.G.C. Renard, 1995. Pectins. In: A.M. Stephen (ed.), Food Polysaccharides and their Applications. Marcel Dekker, Inc., New York, pp. 287-339.

Weinbreck, F., H. Nieuwenhuijse, G.W. Robijn and C.G. De Kruif, 2003. Complex formation of whey proteins: Exocellular polysaccharide EPS B40. Langmuir, 19: 9404-9410.

Valencia orange pectinmethylesterases, charge modification of pectins, and applications to food technology and drug delivery

Y. Kim[1], H. Lee[2], J. Jung[3], J. Rivner[3], R. Lutz[4], R. Arnold[5], N. Garti[4] and L. Wicker[3]
[1]Department of Home Economics Education, Korea University, AnaAnam-dong, Seongbuk-Gu, Seoul, 136-701 Korea
[2]University of Illinois, Department of Food Science, Champagne, IL, USA
[3]University of Georgia, Department Food Science and Technology, Athens, GA, USA
[4]Casali Institute of Applied Chemistry, The Hebrew University, Jerusalem, Israel
[5]University of Georgia, Department of Pharmaceutical and Biomedical Sciences, Athens, GA, USA

Abstract

Fractionation of Valencia orange pulp preparations resulted in PME active fractions containing putative PMEs of 13, 27, and 36 kDa. NMR analysis indicated that pectin desterification by PMEs in these fractions resulted in block wise de-esterification of the substrate. Within the narrow frequency range of 0.57 to 0.76 for the dyad or 0.22-0.55 for the triad, little effect on G' value near 560 Pa is observed. The relative contribution of total charge or distribution of charge cannot be clearly elucidated as the contribution of either depends on achieving a critical limit of de-esterification. The position and shape of this critical limit is influenced by the composition of the dyads and triads of carboxylic acid groups. Applications of charge modified pectins include the ability to interact with proteins, stabilize emulsions for entrapment of addenda and for drug release. Under low pH, cationic milk proteins interact readily with pectins, especially charge modified pectins. At pH 3.8, non fat dry milk, caseinates, αS1,2, β- but not κ-casein precipitated, especially with modified pectin. Modified pectin apparently increased the solubility of sodium caseinate, suggesting a competition between protein-protein and protein-pectin interactions. Further, charge modified pectins reduced the surface tension and interfacial activity of dispersions and reduced the droplet size of emulsions. Finally, modified pectins showed superior entrapment and less release of indomethacin compared to commercial low methoxyl pectins.

Keywords: pectinmethylesterase, charge modification, acidified dairy drinks, gelling, phytochemical delivery

1. Introduction

Pectinmethylesterase (E.C. 1.1.11, PME) initiates a sequence of events that begins with the blockwise de-esterification of pectin to create zones of charge on the pectin backbone. PME activity is implicated in colloidal instability of citrus juices and juice containing beverages. The role of the four main peaks of PME in Valencia orange juice clarification and rates was described (Cameron *et al.*, 1998). PME 2 destabilized cloud slowly (Cameron *et al.*, 1998) and is salt independent; PME 3 was thermostable, destabilized cloud the most rapidly at 4 °C (Cameron *et al.*, 2005); PME 4 had no effect on cloud and was salt dependent (Cameron *et al.*, 2003). Using partially purified PME preparations, separated based on cationic chromatography affinity, Ackerley *et al.* (2002) observed that the least cationic PMEs and/or the presence of a 27 kda peptide clarified juices at the fastest rate. Conversely, PMEs that bound to a cation exchange column and eluted late in the gradient and/or contained a 13 kd peptide were less likely to destabilize cloud (Wicker *et al.*, 2002). These PMEs were later termed U-PME and B-PME, respectively and used to charge modify pectins termed U-Pec and B-Pec (Kim *et al.*, 2005). The degree of esterification (DE) of pectin extracted from the juices decreased from about 27% DE in control juices to about 13% DE in clarified juices. Notably, the DE of pectin from juices that did not clarify also decreased to 19 and 23%. The lower MW peptides have PME activity (Savary *et al.*, 2002) and the presence of the lower MW PMEs affect the pattern of pectin de-esterification (Kim *et al.*, 2005) and calcium gelling ability of pectin modified by the PMEs (Kim and Wicker, 2006).

PME charge modification of pectin creates calcium sensitive pectins that influence functional properties such as gelling, stabilizing or emulsifying. There is conflicting evidence regarding the relative contribution of total charge and the distribution of charge to pectin functionality. Distribution of charge is difficult to quantify and different analytical methods may yield different estimates of distribution of DE (Denes *et al.*, 2000). Although the pattern of de-esterification depends predominantly on PME source, fungal or plant (Willats *et al.*, 2001), but also depends on PME isozyme differences (Kim *et al.*, 2005) and extrinsic factors such as pH (Denes *et al.*, 2000). A number of studies have reported on PME modification of pectins, which has also resulted in better methods to estimate the distribution of DE (Limberg *et al.*, 2000; Denes *et al.*, 2000; Daas *et al.*, 2001; Guillotin *et al.*, 2007).

The ability of low or high methoxy pectins to form stable junction zones, interact with other biopolymers, and form gels lends versatility in many food and pharmaceutical applications (Voragen *et al.*, 1995). Pectin interaction with other biopolymers, such as proteins, influences not only quality attributes such as stability, texture, and flavor, but also influence flavor, phytochemical or drug release if used as an encapsulating agent. As a stabilizer, pectins are used to stabilize acidified dairy drinks and prevent sedimentation via electrostatic interactions. Calcium sensitive pectins tend to impart higher stability than non-calcium sensitive pectins (Glahn and Rolin, 1994; Laurent and Boulenguer, 2003) but low methoxyl pectins reduce stability (Pereyra *et al.*, 1997). Further, only a small portion of pectin directly interacts with the milk proteins and the other portion contributes to stability by structure formation (Maroziene and De Kruif, 2000; Tromp *et al.*, 2004). The nature of pectin interaction with specific milk proteins remains unclear, although interactions are typically attributed to β-lactoglobulin or κ-casein, proteins primarily on the surface of the casein micelle. Charge modified pectins have unique interactions with biopolymers. β-lactoglobulin precipitated with pectin that was PME modified to 67% DE, but formed soluble aggregates with unmodified pectin at 72% DE (Kazmierski *et al.*, 2003). They attributed the difference in interaction to increased charge density in the plant PME modified pectin. More recently, Sperber *et al.* (2008) reported that pectin with localized high charge density interacts with β-lactoglobulin even at high ionic strength and forms soluble complexes. As an emulsifier, pectin-β-lactoglobulin complexes are reported to stabilize oil in water emulsions under wide range of pH and ionic strength (Guzey *et al.*, 2004). While hydrocolloids alone are not typically considered to be good emulsifiers (Leroux *et al.*, 2003), charge modified pectin showed superior reduction of surface tension and interfacial activity (Lutz *et al.*, 2009). Further, the pectin that was modified using U-PME with the 36, 27 kd PME imparted the greatest stability of the pectins evaluated.

Recently, formation of pectin hydrogel beads with calcium and zinc has been investigated as controlled drug delivery agent (Liu *et al.*, 2003). Hydrogel beads can be produced by gradual introduction of pectin solution into calcium or zinc solution (Sriamornsak *et al.*, 1997). Drugs can be entrapped within the gel matrix (Wakerly *et al.*, 1997; Wong *et al.*, 2002; Munjeri *et al.*, 1997). Calcium-pectin hydrogel beads were prepared and release rate in physiological saline decreased as $CaCl_2$ concentration increased during preparation (Murata *et al.*, 2004). The ratio of model drug, indomethacin, content in calcium-induced pectin hydrogel bead to theoretical indomethacin content was used to estimate entrapment efficiency. Calcium-induced pectin hydrogel beads showed high entrapment efficiency (Sriamornsak and Nunthanid, 1998).

Pectin as an ingredient in food products continues to be a versatile, yet elusive biopolymer. While pectin may be an inherent or added component, much of the colloidal instability associated with citrus juice cloud, acidified dairy drinks, and juice added beverages are typically ascribed to changes in the physico-chemical properties and/or interaction of pectin with other biopolymers. Neither Valencia PME activity, nor pectin DE has been directly correlated with cloud instability. However, a PME isozyme extract that resulted in cloud loss also created a more calcium sensitive pectin compared to a PME extract with a different isozyme profile (Kim and Wicker, 2009). While the total charge on the pectin molecule remains important, it is likely that the distribution of esterification plays an important role in use of pectins in foods. The objective of this research is to determine the role of citrus PME isozymes in charge modification of pectin and to evaluate the application of modified pectins as a stabilizer, emulsifier, and as an encapsulating agent. In this paper, we report on the use of Valencia PMEs to modify a series of pectins to the same total charge and to different total charge. Using different PME extracts, we estimate the relative contribution of distribution of DE to gelling ability and present food and pharmaceutical applications of charge modified pectins.

2. Materials and methods

2.1 Materials

Citrus PME was extracted from Valencia orange pulp in alkaline, salt buffer and partially purified using Sepharose SP and Heparin columns, as described by Kim *et al.*, 2005. Citrus pectin was donated by CP Kelco, Copenhagen, DE. Citrus pectin was charge modified by U-PME (36, 27 kd peptides) from about 58%DE to 48%DE. Citrus pectin was also modified by a crude extract of PME (36, 27, 13 kd peptides) to a target DE of 45, 35, and 25%. In a third modification, citrus pectin was modified to 40% DE by different PME extracts. The modifications are described in detail by Kim *et al.* (2005) and Lee *et al.* (2008). Pectin was hydrated in water, the pH was adjusted to 7.5, an aliquot of PME was added and pH was maintained at pH 7.5 until a sufficient time had passed to achieve the desired decrease in degree of esterification. Pectins were precipitated and washed in ethanol, dried and stored at -20 °C until analysis.

2.2 Pectin characterization

The distribution of DE was estimated by NMR as described by Lee *et al.* (2008). Calcium chloride was added at a final concentration of 35 mM to 2% pectin. The G' value was estimated using a Controlled Stress Dynamic Rheometer^{TM} (Rheometrics, Piscataway, NJ) equipped with a cone and plate device (60 mm diameter, 0.0385°) at 20 °C.

2.3 Pectin-protein interactions

The interaction of original and modified pectins with non fat dry milk, sodium caseinates, or individual caseins was evaluated. A 1% dispersion of milk proteins were prepared in sodium acetate buffer (50 mM, pH 3.8) and incubated for 2 hour at 20 °C. O-Pec, U-Pec or B-Pec was added to achieve a ratio of 1:10 pectin: protein. The particle size distribution was determined by laser diffraction using a Mastersizer S, (Malvern Instruments, Southborough, MA) as described by (Ackerley and Wicker, 2003). Size distributions of 1% pectin/protein (1:10) dispersions were calculated on the volume fractions against particle size and the weight-average size were expressed as $d_{3,2}=(\Sigma\, n_i d_i^3/\Sigma\, n_i d_i^2)$ and $d_{4,3}=(\Sigma\, n_i d_i^4/\Sigma n_i d_i^3)$, where n_i is the number of particles of diameter $d_{i.}$

2.4 Entrapment and release of indomethacin

Pectin hydrogels were prepared using commercial pectins of 10 or 35%DE (TIC Gums, MD) or PME modified pectin (38% DE) by mixing 1% pectin with a model drug indomethacin. The pectin/indomethacin was dispersed into 300 mM calcium chloride, equilibrated for 4 hr at room temperature. The calcium chloride was decanted and tested for indomethacin content by absorption at 318 nm. The beads were dried at 50 °C for 12hr under 28 psi vacuum. Entrapment efficiency and release under simulated gastrointestinal conditions were determined as described by Jung *et al.* (2008).

3. Results and discussion

Pectins with different patterns of carboxylic acid distributions are produced by PME isozymes and differ in calcium sensitivity. Conventionally, a 50% ester content is the partition point between non-calcium and calcium sensitive pectin (Voragen *et al.*, 1995). Low ester pectins will readily form calcium pectate gels and block charge modified high ester pectins can also form gels. Citrus PME modified pectins were prepared and the relationship between gelling and % DE was described by Lee *et al.* (2008). It is noted that a minimal DE of about 48% DE is needed before a gel structure is formed under the conditions of this study (2% pectin, 35 mM calcium chloride) and the gel strength increases about 50 fold with a decrease from 58% to about 48% DE. There is no marked change in the gel strength until a DE of about 37% is reached. Below 37% DE, a two fold increase in gel strength is observed (Figure 1).

The relationship between probability for the triads GGE, EGG, and GGG and gel strength are plotted in Figure 2, where G and E represent the carboxylic acid and methyl ester groups, respectively. Note that the probability of three contiguous carboxylic acid groups must be near about 0.22 in order to form a gel and only when the probability for GGG is greater than 0.22 is there a marked increase in gel strength (Lee *et al.*, 2008). Between a probability between 0.22 and 0.50, there is little change

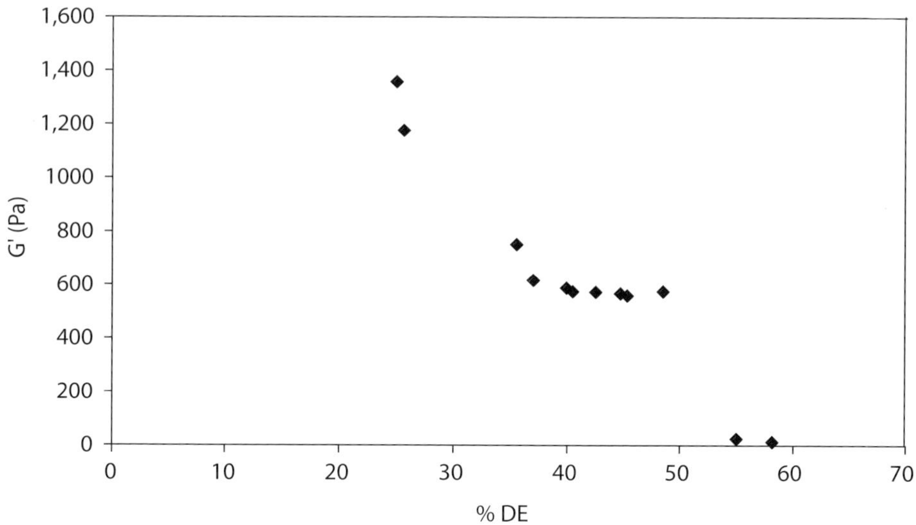

Figure 1. Correlation of G' (Pa) and % DE of 2% pectins in the presence of 35 mM calcium at 0.4 Hz (Adapted from Lee et al., *2008).*

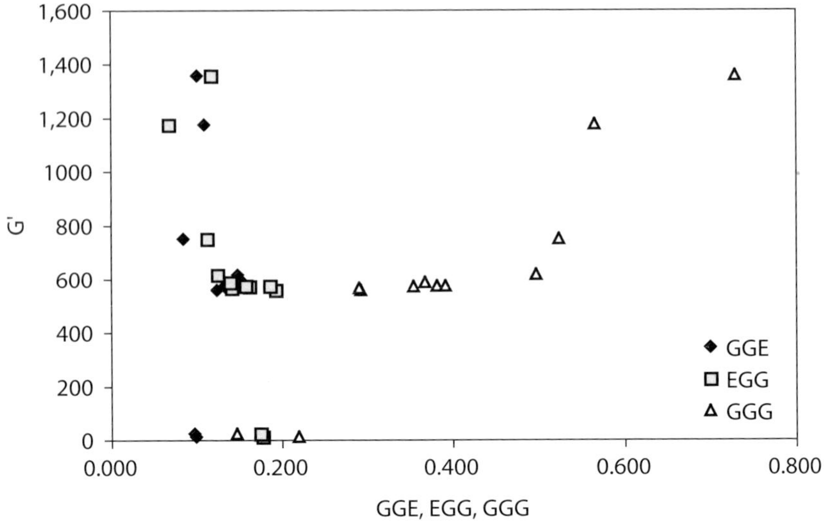

Figure 2. Correlation of G' (Pa) of 2% pectins in the presence of 35 mM calcium at 0.4 Hz and 1 Pa with % DE (a), probabilities of GGE, EGG, and GGG. G' values were measured in duplicate. Average values were used in the plots. Probability of GGG was adapted from Lee et al.*, 2008 (Figure 5).*

in the gel strength of about 560 Pa. Near a probability of 0.50 for three contiguous carboxylic acid groups, there is a second inflection and increase in gel strength. The presence of an ester group in the triad, regardless of position, dramatically influences the relationship to gelling ability. As the probability of GGE or EGG increases, the gelling ability decreases markedly. A similar influence on gel strength is observed when the relationship for probability for the dyad GG is considered (Figure 3). There is a dramatic increase in gelling when the probability for GG reaches about 0.57 and no significant change until the probability for GG is near 0.76. The extent of a change in probability of contiguous triads or diads of the carboxylic acid groups without a corresponding increase in G' varies with triads or diads or the presence of an ester in the triad or diad. Again, the consecutive carboxylic acid groups are required for gelling. Considering the impact of total charge as estimated by % DE, distribution of charge as estimated by the frequency of the dyad, GG, or triad, GGG, it appears the relative contribution to gelling potential varies depending on the position.

Clearly, the total degree of esterification and distribution of ester content are important in terms of calcium senstivity and gelling of charge modified pectins. However, there are three distinct regions of cation sensitivity. The relative contribution of charge or charge distribution cannot be clearly distinguished unless the position or region of cation sensitivity is known.

Because modified pectins are more sensitive to cations, such as calcium, they are likely to be more reactive with positively charged proteins as well. Pectins are commonly

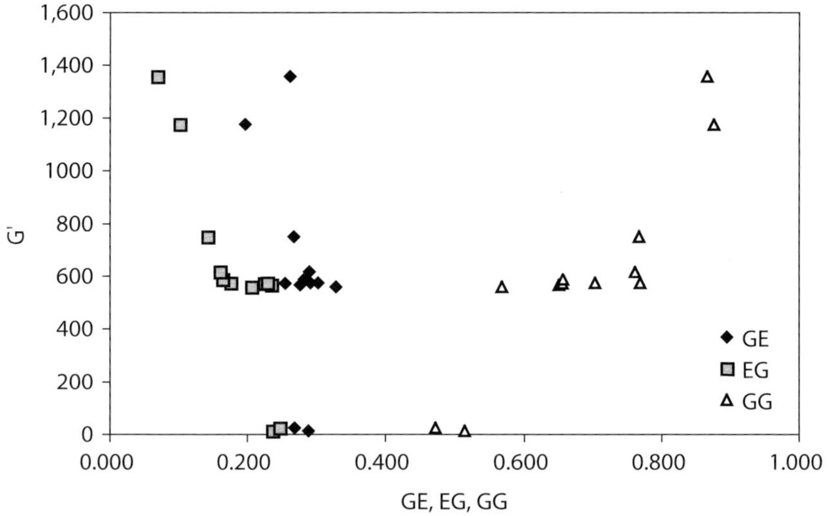

Figure 3. Correlation of G' (Pa) of 2% pectins in the presence of 35 mM calcium at 0.4 Hz and 1 Pa with % DE (a), probabilities of GGE, EGG, and GGG. G' values were measured in duplicate. Average values were used in the plots. Probability of GG was adapted from Lee et al.*, 2008 (Figure 5).*

used to stabilize acidified dairy drinks, most likely due to electrostatic interactions between pectins and milk proteins. The nature of the interaction and contribution of calcium sensitive pectins is not clearly understood. In this study, milk proteins were acidified in acetate buffer to a pH below the pI of the caseins, and mixed with unmodified and charge modified pectin in an effort to determine the contribution of the individual milk proteins. The visual appearance of the dispersions varied with the type of milk protein and pectin (Figure 3). Non-fat milk, casein, $\alpha S_{1,2}$- and β-casein fractions made cloudy dispersions with evidence of precipitation in acetate buffer, pH 3.8. Only the κ-casein dispersion gave a clear appearance with no evidence of precipitation at pH 3.8. Addition of pectin to the milk fractions usually increased the extent of sedimentation (Figure 4a, c, d, e), but pectin increased the apparent stability and opalescent appearance of sodium caseinate (Figure 4b). A greater precipitate was observed in non-fat milk at pH 3.8 in the presence of O-Pec, B-Pec, or U-Pec. On the other hand, the opalescent appearance of caseinate at pH 3.8 increased in the presence of O-Pec, B-Pec, or U-Pec with U-Pec showing the greatest effect. The different effects between non-fat milk and casein in the presence of pectin suggests that non-casein components in non-fat milk contribute to greater instability. The major non-casein protein in NFDM is β-lactoglobulin, which was more reactive to charge modified pectin than unmodified pectin (Kazmierski *et al.*, 2003). Of the fractionated caseins, O-Pec, B-Pec, or U-Pec had the least effect on appearance of $\alpha S_{1,2}$-casein. The dispersion was slightly more opalescent and had slightly larger precipitate, regardless of type of pectin. The opalescence of β-casein was minimally affected by pectins, but the size of the precipitate increased with pectin, especially U-Pec. Although κ-casein

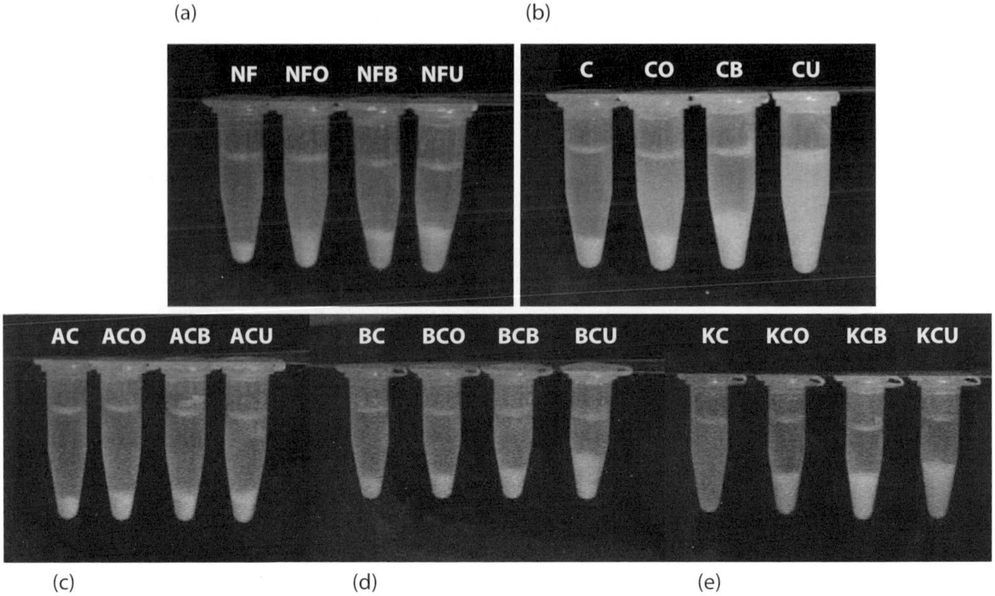

Figure 4. Photograph of 1% milk dispersion with pectins (10:1) following equilibration at 20°C for 2hour. a) NF: non-fat milk, NFO: NF in O-Pec, NFB: NF in B-Pec, NFU: NF in U-Pec; b) C: Casein, CO: C in O-Pec, CB: C in B-Pec, CU: C in U-Pec; c) AC: $\alpha S_{1,2}$-Casein, ACO: AC in O-Pec, ACB: AC in B-Pec, ACU: AC in U-Pec; d) BC: β-Casein, BCO: BC in O-Pec, BCB: BC in B-Pec, BCU: BC in U-Pec; e) KC: κ-Casein, KCO: KC in O-Pec, CB: KC in B-Pec, KCU: KC in U-Pec.

had no precipitate initially, addition of pectin resulted in sedimentation, and U-Pec resulted in the greatest precipitate.

Particle size distributions of milk protein and pectin dispersions are presented (Table 1). No significant differences (P>0.05) were observed in volume to surface average diameter, $D_{3,2}$, except for κ-casein and U-Pec. $D_{4,3}$ values were larger in pectin added dispersions, except $\alpha S_{1,2}$ -casein, where a decrease was observed with all pectin types. In non-fat milk and κ-casein, $D_{4,3}$ values were higher in pectin added dispersions and U-Pec was most reactive. The particle size of β-casein also increased following the same tendency of non-fat milk dispersion and κ-casein, but U-Pec added dispersion had only significantly different particle size comparing O-Pec, B-Pec – added or no added dispersion. In contrast, $\alpha S_{1,2}$ -casein dispersions with pectin had smaller particle size regardless of type of pectin.

These results indicate that pectins interact with milk proteins individually and the micellular structure is not essential for milk protein-pectin interaction. Our results show that unmodified and modified pectin, interact with milk proteins strongly in the order of UPec, BPec and OPec. However, the presence of β-lactoglobulin or other milk proteins in NFDM or collective caseins in sodium caseinate influences the reaction.

Table 1. The weight-average size of 1% dispersion milk protein with pectins (10:1) in acetate buffer, pH 3.8.

Sample		Weight average size[1]	
		$D_{3,2}$ (μm)	$D_{4,3}$ (μm)
Non-fat milk	non-fat milk	$7.74^a \pm 0.11$	$25.8^a \pm 1.63$
	NF in O-Pec	$11.28^a \pm 1.07$	$56.74^b \pm 13.33$
	NF in B-Pec	$15.27^a \pm 0.50$	$102.4^c \pm 14.13$
	NF in U-Pec	$21.5^a \pm 2.19$	$183.6^d \pm 13.91$
$\alpha S_{1,2}$-Casein	$\alpha S_{1,2}$-Casein	$5.87^a \pm 0.25$	$44.63^a \pm 14.21$
	$\alpha S_{1,2}$-Casein in O-Pec	$6.51^a \pm 0.20$	$24.18^b \pm 2.99$
	$\alpha S_{1,2}$-Casein in B-Pec	$7.52^a \pm 0.04$	$34.55^b \pm 5.00$
	$\alpha S_{1,2}$-Casein in U-Pec	$7.78^a \pm 0.35$	$33.68^b \pm 1.22$
β-Casein	β-Casein	$5.67^a \pm 0.12$	$24.00^a \pm 4.71$
	β-Casein in O-Pec	$13.37^a \pm 1.15$	$29.12^a \pm 1.11$
	β-Casein in B-Pec	$10.28^a \pm 0.65$	$28.71^a \pm 4.07$
	β-Casein in U-Pec	$14.71^a \pm 1.43$	$43.23^b \pm 6.23$
κ-Casein	κ-Casein	not detected	not detected
	κ-Casein in O-Pec	$35.94^a \pm 6.22$	$135.2^a \pm 7.34$
	κ-Casein in B-Pec	$39.13^a \pm 7.13$	$215.3^b \pm 43.93$
	κ-Casein in U-Pec	$69.29^b \pm 44.87$	$258.3^c \pm 5.91$

[1]Under same milk protein, means with the same superscript in a column are not significantly different at $p = 0.05$.

In NFDM, only an increase in particle size is observed with pectin, suggesting that β-lactoglobulin or other milk proteins in NFDM contribute to pectin interactions. There was less precipitate in sodium caseinate dispersions and smaller particle size (data not shown) with modified pectins. The results suggest that there is competition between different caseins for casein-casein interaction or casein-pectin interaction. In acidified dairy drinks, the micellular structure is not destroyed by acidification but it differs in composition and αS1, αS2, β-caseins migrate from the micelle. Tromp *et al.* (2004) noted that less than 20% of pectin interacts with proteins in the micelle of acidified protein and that the balance of the pectin contributes to stability by forming a network structure. Potentially, these proteins interact with pectin in the bulk phase to enhance the formation of a network structure.

Charge modified pectins also have potential as a stabilizer for emulsions and has great high potential impact. While hydrocolloids are not typically considered to be good emulsifiers, gum acacia and pectin are unique in that both have shown to be surface active. In the case of pectin, smaller MW and more highly branched pectin tends to be more surface active (Leroux *et al.*, 2003). PME modified pectins reduce surface tension more than unmodified pectin and the ability was related to block wise de-esterification which allowed better adsorption to the oil interface. Lutz *et al.* (2008) compared commercial apple pectin, PME modified pectin by U-PME or crude PME enzyme, UPec and C-Pec) for the ability to reduce surface tension. They observed values of 63, 58 and 55 mN/m, respectively for commercial apple pectin, C-Pec and U-Pec, respectively. When proteins and hydrocolloids, such as pectin interact, the resulting complex often serves as good emulsifying activity. Often the complex is formed by electrostatic interaction between the anionically charged pectin and the cationically charged protein. After adsorption of the protein at the oil interface, stability of the emulsion is enhanced by protein-pectin interactions. In the case of double emulsions, the droplet size is decreased from about 22-35 μm if stabilized by whey protein isolate alone to 10-13 μm if stabilized by a complex of whey protein isolate-charge modified pectin (Figure 5).

A final potential application of charge modified pectin is the development of encapsulating techniques to deliver addenda such as phytochemicals or drugs to the colon for release. For acid or hydrolytic enzyme sensitive addenda that may be altered by the low pH of the stomach or degraded by proteolytic or lipolytic enzymes in the small intestine, such as peptides, protection by encapsulation in a pectin hydrogel offers the possibility for enhanced absorption downstream. We have used indomethacin as a model drug and compared the entrapment efficiency and release under simulated gastrointestinal conditions (Table 2). In this case, modified pectin had the highest entrapment efficiency of indomethacin and the lowest amount of indomethacin in the calcium chloride equilibration medium. Furthermore, the release of indomethacin in HCl, pH 1.2 or phosphate buffer, pH 7.4, was less than 1% and less than 15%, respectively (data not shown).

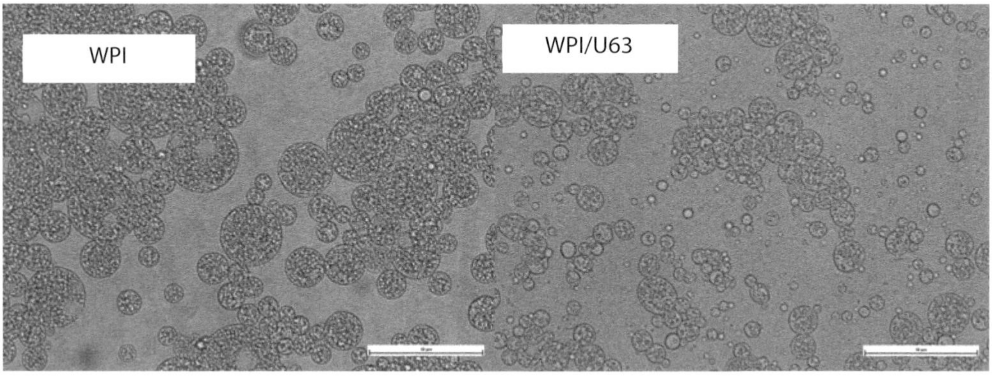

Figure 5. Water in oil in water double emulsions stabilized by whey protein isolate (WPI) or a complex of WPI and charge modified pectin. Adapted from Lutz et al.*, 2008.*

Table 2. Bead formulation and indomethacin entrapment efficiency in dried beads.[1]

Pectin type	Drug:pectin ratio	Release into calcium chloride	Entrapment efficiency (5)
DM 10	0.5	0.75	53.6
	1.0	0.17	28.2
DM 35	0.5	0.38	66.8
	1.0	0.15	40.5
MP 38	0.5	0.06	82.1
	1.0	0.01	43.5

[1]Calcium chloride concentration was 300 mM. DM10 and DM35 refer to commercial low methoxyl pectins and MP 38 is PME charge modified pectin. Adapted from Jung *et al.*, 2008.

4. Conclusion

In this manuscript, we have described results of multiple charge modifications of citrus pectin by citrus PME to targeted DE values. Clearly, PME isozymes and presence of multiple isozymes under the same conditions of modification, influence the charge distribution. The relative contribution of total charge or distribution of charge cannot be clearly elucidated as the contribution of either depends on achieving a critical limit of de-esterification. The position and shape of this critical limit is influenced by the composition of the dyads and triads of carboxylic acid or ester groups. Charge modified pectins have great potential applications and offers the possibility to create tailored pectins that have specific functional attributes of gelling, emulsifying, stabilizing or encapsulation.

Acknowledgement

This research was partially supported by Research Grant Award No. US 3655-04 from BARD, The United States – Israel Binational Agricultural Research and Development Fund.

Reference

Ackerley, J. and L. Wicker, 2003. Floc formation and charges in serum soluble cloud components of fresh valencia orange juice. Journal of Food Science 68: 1169-1174.

Ackerley, J., M. Corredig and L. Wicker, 2002. Clarification of citrus juice is influenced by specific activity of thermolabile pectinmethylesterase and inactive PME-pectin complexes, *Journal of Food Science,* 67: 2529-2533.

Cameron, R.G., R.A. Baker and K. Grohmann, 1998. Multiple forms of pectinmethylesterase from citrus peel and their effects on juice cloud stability. Journal of Food Science, 63: 253-256.

Cameron, R.G., B.J. Savary, A.T. Hotchkiss, M.L. Fishman, H.K. Chau, R.A. Baker and K. Grohmann, 2003. Separation and characterization of a salt-dependent pectin methylesterase from *Citrus sinensis* var. Valencia fruit tissue. Journal of Agricultural and Food Chemistry, 51: 2070-2075.

Cameron, R.G., B.J. Savary, A.T. Hotchkiss and M.L. Fishman, 2005. Isolation, characterization, and pectin-modifying properties of a thermally tolerant pectin methylesterase from *Citrus sinensis* var. Valencia. Journal of Agricultural and Food Chemistry, 53: 2255-2260.

Daas, P.J.H., A.G.J. Voragen and H.A. Schols, 2001. Study of the methyl ester distribution in pectin with Endo-polygalacturonase and high-performance size-exclusion chromatography. Biopolymers, 58: 195-203.

Denes, J.-M., A. Baron, C.M.G.C. Renard, C. Pean and J.-F. Drilleau, 2000. Different action patterns for apple pectin methylesterase at pH 7.0 and 4.5. Carbohydrate Research, 327: 385-393.

Glahn, P.E. and C. Rolin, 1994. Casein-pectin interaction in sour milk beverages. Food Ingredients Europe, 1: 1-5.

Guillotin, S.E., E.J. Bakx, P. Boulenguer, H.A. Schols and A.G.J. Voragen, 2007. Determination of the degree of substitutin, degree of amidation and degree of blockiness of commercial pectins by using capillary electorphoresis. Food Hydrocolloids 21: 444-451.

Guzey, D., H.J. Kim and D.J. McClements, 2004. Factors influencing the production of o/w emulsions stabilized by [beta]-lactoglobulin-pectin membranes. Food Hydrocolloids, 18: 967-975.

Jung, J., R. Arnold and L. Wicker, 2008. Pectin and modified pectin hydrogel beads as a colon-targeted drug delivery carrier. Annual Meeting of Institute of Food Technologists, June 28-July 2, New Orleans, LA.

Kazmierski, M., L. Wicker and M. Corredig, 2003. Thermally induced aggregation of whey proteins; characterization of protein isolates and beta-lactoglobulin/pectin interactions. Journal Food Science 68: 1673-1679.

Kim, Y.K. and L. Wicker, 2006. Pectinmethylesterase Modification to Produce Charge Modified Pectin. In: M.L. Fishman, P.X. Qi and L. Wicker (ed.) Advances in Biopolymers, Molecules, Clusters, Networks and Interactions, Vol. ACS Symposium Series 935, American Chemical Society, Washington, D.C., 2006.

Kim, Y.K. and L. Wicker, 2009. Valencia PME isozymes create charge modified pectins with distinct calcium sensitivity and rheological properties. Food Hydrocolloids, 23(3): 957-963.

Kim, Y.K., Q. Teng and L. Wicker, 2005. Action pattern of Valencia orange PME de-esterification of high methoxyl pectin and characterization. Carboydrate Polymers 340: 2620-2629.

Laurent, M.A. and P. Boulenguer, 2003. Stabilization mechanism of acid dairy drinks (ADD) induced by pectin. Food Hydrocolloids 17: 445-454.

Lee, H., J. Rivner, J. Urbauer and N. Garti, 2008. De-esterification Pattern of Valencia Orange Pectinmethylesterases and Characterization of Modified Pectins. Journal Science Food Agriculture 88: 2102-2110.

Leroux, J., V. Langendroff, G. Schick, V. Vaishnav and J. Mazoyer, 2003. Emulsion stabilizing properties of pectin. Food Hydrocolloids 17: 455-462.

Limberg, G., R. Korner, H.C. Buchholt, T.M.I.E. Christensen, P. Roepstorff and J.D. Mikkelsen, 2000. Quantification of the amount of galacturonic acid residues in block sequences in pectin homogalacturonan by enzymatic fingerprinting with exo- and endo-polygalacturonase II from *Aspergillus niger*. Carbohydrate Research, 327: 321-332.

Liu, L., M.L. Fishman, J. Kost and K.B. Hicks, 2003. Pectin-based systems for colon-specific drug delivery via oral route. Biomaterials. 24: 3333.

Lutz, R., A. Aserin, L. Wicker and N. Garti, 2009. Structure and Physical Properties of Pectins with Block-Wise Distribution of Carboxylic Acid Groups. Food Hydrocolloids 23(3): 786-794.

Maroziene, A. and C.G. De Kruif, 2000. Interaction of pectin and casein micelles. Food Hydrocolloids. 14: 391-394.

Munjeri, O., J.H. Collett and J.T. Fell, 1997. Hydrogel beads based on amidated pectins for colon-specific drug delivery: the role of chitosan in modifying drug release. Journal of Controlled Release. 46: 273.

Murata, Y., M. Miyashita, K. Kofuji, E. Miyamoto and S. Kawashima, 2004. Drug release properties of a gel bead prepared with pectin and hydrolysate. Journal of Controlled Release. 95: 61.

Pereyra, R., K.A. Schmidt and L. Wicker, 1997. Interaction and stabilization of acidified casein dispersions with low and high methoxyl pectins. J. Agric. Food Chem. 45: 3448-3451.

Savary, B.J., A.T. Hotchkiss and R.G. Cameron, 2002. Characterization of a salt-independent pectin methylesterase purified from Valencia orange peel. Journal of Agricultural and Food Chemistry, 50: 3553-3558.

Sperber, B.L.H.M., H.A. Schols, M.A. Cohen Stuart, W. Norde and A.G.J. Voragen, 2008. Influence of the overall charge and local charge density of pectin on the complex formation between pectin and β-lactoglobulin. Food Hydrocolloids, 23 (3): 765-772.

Sriamornsak, P., S. Puttipipatkhachorn and S. Prakongpan, 1997. Calcium pectinate gel coated pellets as an alternative carrier to calcium pectinate beads. International Journal of Pharmaceutics. 156: 189-194.

Sriamornsak, P. and J. Nunthanid, 1998. Calcium pectinate gel beads for controlled release drug delivery: I. Preparation and in vitro release studies. International Journal of Pharmaceutics. 160: 207.

Tromp, H.R., C.G. de Kruif, M. van Eijk and C. Rolin, 2004. On the mechanism of stabilization of acidified milk drinks by pectin. Food Hydrocolloids 18: 565-572.

Voragen, A.G.J., W. Pilnik, J.-F. Thibault, M.A.V. Axelos and C.M.G.C. Renard, 1995. Pectins. In: A.M. Stephen (ed.) Food Polysaccharides and Their Applications, Vol. 10, Marcel Dekker, New York, pp. 287-339.

Wakerly, Z., J.T. Fell, D. Attwood and D. Parkins, 1997. Studies on drug release from pectin/ethylcellulose film-coated tablets: a potential colonic delivery system. International Journal of Pharmaceutics. 153: 219-224.

Wicker, L., J.L. Ackerley and M. Corredig, 2002. Clarification of juice by thermolabile Valencia pectinmethylesterase is accelerated by cations. Journal of Agricultural and Food Chemistry, 50: 4091-4095.

Willats, W.G.T., C. Orfila, G. Limberg, H.C. Buchholt, G.J.W.M. van Alebeek, A.G.J. Voragen, S.E. Marcus, T.M.I.E. Christensen, J.D. Mikkelsen, B.S. Murray and J.P. Knox, 2001. Modulation of the degree and pattern of methyl-esterification of pectic homogalacturonan in plant cell walls: implications for pectin methyl esterase action, matrix properties, and cell adhesion. Journal of Biological Chemistry 276: 19404-19413.

Wong, T.W., H.Y. Lee, L.W. Chan and P.W.S. Heng, 2002. Drug release properties of pectinate microspheres prepared by emulsification method. International Journal of Pharmaceutics. 242: 233.

The use of pectin modifying treatments to minimize texture and structure degradation of frozen fruits and vegetables

S. Van Buggenhout, A. Van Loey and M. Hendrickx
Laboratory of Food Technology and Leuven Food Science and Nutrition Research Centre (LFoRCe), Department of Microbial and Molecular Systems (M2S), Katholieke Universiteit Leuven, Kasteelpark Arenberg 22, 3001 Leuven, Belgium

Abstract

Pectinmethylesterase (PME) is an enzyme that can be used in the context of texture preservation during processing of plant-based foods: a controlled action of this enzyme is sometimes desired because the enzyme catalyzed de-esterification of pectin creates the possibility of intermolecular interactions between Ca^{2+} ions and low-methyl esterified pectin. As a result of the strengthened cell-cell adhesion, pectin modifying treatments have been shown useful to reduce softening of for example thermally treated plant-based foods. In this work it was investigated whether pectin modifying treatments could reduce textural changes of frozen fruits and vegetables. To evaluate structural changes, a new microscopic methodological approach including detailed image analysis and tissue damage quantification in case of cell wall breakage, was evolved. Two case studies were selected: (1) carrots, a system where mainly endogenous (plant) PME is responsible for pectin conversions and (2) strawberries, a system suitable for infusion of fungal PME. Pectin in carrots could be modified using a thermal (60 °C) or a high pressure (60 °C/300 MPa) treatment. The degree of methyl esterification (DM) of strawberry pectin on the other hand was decreased using vacuum infusion (VI) with PME and Ca^{2+}. Results showed that the effect of these pectin modifying treatments on the final textural and structural quality of frozen products depended on the freezing conditions and on the product. When combined with slow freezing, the pectin modifying treatments had no significant effect on the resulting textural quality. When combined on the other hand with rapid and cryogenic freezing, the impact of the pretreatments was significant but product dependent: a negative effect was observed for carrots, whereas a positive effect on the hardness of strawberries was noticed. When pretreatments were applied prior to high pressure shift freezing (HPSF), a desired hardness improvement was remarked.

Keywords: texture, microstructure, fruit and vegetable processing, freezing

1. Introduction

To food scientists water is the medium that causes the most deleterious reactions, and its removal or immobilization, when possible, is highly desirable. It is known that the rate of deleterious reactions decreases with temperature. Freezing, involving simultaneous immobilization of water as ice and temperature reduction, seems the perfect food preservation technique. This would be true if it were not for the extensive microstructural changes inside organized tissue caused by ice crystallization. As a consequence, the preservation of a desirable texture in frozen fruits and vegetables is very complex. However, because freezing in general preserves the taste and nutritional value of foods better than other preservation techniques, freezing preservation is one of the most important methods for retaining quality in agricultural products over long-term storage periods.

The most important factor affecting the final quality of frozen fruit and vegetables is the control of temperature during and subsequent to the freezing process (Brown, 1976). In most cases, slow freezing will lead to significant cell dehydration and further extracellular ice formation (Pearce, 2001). Formation of large extracellular ice crystals can cause cell separation and eventually cell rupture (Munoz-Delgado, 1977) and very extensive cell dehydration may lead to destruction of membrane compartmentalization (Stepkonus, 1999). On the other hand, rapid freezing will lead to limited cell dehydration and, as a consequence, excessive supercooling of the intracellular solution will increase the probability of intracellular freezing. Although intracellular freezing is an inherently lethal event, intracellular freezing is advantageous for the overall eating quality of frozen products as the gross tissue structure is maintained (Reid, 1999).

Another factor that might be important for the final textural quality of frozen fruit and vegetables is the structure of cell wall pectin. From several studies in the thermal processing domain, it is known that a controlled activity of pectinmethylesterase (PME) can improve the textural quality of processed plant-based products (Van Buren, 1979, Waldron *et al.*, 1997, 2003). PME catalyzes the de-esterification of pectin and creates in this way the possibility of intermolecular interactions between Ca^{++} ions and low esterified pectin. Consequently, cell-cell adhesion is reinforced. In this work it was evaluated to what extent textural and structural changes during freezing of fruit and vegetables is related to pectin conversions and thus to what extent the suitability of fruit and vegetables to be frozen can be enhanced by the use of pectin modifying (pre-)treatments. In this context, two case studies were selected: (1) carrots were chosen as a model product for hard, pectin-rich plant tissues whose texture is influenced considerably by pectin modifications catalyzed by the endogenous (plant) PME and (2) strawberries were chosen as a case study for soft plant tissues whose texture depends noticeably on the turgor pressure and that as a result of their porous structure allow addition of exogenous enzymes like (fungal) PME. In order to better understand textural freezing damage, structural changes were also studied in detail. Hereto, a new microscopic methodological approach including detailed image analysis and tissue damage quantification in case of cell wall breakage, was evolved.

2. Material and methods

Carrots (*Daucus carota* L., cultivar Nerac) were bought from a local wholesale trade and stored at 4 °C for maximum 5 days. The cores of the carrots were cut with a sharp stainless-steel cork borer and chopped using a sharp knife to obtain small cylinders of 12 mm diameter and 10 mm height. A single batch of carrots was used and a maximum randomization of samples was established.

Fresh Belgian strawberries (*Fragaria ananassa*, cultivar Elsanta) were bought from a local grocer and stored at 2 °C for 3 days maximally. Strawberries were stemmed and cut into halves just before infusion or freezing.

In case pectin modifying pretreatments were applied, carrot cylinders were soaked in a $CaCl_2$ solution (0.5%) for 1 h and packed under vacuum. This step was followed by thermal treatment for 30 min at 60 °C in a thermostated water bath (Julabo UC, Merck Belgolabo, Belgium) or high pressure treatment for 15 min at 60 °C and 300 MPa using a single vessel pressure unit (EPSI, Belgium).

In order to infuse the strawberry fruits, strawberry halves were immersed in a solution containing 0.12% (v/v) *Aspergillus aculeatus* PME (EC 3.1.1.11, Carbohydrate Esterase Family 8, Novoshape, Novozymes, Denmark) and 0.5% (w/w) $CaCl_2$. A vacuum pulse (10 hPa) was applied for 5 min at room temperature (Duvetter *et al.*, 2005). After releasing the vacuum very slowly, strawberry halves were allowed to drain for 30 min. 150 ml of infusion solution was used per 125 g fruits. Non-infused and water-infused strawberry halves were used as controls. The weight gain of the fruits upon infusion (%) was determined by weighing the samples.

The Alcohol Insoluble Residue (AIR) was extracted (McFeeters and Armstrong, 1984) from carrots and strawberries and hydrolyzed for methanol (Ng and Waldron, 1997) and galacturonic acid (Ahmed and Labavitch, 1977) analysis. Hydrolysis was done in duplicate. The amount of methanol formed was determined colorimetrically according to the method of Klavons and Bennett (1986). The hydroxydiphenyl method of Blumenkrantz and Asboe-Hansen (1973) was used to determine the galacturonic acid content of the hydrolyzed samples. Both determinations were done in triplicate. The DM was defined as the ratio of the amount of methanol and the amount of galacturonic acid:

$$DM = \frac{(N_{methanol} / AIR)}{(N_{galac.\ acid} / AIR)} \times 100$$

where $N_{methanol}$ is the number of moles of methanol, $N_{galac.\ acid}$ is the number of moles of galacturonic acid and AIR is the amount (g) of AIR.

Slow freezing was performed in stagnant air at -18 °C for one night. To freeze the carrots and strawberries rapidly, samples packed in double film bags were kept in a pre-cooled cryostat bath (Heto CBN18-50, Heto-Holten, Denmark), at -18 °C and at -25 °C respectively, until the center temperature of the sample decreased to -18 ± 1 °C. Subsequently, samples were transferred to cold storage (overnight) at -18 °C until further analysis. A programmable cryogenic freezer (Nicool PC Plus, Air Liquide, France) was used to freeze the carrot and strawberry samples cryogenically. Packed samples were placed in a ventilated pre-cooled freezing chamber at -80 °C. When the temperature at the center of the samples reached -18 ± 1 °C, samples were transferred to cold storage at -18 °C until further analysis. High pressure shift freezing (HPSF) experiments were performed in a pilot-scale high pressure apparatus (EPSI, Belgium). Samples packed in a double film polyethylene bag were put in the pre-cooled vessel at 0 °C. After equilibration of the sample temperature to the temperature of the pre-cooled vessel, pressure was built up to 200 MPa. Then the system was cooled down until the temperature at the center of the samples reached -15 ± 1 °C. At this point, the product was still in the non-frozen state. Then, pressure was suddenly released initiating the actual freezing process. The freezing process was completed at atmospheric pressure and continued until the temperature at the center of the sample reached -18 ± 1 °C. Samples were transferred to cold storage at -18 °C until further analysis.

The nominal freezing time (Int. Inst. Refrig., 1986) was determined as the time to decrease the temperature of the thermal center from 0 °C to 10 °C below the initial freezing point.

After thawing at room temperature, the hardness of carrot and strawberry samples, defined as the maximal force during a compression (Van Buggenhout *et al.*, 2006a) or a penetration test (Van Buggenhout *et al.*, 2006b), was determined using a 'TA-XT2i Texture Analyser' (Stable Micro Systems, Surrey, U.K.). Each hardness value reported was the mean of 10 (carrots) or 16 (strawberry halves) measurements and expressed relatively to the hardness of the fresh material. Texture results were assayed statistically using multiple way ANOVA (SAS v9.1, Cary, U.S.A.). Significant differences among texture means were analyzed using the post-hoc Tukey test. The significance level was set at α=0.05.

After thawing, also the drip loss of strawberries was determined (Van Buggenhout *et al.*, 2006b).

Samples were prepared for microscopy analysis: samples were partially thawed before slices were cut. Duplicate transverse slices, from two different carrot cylinders and duplicate transverse and duplicate longitudinal slices, from 4 different strawberry halves were prepared. Fixation, dehydration, embedding and micro-sectioning were performed according to the method of Van Buggenhout *et al.* (2006b). Sections were stained with 0.1% toluidin blue for 5 min followed by washing with water. Fixed and dyed slices were examined with a light microscope (Olympus BX50 and BX41, Optical Co. Ltd, Tokyo, Japan).

Image analysis software (analySIS pro 3.1 and 5, Soft Imaging System GmbH, Bensheim, Germany) was used to analyze the white areas in the micrographs, representing cells or intercellular spaces and further called 'particles'. Size and shape factor parameters were determined. Shape factor ($4 \cdot \pi \cdot$ size/perimeter^2) was considered as an indication of the 'roundness' and the 'smoothness' of the particle: it is equal to 1 for perfectly round and smooth particles and decreases to 0 for more elongated or more irregular particles. Histograms of all measured parameters were constructed using the statistical software package SAS (v9.1, Cary, U.S.A.). The average tissue area, which was studied per sample during image analysis, was 56 mm^2 for carrot samples and 36 mm^2 for strawberry samples. In this way analyses of around 40,000 and 17,000 particles respectively were performed.

Tissue damage in carrots was defined as the percentage of the total area occupied by particles larger than the largest particle in the fresh tissue. Hereto, tissue particles were classified according to their size and marked with different colors: red (100-500 μm^2), green (500-1000 μm^2), dark blue (1000-5,000 μm^2), yellow (5,000; approximate size of largest particle in fresh carrot tissue) and light blue (approximate size of largest particle in fresh carrot tissue - 2,000,000 μm^2). To identify the approximate size of the largest particles in fresh carrot tissue, the fresh carrot tissue was studied in detail. For the batch used in this experiment, particles larger than 10,000 μm^2 that occupied less than 0.14% of the total analyzed area were used to define tissue damage (Van Buggenhout *et al.*, 2006a).

3. Results and discussion

3.1 Case 1: carrots

Pectin in fresh carrots was highly esterified (74%). Pretreatments of carrots clearly affected the endogenous pectin: thermal and high pressure pretreatment reduced the DM to respectively 63% and 56%. Moreover, carrot pectin DM was not remarkably affected by freezing and thawing. These results confirmed that the selected pretreatment conditions are suitable to enhance carrot PME activity which was also pointed out by Sila *et al.* (2007) and suggest that both thermal treatment at 60 °C and high pressure treatment and 300 MPa and 60 °C could be used as a pre-freezing treatment to alter the carrot pectin and to possibly affect the resistance of carrots to freezing damage.

Statistical analysis revealed a significant interaction between these pretreatments and the different freezing conditions in their influence on the texture results (F-value 32.66; $P<0.0001$). Therefore, the effect of pretreatments (F-value 24.85; $P<0.0001$) was considered for each freezing method separately.

Slow freezing (nominal freezing time 2h 28 min ± 8 min) of carrots induced severe hardness loss. After freezing and thawing, only about 10% of the original hardness could be retained. Severe texture loss was correlated with serious radial tissue damage: around 20% of the carrot tissue was damaged. Nor calcium soaking followed by thermal pretreatment, neither calcium soaking followed by high pressure pretreatment, could reduce the hardness loss and tissue damage of slowly frozen carrot samples. This conclusion was confirmed by the absence of striking differences between the distribution curves of both particle size and particle shape factor between the untreated, slowly frozen sample and the pretreated, slowly frozen sample (Van Buggenhout *et al.*, 2006a).

Textural and microstructural freezing damage was significantly minimized by increasing the freezing rate (the nominal freezing time for rapid and cryogenic freezing was determined respectively at 26 min ± 2 min and 10 min 17 s ± 48 s). Rapidly and cryogenically frozen carrots could retain about 35-40% of their original hardness values and their microstructure resembled the fresh carrot tissue microstructure. For rapidly and cryogenically frozen carrot cylinders, pretreatment had a significantly negative impact on the hardness of the frozen-then-thawed samples (Van Buggenhout *et al.*, 2006a). Carrots that were pretreated and subsequently frozen by rapid freezing or by cryogenic freezing lost about 90% and 75-85% of their hardness respectively. The negative effect of pretreatment on the texture of rapidly and cryogenically frozen carrots was correlated to a small negative effect of pretreatment on the tissue damage of the carrot samples. Image analysis revealed that thermal and high pressure pretreatment caused respectively 6% and 8% of tissue damage when combined with rapid freezing. Also a shift in tissue particle shape factor due to the application of pretreatments prior to rapid freezing was observed. The mean shape factor of untreated, rapidly frozen tissue particles (0.71) was decreased to 0.61 for both thermal and high pressure pretreatment. Tissue damage of the thermally and high pressure pretreated, cryogenically frozen carrot tissue was negligible (1% and 2% respectively). Mean shape factor of untreated, cryogenically frozen carrot samples (0.70) was reduced to 0.65 for thermal pretreatment and to 0.63 for high pressure pretreatment. The negative effect of pretreatment was thus smaller for cryogenic freezing than for rapid freezing. Therefore, it was concluded that the negative impact of pretreatment on the textural and microstructural quality of frozen carrots depended on the freezing rate. The above results also implied that the positive effect that fast freezing had on the hardness and tissue damage of untreated carrot tissue compared to slow freezing conditions disappeared when samples were pretreated. Possibly, membrane disruption during thermal (Kidmose and Martens, 1999) and high pressure pretreatment (Prestamo and Arroyo, 1998, Tangwongchai *et al.*, 2000), and thus turgor loss during pretreatments prior to the freezing step play an important role in these findings. Unfortunately, the microscopy based method used did not allow detecting cell membrane damage what declared the missing correlation between the tissue damage and texture results for pretreated carrots frozen in different ways.

The application of HPSF (nominal freezing time: 15 min 56 s ± 48 s, high and homogeneous degree of supercooling prior to nucleation) only had a small positive, but non-significant effect on the hardness of the frozen-then-thawed carrots (residual hardness 17%, see Figure 1 1st row) compared to slow freezing (residual hardness

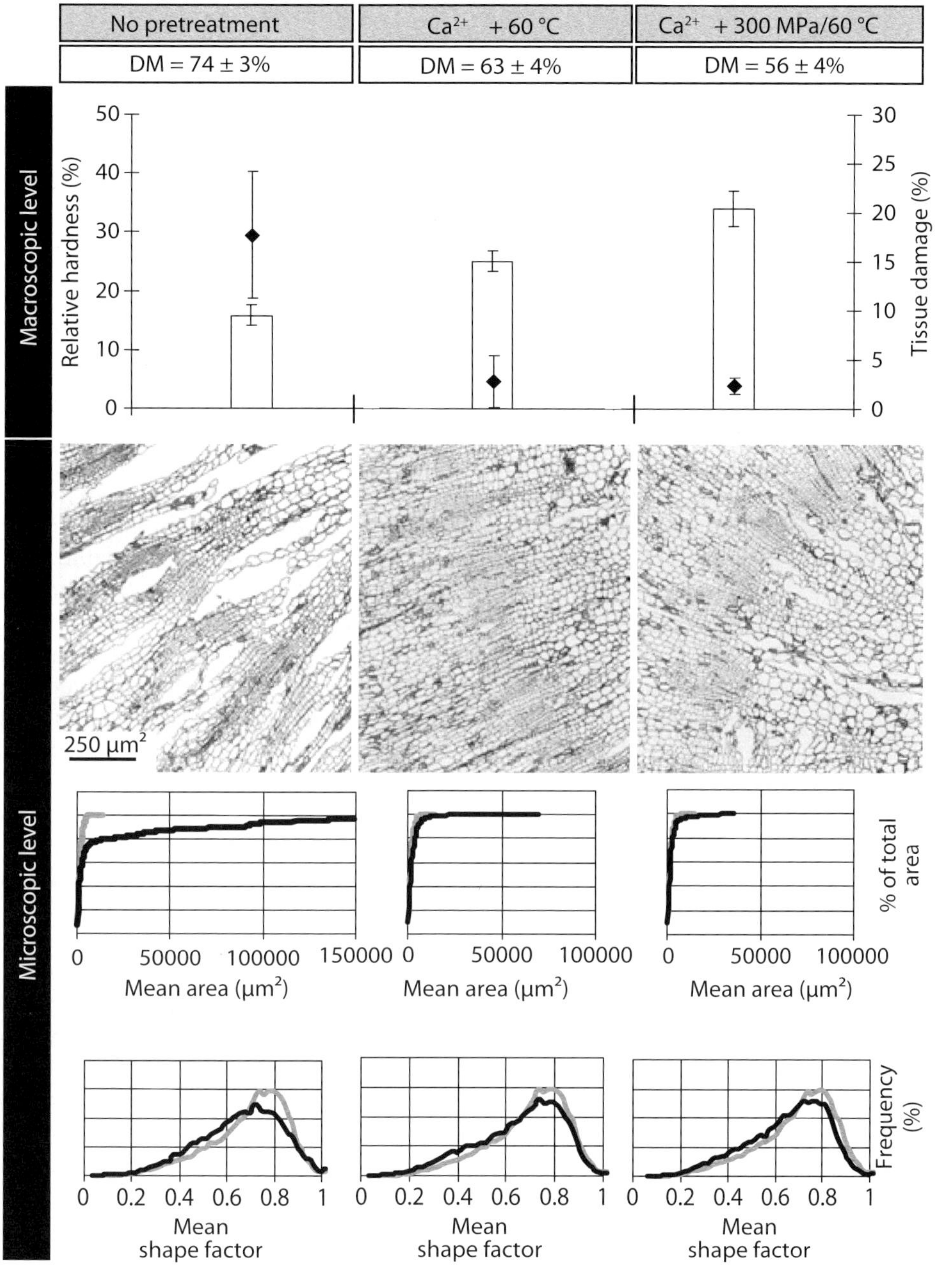

Figure 1. Texture ((bars - 1st row), tissue damage (symbols - 1st row), microstructure (2nd row) and tissue particle results (3rd and 4th row) for non-pretreated and pretreated (calcium soaking followed by a thermal treatment at 60 °C or followed by a high pressure treatment at 300MPa/60 °C) carrot cylinders frozen by HPSF. Tissue particle results: individual particle area in function of the percentage of total area (3rd row) and shape factor distribution curves (4th row) for (—) fresh carrots and (—) frozen carrots.

10%). Like for slowly frozen carrot, also for high pressure frozen carrots, severe hardness loss was correlated with serious radial tissue damage which was esteemed around 20% (Figure 1, 2nd, 3rd and 4th row). When the individual mean particle area was represented in function of the total particle area, the curve for the high pressure frozen sample clearly deviated from the curve of the fresh tissue particles from around 80% of the total area. This way of representing the tissue particle results confirmed that around 20% of the tissue was damaged. Most frequent (0.77) and mean shape factor (0.69) of fresh carrot tissue particles changed to respectively 0.72 and 0.66 for HPSF. This shift towards smaller shape factor values for particles in high pressure frozen carrot tissue revealed that carrot tissue frozen under these conditions showed more elongated and more irregular particles when compared to fresh particles and particles in tissue frozen by rapid (most frequent and mean shape factor values: 0.79-0.71) and cryogenic freezing (most frequent and mean shape factor values: 0.83-0.70).

Texture results (Figure 1, 1st row) clearly showed that high pressure frozen carrot cylinders were significantly harder when they were submitted to calcium soaking followed by a thermal pretreatment or calcium soaking followed by high pressure pretreatment prior to the freezing step. This hardness increase was correlated with a remarkable decrease of about 84-87% in tissue damage: the pretreated, high pressure frozen tissue was only damaged for about 2% (Figure 1, 2nd row). Tissue particle results illustrated the obvious differences between the untreated and pretreated high pressure frozen carrot tissue: while the tissue particle area curve of untreated, high pressure frozen carrot clearly deviated from the curve of fresh carrot tissue particles, the curves of pretreated, high pressure frozen carrot tissue particles coincided the curve of the fresh carrot tissue particles. Tissue particle shape factor distribution of high pressure frozen samples was not noticeable affected by pretreatment, but the distribution curves of pretreated, high pressure frozen carrot tissue coincided more the curve of the unfrozen tissue when compared to the untreated, high pressure frozen carrot tissue.

The results of this case study suggested that increased tissue strength, as a result of pretreatment, seem to improve the hardness and to minimize the tissue damage of HPS frozen carrots while this was not the case when pretreatment was combined with slow, rapid or cryogenic freezing. Because the actual freezing step of HPSF was longer than the cryogenic freezing process, the uniformity of the rapid nucleation step during HPSF was thought to be the key factor in the effectiveness of pectin modifying pretreatments to improve the hardness and limit the tissue damage of frozen samples. According to Ormerod *et al.* (2004), firming pretreatments are the most effective when combined with a controlled freezing regime that assures a high degree of supercooling throughout the whole vegetable product.

3.2 Case 2: strawberries

Strawberry halves were successfully infused with commercial fungal PME and calcium chloride under vacuum. The infusion process affected the overall appearance of the fruit as fruit halves had a soaked appearance. Visual observations confirmed that infusion proceeded from the bundle zone and not from the strawberry skin (Banjongsinsiri *et al.*, 2004). On average, VI with PME/Ca decreased the DM of strawberry pectin from 87% to 56%. Freezing and thawing did not remarkably affect the endogenous pectin. These observations made clear that VI using PME/Ca-solutions could be used to de-esterify the strawberry pectin and might be used as a pre-freezing treatment.

Hardness of PME/Ca-infused fruit was about twice that of fresh fruit and about four times that of water-infused control samples. The hardness increase of PME/Ca-infused fruits was accompanied by a weight increase of about 30%, whereas water-infused fruits only gained around 18% weight upon infusion. According to Duvetter *et al.* (2005), texture improvement of PME/Ca-infused strawberries is (partly) caused by the action of the infused PME, which was also indicated by the DM of the infused strawberries in this study.

Microstructural studies of untreated and vacuum infused strawberry tissue indicated that the cell wall integrity of the isodiametric cells of the parenchymatous strawberry tissue was not affected by VI with PME/Ca. This is in agreement with previous work of Muntada *et al.* (1998), which showed that the size of the cells in infused kiwi fruit and their arrangement were preserved although minor rupture in cell walls was observed. Possibly, the moderated loss of integrity as a consequence of the vacuum treatment can be compensated to a large extent by the active role of the transferred solutes (Saurel, 2002). The quantitative characterization of the strawberry tissue in terms of particle size and particle shape confirmed that nor the size, neither the shape of the particles was influenced by the infusion process. Although the integrity of the tissue and the size/shape of tissue particles were not influenced by VI, a striking difference was observed between the micrographs of the untreated and the vacuum infused strawberry tissues. In the untreated sample, cell walls were poorly stained compared to the infused sample. In infused strawberry tissue, cell walls were denser and strongly stained. Moreover, spaces in between different cells seemed to be filled up. Since it is known that VI with PME /Ca affects the DM of strawberry pectin, it was therefore suggested that cell walls colored stronger after infusion because of the lower DM of pectin in the middle lamella of the cell walls and because of the resulting pectin-calcium-network. Suutarinen *et al.* (2000) also observed that pectin in Ca and/or PME treated strawberries was colored (using ruthenium red) stronger compared to the untreated reference sample.

Hardness and drip loss of non-infused strawberries frozen by different conditions were analyzed after overnight storage at -18 °C and thawing at room temperature (Van Buggenhout *et al.*, 2006b). Strawberry halves frozen under slow freezing conditions (nominal freezing time: 5 h 47 min ± 3 min) lost about 87% of their initial hardness value and 40% of their weight upon thawing. Although rapidly (nominal freezing time: 42 min ± 1 min) and cryogenically (nominal freezing time 10 min ± 4 min) frozen strawberries were slightly harder (and only significantly harder in case of rapid freezing) than strawberries frozen by other freezing methods, quality loss (both texture loss and drip loss) of frozen strawberries was not strikingly prevented by reducing the freezing time. Neither the application of HPSF (nominal freezing time: 41 min ± 1 min, high and homogeneous degree of supercooling prior to nucleation) did result in a positive effect on the texture or on the drip loss of frozen berries.

The micrographs of fresh and untreated, frozen strawberry tissue illustrated a large difference between the intact, isodiametric and convex particles in non-frozen samples and the damaged, irregular and distorted particles in frozen samples (Van Buggenhout *et al.*, 2008). Whatever the freezing process, ice crystal formation seemed to damage the delicate strawberry tissue. Untreated strawberry tissue was distorted most when it was frozen under slow freezing conditions. When untreated strawberry tissue was frozen under cryogenic freezing conditions (see micrograph in Figure 2, 2nd row), the tissue seemed to be more intact as compared to the other applied freezing methods. These micrographs clearly illustrated that ice crystals formed during slow freezing damaged the tissue more compared to rapid/cryogenic freezing conditions. Therefore, it is suggested that increasing the freezing rate prevented the formation of large extracellular ice crystals, usually formed during slow freezing processes. Although the cell walls in untreated strawberry tissue frozen by HPSF were rather intact, the tissue was damaged seriously because of air bubbles in the sample, presumably because air was pushed into the tissue during pressure build-up of the freezing process. The negative effect of HPSF on the microstructural integrity of the strawberry tissue is in agreement with the negative effect that HPSF had on the texture of fresh strawberry halves and can be explained by the high porosity, and thus the high air content, of the fruits. In the distorted, untreated frozen tissue it was not possible to distinguish cell walls from cell wall material that occurred outside the cell walls. Therefore, threshold-based segmentation of different cells, and thus image analysis, of these tissues was not feasible, and consequently not performed.

Hardness results showed that the applied infusion treatment could not prevent the softening of slowly frozen strawberries. Like for the untreated, frozen strawberry tissue, the PME/Ca-infused, slowly frozen sample appeared to be less intact compared to the infused tissues frozen by the other freezing methods. For infused, slowly frozen tissue, particles were very irregular and damaged and the contact between cells was weak. Image analysis of these micrographs was not feasible.

For rapid freezing, a remarkable texture improvement was noticed due to infusion with PME and calcium: 55% of the original hardness value was maintained when strawberry halves were infused with PME/Ca and subsequently frozen by rapid freezing conditions (Van Buggenhout *et al.*, 2008). When combined with rapid freezing, VI with PME and calcium also seemed to be effective in preserving the structural integrity of the tissue. In infused strawberry tissue frozen by rapid freezing conditions, cells were more or less intact and there was a good contact between neighboring cells. Cell wall material was also found in the extracellular spaces.

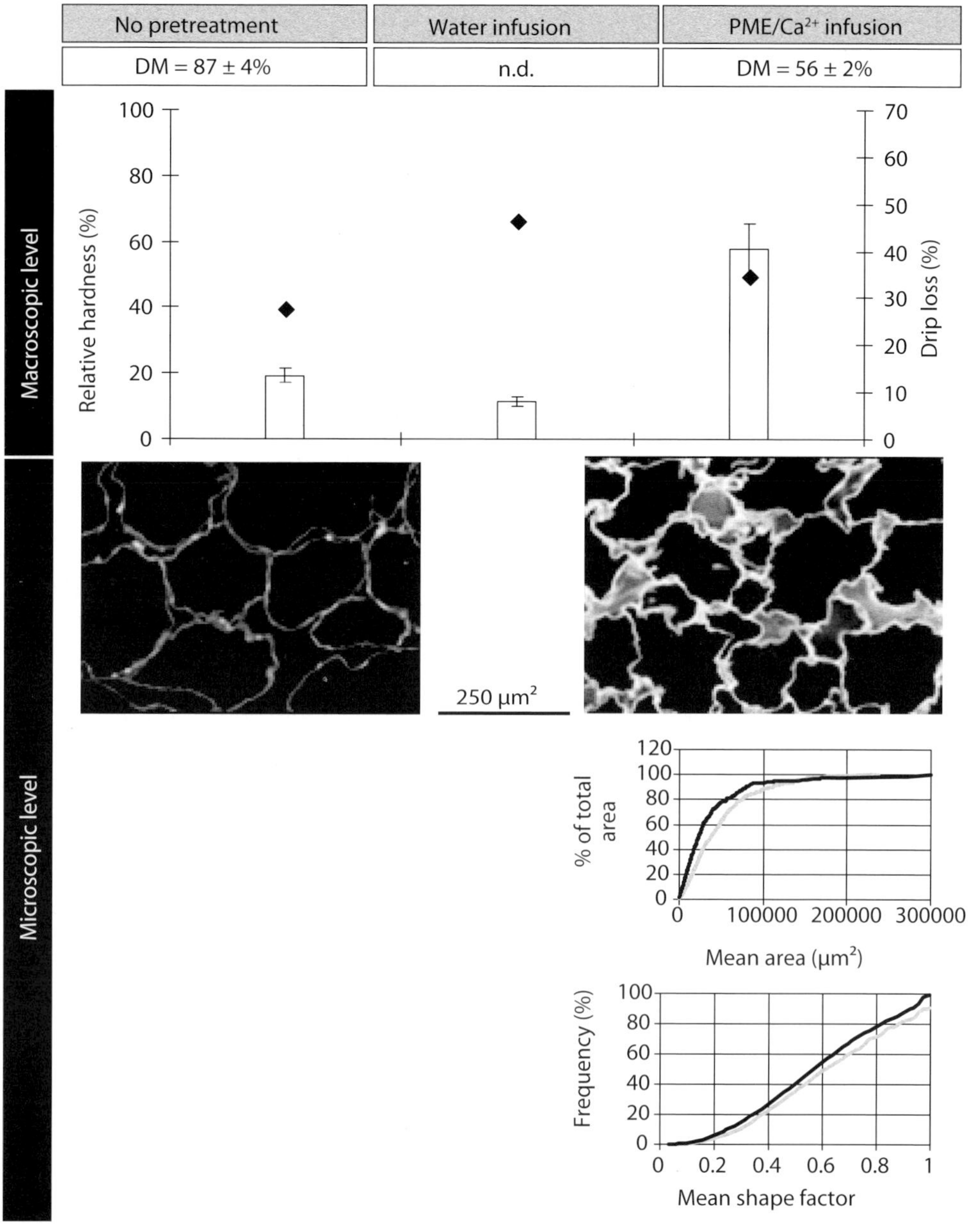

Figure 2. Texture ((bars - 1st row), drip loss (symbols - 1st row), microstructure (2nd row) and tissue particle results (3rd and 4th row) for non-pretreated and PME/Ca-infused strawberries frozen by cryogenic freezing. Tissue particle results: individual particle area in function of the percentage of total area (3rd row) and shape factor distribution curves (4th row) for (—) infused strawberries and (—) infused, frozen strawberries.

Also for cryogenically frozen strawberry halves, a positive effect of VI with PME/Ca on the hardness of the frozen-then-thawed fruits was noticed (Figure 2, 1st row). PME/Ca-infused fruits was significantly harder (relative hardness 58%) than the control strawberries (relative hardness 19%) and that of fruits infused with water (relative hardness 11%). Micrographs (Figure 2, 2nd row) revealed that, like for rapidly frozen strawberries, VI with PME/Ca was effective in preserving the structural integrity of cryogenically frozen strawberry tissue which corresponded with the positive effect of this treatment on the hardness of the frozen-then-thawed fruits.

A prominent advantage of VI with PME/Ca on the hardness retention of high pressure frozen strawberries was also noticed. In fact, PME/Ca-infused strawberries frozen by HPSF were harder (relative hardness 112%) than fresh fruits. It seemed that this hardness improvement was caused by the combined effect of the infused PME and the high degree of supercooling during HPSF. Moreover, it seemed that liquid retention upon thawing was also maximized by the action of PME and calcium. Like for rapidly and cryogenically frozen strawberry tissue, structural damage upon HPSF was limited by the application of VI with PME/Ca prior to freezing.

When feasible (for samples that were infused prior to rapid, cryogenic and HPSF), image analysis was performed. The area of the particles was not strikingly different for the different samples: the non-frozen, infused tissue had slightly larger particles compared to the infused tissue that was frozen by HPSF and the latter had slightly larger particles as compared to the infused tissue frozen under rapid and cryogenic freezing conditions (Figure 2, 3rd row). However, the differences in particle area of these tissues remained small. Slight differences in particle area could be explained by a slightly different effect of the different freezing processes on the cells and intercellular spaces in strawberry tissue. In infused, non-frozen tissue, cells looked turgid and intercellular spaces were small. Moreover, most of the latter spaces were filled up with cell wall material, thus strongly colored during preparation, and thus not analyzed as particles during image analysis. When the infused tissue was frozen, cells were deformed and less turgid, possibly due to loss of turgor as a consequence of ice formation and intercellular spaces were expanded. Changes of tissue convexity/smoothness due to freezing of the infused tissue were indicated by differences in particle shape factor (Figure 2, 4th row). Mean shape factor values (0.64 for infused, unfrozen tissue; 0.58 for infused, high pressure frozen tissue; 0.57 for infused, cryogenically frozen tissue; 0.50 for infused, rapidly frozen tissue) showed that freezing decreased the smoothness of tissue particles. The shape factor distributions showed that particles were most smooth in non-frozen, infused tissue, followed by infused tissue frozen by HPSF or under cryogenic freezing conditions. Strawberry tissue infused and then frozen under rapid freezing conditions had the most irregular particles.

The above comparison between non-infused and infused strawberry tissue frozen under particular freezing conditions revealed that infusion with PME and calcium prior to freezing had a positive effect on the preservation of the frozen strawberry tissue. However, this positive effect depended on the freezing method applied: the lower the freezing rate, the smaller the positive effect. Although infusion had a positive effect on the preservation of frozen strawberry tissue, it must be pointed out that damage to the strawberry tissue due to freezing could not be avoided completely. For all freezing methods applied, particles of infused, frozen tissue were less smooth and cells were less adhered to each other when compared to the vacuum infused tissue before freezing. Nevertheless, structural integrity of strawberries could be preserved to a maximal extent when the PME/Ca-infused tissue is frozen rapidly or by HPSF.

4. Conclusions

Although pectin changes only play a minor role in freezing-induced texture degradation when compared to thermal texture degradation, pectin alterations that strengthen the middle lamella complex can result in reduced freezing-induced structural and thus reduced textural damage. In this work it became clear that the use of pectin modifying pretreatments to enhance the product suitability to be frozen depends on the freezing process and on the product. For the optimization of the textural and structural quality of specific frozen products, each case thus should be considered separately. Based on the results of the present study and under the assumption that textural quality of frozen fruit and vegetables is optimal when hardness retention is maximal, rapid conventional and cryogenic freezing conditions are recommended to freeze carrots. Pectin modifying treatments for carrots, it are calcium soaking followed by thermal treatment at 60 °C and calcium soaking followed by high pressure treatment at 300 MPa and 60 °C, have a negative effect the textural and structural quality parameters of rapidly and cryogenically frozen carrots. Only the texture and tissue integrity of high pressure frozen carrots can be optimized by the use of the above mentioned pectin modifying treatments. For strawberries, on the other hand, a combination of VI with PME and Ca^{2+} and rapid, cryogenic or high pressure freezing is advised to freeze strawberries. From an industrial point of view, it can be stated that infusion with exogenous PME has some promising perspectives to create frozen plant-based products with novel properties meeting consumer quality demand.

Acknowledgements

Sandy Van Buggenhout is a post-doctoral researcher funded by the Fund For Scientific Research Flanders (FWO).

References

Ahmed, A.E.R and J.M. Labavitch, 1977. A simplified method for accurate determination of cell wall uronide content. Journal of Food Biochemistry 1:361-365.

Banjongsinsiri, P., J. Kenney and L. Wicker, 2004. Detection of vacuum infusion of pectinmethylesterase in strawberry by activity staining. Journal of Food Science 69:FCT179-183.

Blumenkrantz, N. and G. Asboe-Hansen, 1973. New method for quantitative determination of uronic acids. Analytical Biochemistry 54:484-489.

Brown, M.S, 1976. Effects of freezing on fruit and vegetable structure. Food Technology 106-109.

Duvetter, T., I. Fraeye, T. Van Hoang, S. Van Buggenhout, I. Verlent, C. Smout, A. Van Loey and M. Hendrickx M, 2005. Effect of pectinmethylesterase infusion methods and processing techniques on strawberry firmness. Journal of Food Science 70:S383-S388.

International Institute of Refrigeration, 1986. Recommendations for the processing and handling of frozen foods. 3rd edition. Paris.

Kidmose, U. and H. Martens, 1999. Changes in texture, microstructure and nutritional quality of carrot slices during blanching and freezing. Journal of the Science of Food and Agriculture 79:1747-1753.

Klavons, J.A. and R.D. Bennett, 1986. Determination of methanol using alcohol oxidase and its application to methyl ester content of pectins. Journal of Agricultural Food Chemistry 34:597-599.

McFeeters, R.F. and S.A. Armstrong, 1984. Measurement of pectin methylation in plant cell walls. Analytical Biochemistry 139:212-217.

Munoz-Delgado, J.A, 1977. Effects of freezing, storage and distribution on quality and nutritive attributes of foods, in particular fruit and vegetables. In: W.K. Downey (ed.) Food Quality and Nutrition. Applied Science Publishers, London, 353-384.

Muntada, V., L.N. Gerschenson, S.M. Alzamora and M.A. Castro, 1998. Solute infusion effects on texture of minimally processed kiwifruit. Journal of Food Science 63: 616-620.

Ng, A. and K.W. Waldron, 1997. Effect of cooking and pre-cooking on cell wall chemistry in relation to firmness of carrot tissues. Journal of the Science of Food and Agriculture 73: 503-512.

Ormerod, A.P., J.D. Ralfs and C.M. Sidebottom, 2004. Freezing vegetables. Patent US20040065094 A1. Unilever Bestfoods North America, division of Conopco.

Pearce, R.S., 2001. Plant freezing and damage. Annals of Botany 87: 417-424.

Prestamo, G. and G. Arroyo, 1998. High hydrostatic pressure effects on vegetable structure. Journal of Food Science 63: 878-881.

Reid, D.S, 1999. Dehydration and freezing stresses. In: Y.H. Roos, R.B. Leslie and P.J. Lillford (eds.) Water Management in the Design and Distribution of Quality Foods. Technomic Publishing Company, Lancaster, Pennsylvania. 87-105.

Saurel, R., 2002. The use of vacuum technology to improve processed fruit and vegetables. In: W. Jongen (ed.) Fruit and Vegetable Processing. Improving Quality. Woodhead Publishing Limited, Cambridge, 363-380.

Sila, D.N., C. Smout, Y. Satara, V. Truong, A. Van Loey and M. Hendrickx, 2007. Combined thermal and high pressure effect on carrot pectinmethylesterase stability and catalytic activity. Journal of Food Engineering 78: 755-764.

Stepkonus, P.L., 1999. Freeze-induced dehydration and the cryostability of biological membranes. In: Y.H. Roos, R.B. Leslie and P.J. Lillford (eds.) Water Management in the Design and Distribution of Quality Foods. Technomic Publishing Company, Lancaster, 53-85.

Suutarinen, J., K. Honkapää, R.L. Heiniö, K. Autio and M. Mokkila, 2000. The effect of different prefreezing treatments on the structure of strawberries before and after jam making. Lebensmitteln Wissenschaft und Technologie 33: 188-201.

Tangwongchai, D.A., D.A. Ledward and J.M. Ames, 2000. Effect of high-pressure treatment on the texture of cherry tomato. Journal of Agricultural Food Chemistry 48: 1434-1441.

Van Buggenhout, S., M. Lille, I. Messagie, A.Van Loey, K. Autio and M. Hendrickx, 2006a. Impact of pretreatment and freezing conditions on the microstructure of frozen carrots: quantification and relation to texture loss. European Food Research and Technology 222: 543-553.

Van Buggenhout, S., I. Messagie, V. Maes, T. Duvetter, A. Van Loey and M. Hendrickx, 2006b. Minimizing texture loss of frozen strawberries: effect of infusion with pectinmethylesterase and calcium combined with different freezing conditions and effect of subsequent storage/thawing conditions. European Food Research and Technology 223: 395-404.

Van Buggenhout, S., T. Grauwet, A. Van Loey and M. Hendrickx, 2008. Structure/processing relation of vacuum infused strawberry tissue frozen under different conditions. European Food Research and Technology 226(3): 437-448.

Van Buren, J.P., 1979. The chemistry of texture in fruits and vegetables. Journal of Texture Studies 10:1-23.

Waldron, K.W., A.C. Smith, A.J. Parr, A. Ng and M.L. Parker, 1997. New approaches to understanding and controlling cell separation in relation to fruit and vegetable texture. Trends in Food Science and Technology 8: 213-221.

Waldron, K.W., M.L. Parker and A.C. Smith, 2003. Plant cell walls and food quality. Comprehensive Reviews in Food Science and Food Safety 2: 101-119.

Texture improvement of processed carrots by modifying pectin

A. De Roeck, D.N. Sila, T. Duvetter, A. Van Loey and M. Hendrickx
Laboratory of Food Technology, Centre for Food and Microbial Technology, Department of Microbial and Molecular Systems, Katholieke Universiteit Leuven, Kasteelpark Arenberg 22, 3001 Leuven, Belgium

Abstract

Heat sterilization of fruits and vegetables is accompanied by a pronounced texture loss, mainly due to beta-eliminative depolymerization of the cell wall pectin resulting in reduced intercellular adhesion. Food technologists are trying to minimize this texture loss. On the one hand, they focus on optimizing the existing thermal methods. On the other hand, they search for novel, less degradative processing technologies. High pressure processing has shown potential and is currently successfully applied on a commercial scale for pasteurization. To explore the potential of high pressure sterilization, the effect of combined high pressure/high temperature (HP/HT) processing on the texture of carrots and on the related pectin components was compared with that of high temperature treated samples. Carrot disks (*Daucus carota* var. Yukon) were subjected to three different treatments (80 °C-0.1 MPa, 100 °C-0.1 MPa, 80 °C-600 MPa). Subsequently, the residual hardness and microstructure were investigated. The pectin, isolated as alcohol insoluble residue, was fractionated into water (WSP), chelator (CSP) and sodium carbonate (NSP) soluble pectin. Thermal treatments at 0.1 MPa caused enormous tissue softening. This was marked by reduced cell-cell adhesion and an increase in WSP paralleled by a decrease in CSP and NSP. These results confirm that an extensive solubilization and/or degradation of pectin occurs during thermal processing. In contrast, HP/HT treated carrots showed minimal softening and negligible changes in intercellular adhesion. This was accompanied by a significant lower degree of methyl esterification of the pectin, dissimilar pectin fractions ratios and minor inter-conversions in the pectin fractions during treatment, indicating less solubilization. These results suggest an inhibition of the beta-elimination at HP/HT. So, there was a clear difference between pectin in HP/HT and thermally processed carrots; HP/HT treated carrots showing pronounced texture preservation. Concerning the quality parameter texture, high pressure sterilization offers an interesting alternative for heat sterilization.

Keywords: carrot, pectin, texture, high pressure sterilization, beta-elimination

1. Introduction

Consumers are demanding high quality, microbiologically safe foods with an extended shelf-life. Consequently, processing of food is necessary to inactivate enzymes and micro-organisms present. Until now, the most commonly used preservation method has been thermal processing, such as pasteurization and sterilization. However, the high temperatures used result in organoleptic and nutritional quality losses. Food scientists are therefore continuously searching for novel, less degradative processing technologies (Mertens and Knorr, 1992). High pressure technology has shown potential to meet consumer demands (Ludikhuyze *et al.*, 2001). High pressure processing applies pressures (in the range of 400-600 MPa in combination with 0-50 °C for pasteurization and in the range of 500-800 MPa / 95-110 °C for sterilization) to inactivate enzymes and micro-organisms. At the same time, it offers an advantage in minimal deleterious effects on food quality attributes (e.g. colour, flavour, and nutritional value) as it does not affect covalent bonds. Pressure is transmitted uniformly and instantaneously throughout the food, which allows very homogeneous products to be obtained. Currently, high pressure processing is successfully applied on a commercial scale for pasteurization of a whole range of food products, e.g. fruit juices, guacamole, oysters and ham. Given the technical progress made in high pressure equipment, commercial high pressure sterilization may be feasible in the near future (Hoogland *et al.*, 2001). This process would involve both elevated pressure and elevated temperature contributing to sterilization by inactivating spores and enzymes. Before commercial introduction, the influence of combined high pressure/ high temperature processing on different quality attributes has to be investigated to evaluate whether it meets consumer demands.

A very important quality attribute of fruits and vegetables is texture. Texture is a characteristic at the plant organ level, which is dependent on several features of plant tissues, cells, and ultimate basically on structures and interactions of molecules and ions (Waldron, 2004). So, texture can be looked at at different scales. At macroscopic level, the hardness of a fruit or vegetable can be determined, either instrumentally, either by sensory analysis. Mesoscopically, the plant tissue can be examined by microscopy. In this regard, the degree of cell-cell adhesion is of particular importance. At molecular scale, the different cell wall components (cellulose, hemicellulose, and pectin) can be investigated. Most attention is given to pectin. Pectin is a cell wall polysaccharide which is especially abundant in the middle lamella of plant cells and which plays a crucial role in cell-cell adhesion and, consequently, in providing mechanical strength to plant tissues. Pectin is brought into solution more easily than the other cell wall polymers and is more chemically reactive. Generally, pectin consists of (1→4)α-D-galacturonan regions, with varying degrees of methyl esterification, and rhamnogalacturonan regions substituted by side chains rich in neutral sugars, for example arabinose and galactose (Ridley *et al.*, 2001). Specific biochemical and/ or chemical transformations occur in the homogalacturonan part of pectin during processing, leading to texture changes of the resulting products (Figure 1; Van Buren, 1979). At high temperature and a pH above 4.5, highly methoxylated pectin undergoes beta-eliminative depolymerization. This degradation of pectin results in reduced intercellular adhesion and consequently in increased softening. In most cases, extensive softening is not desired by the consumer. The reaction rate of beta-elimination is strongly dependent on the degree of methyl esterification of pectin,

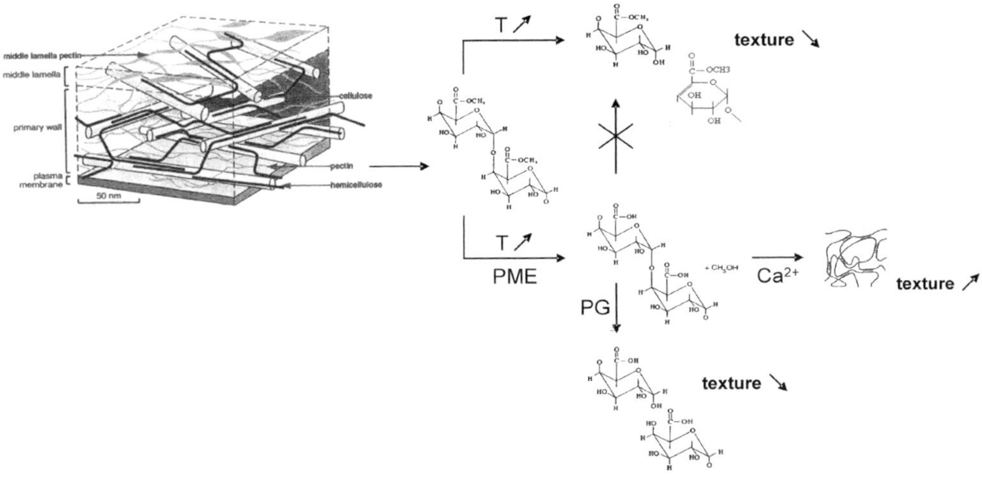

Figure 1. (Bio-)chemical transformations in the homogalacturonan part of cell wall pectin resulting in texture changes (T: temperature, PME: pectin methylesterase, PG: polygalacturonase).

pH and presence of ions. Pectin can also undergo demethoxylation, catalyzed by the enzyme pectin methylesterase (PME) or chemically at high temperature. This reaction is beneficial for texture conservation because demethoxylated pectin is less susceptible to beta-elimination. Moreover, the demethoxylated pectin can be cross-linked in the presence of divalent ions, such as calcium, forming fortifying networks. However, the enzyme polygalacturonase (PG) (if endogenously present) can depolymerize the lowly methoxylated pectin, resulting in tissue softening.

Traditional heat sterilization of fruits and vegetables is accompanied by a pronounced texture loss due to beta-eliminative depolymerization of the pectin resulting in reduced intercellular adhesion. This texture loss can, on the one hand, be minimized by optimizing the existing thermal methods. It has been established that minimizing texture loss is possible by applying pre-treatments that make the plant tissue less vulnerable to beta-elimination during a subsequent thermal treatment (Smout *et al.*, 2005). During these pre-treatments (low temperature blanching, high pressure pre-treatment) PME is activated, resulting in pectin with a lower degree of esterification which is more stable during subsequent thermal processing. Additional texture improvement can be achieved by enhancing cross-links between cell wall polymers, e.g. cross-links between free carboxyl groups of pectin and calcium ions, as this results in a stronger cell-cell adhesion. A possible alternative could be ferulic acid cross-linking. A small percentage of the sugars in pectin carry ferulic acid, which may be cross-linked by the action of the enzyme peroxidase and H_2O_2 (Waldron *et al.*, 1997). On the other hand, texture loss can be minimized by applying novel processing technologies with less adverse effects on food quality. In this research, attention was given to the novel high pressure technology mentioned above. The aim was to explore the potential of high pressure sterilization, as alternative for heat sterilization, regarding texture preservation of processed fruits and vegetables. For that, the effect of combined HP/HT treatments on the texture of carrots was compared with that of high temperature (HT) treated samples. A detailed picture was obtained by abstracting information at the macro-, mesoscopic and molecular levels.

2. Materials and methods

Figure 2 gives an overview of the experimental set-up.

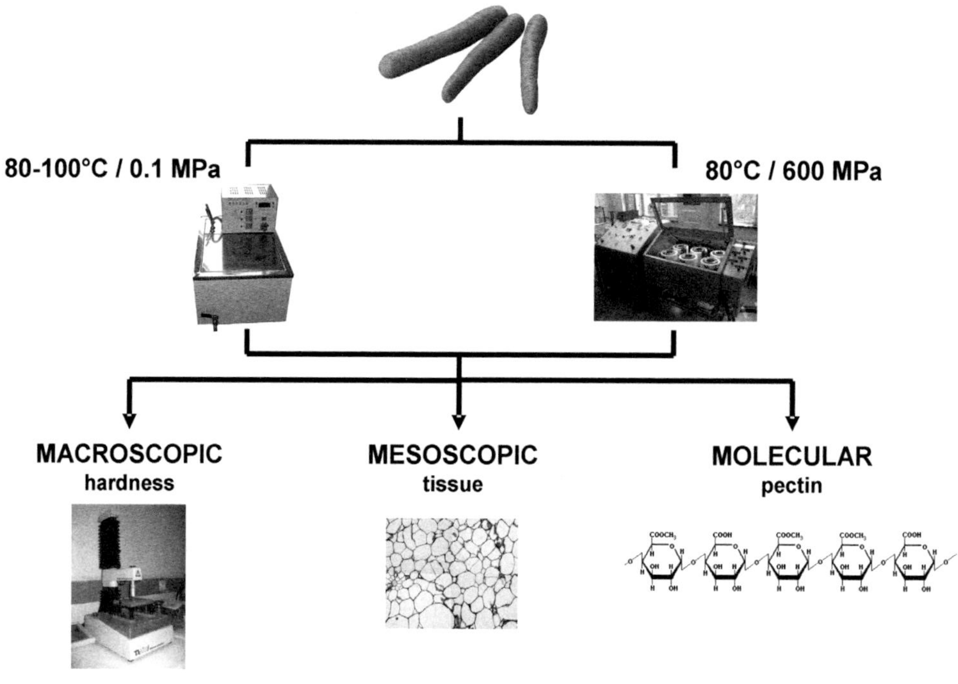

Figure 2. Schematic overview of the experimental set-up.

2.1 Carrots

Carrots (*Daucus carota* var. Yukon) were obtained from a local shop in Belgium and stored at 4 °C before use. Disks of 10 mm height and 12 mm diameter were cut out of the core of the carrots before processing.

2.2 Treatments

Carrot disks were on the one hand treated at 80 or 100 °C and atmospheric pressure, on the other hand at 80 °C in combination with 600 MPa, and this for different times. Thermal treatments were carried out in a temperature controlled oil bath. Carrot disks were encapsulated in stainless steel tubes already filled with demineralized water as a brine. After an equilibration time of 5 min, a first sample (time 0 min) was withdrawn from the oil bath. After preset time intervals, the rest of the samples were removed. High pressure/high temperature treatments were conducted in a laboratory scale multi-vessel high pressure equipment (Resato, Roden, the Netherlands). Carrot disks were, together with demineralized water, enclosed in polyethylene bags. Pressure was built up slowly using a pressurization rate of 100 MPa/min. A slow pressurization rate allowed the heat, inextricably bound up with compression, to dissipate to the surroundings. Despite this, the temperature rose above 80 °C. Therefore, an equilibration period of 5 min allowing the temperature to decrease to 80 °C was taken into account before the first sample (time 0 min) was removed.

2.3 Texture measurement

The residual hardness of the carrots was evaluated by a compression test using a TA-XT2i Texture Analyzer (Stable Micro Systems, Surrey, UK) equipped with a 25 kg load cell and a cylindrical flat head aluminium probe of 25 mm diameter. The peak force required to deliver a constant strain of 30% was measured at a compression rate of 1 mm/s.

2.4 Microstructure analysis

Carrot slices were fixed using glutaraldehyde and subsequently washed and dehydrated in a series of ethanol solutions (50, 70 and 95%). The dehydrated slices were infiltrated with a Historesin Embedding Kit (Leica, Bensheim, Germany) for 5 days, polymerized and cut into thin sections (5 μm) with a Microm HM355 microtome (Microm Laborgeräte GmbH, Walldorf, Germany). Sections were stained with 0.1% toluidin blue for 5 min followed by washing with water. Dyed slices were examined with a light microscope (Olympus BX-41, Optical Co. Ltd, Tokyo, Japan). Micrographs were taken using image analysis software (analySIS pro 3.1 and 5, Soft Imaging System GmbH, Bensheim, Germany).

2.5 Determination of the degree of methyl esterification of pectin (DM)

Cell wall material was isolated as alcohol insoluble residue (AIR) as described by McFeeters and Armstrong (1984). The DM of the pectin was estimated by taking the ratio of moles of methoxyl groups to the moles of anhydrous galacturonic acid in the AIR. The methoxyl content was estimated by saponifying the ester bonds with NaOH (Ng and Waldron, 1997) and colorimetric quantification of the released methanol as described by Klavons and Bennett (1986). The galacturonic acid content was determined by hydrolysis with concentrated sulphuric acid (Ahmed and Labavitch, 1977), followed by colorimetric quantification of the galacturonic acid residues (Blumenkrantz and Asboe-Hansen, 1973).

2.6 Fractionation of pectin

The AIR was fractionated into various pectin fractions as described by Sila *et al.* (2006). By incubating the AIR in demineralized water for 5 min at 100 °C the water soluble pectin was extracted. This is rather highly methoxylated pectin neither covalently, nor ionically bound to the cell wall. The residue was incubated with cyclohexane-trans-1,2-diamine tetra-acetic acid to obtain the chelator soluble pectin. This is lowly methoxylated pectin ionically bound to the cell wall. By using Na_2CO_3 pectin covalently bound to the cell wall was extracted (NSP). The residue that remained contains mainly cellulose and hemicellulose. The pectin in the fractions was quantified by determining the galacturonic acid content.

2.7 Molar mass distribution

Changes in molar mass distribution of carrot pectin during processing were studied using high performance size exclusion chromatography, after lyophilization and dialysis of the pectin fractions. This was performed using an Äkta Purifier (GE Healthcare, Uppsala, Zweden) equipped with a mixed bed column of Bio-Gel TSK (dimensions = 300 mm L × 7.5 mm Ø, pore size = 100-1000 Å, particle size = 13 µm, theoretical plates/column ≥ 7000, pH range = 2-12, maximum pressure = 300 psi; Bio-Rad Labs, Richmond, CA, USA) in combination with a TSK guard column. A 20 µl injection loop was used. Elution was executed at 35 °C with 0.05 M $NaNO_3$ at a flow rate of 0.7 ml/min. The eluent was monitored using a Shodex R101 refractive index detector (Showa Denko, K.K., Tokyo, Japan).

3. Results and discussion

3.1 Influence of processing on the hardness of carrots

Figure 3 shows the changes in carrot disks' hardness during processing. Both high temperature and high pressure/high temperature treated carrots showed an initial loss of texture due to membrane damage and turgor pressure loss (Greve *et al.*, 1994b; Basak and Ramaswamy, 1998). In contrast to the thermally treated carrots which underwent further softening with treatment time, the high pressure treated carrots did not. Texture loss of thermally treated carrots is mostly explained by the beta-eliminative depolymerization of pectin (Greve *et al.*, 1994a; Sila *et al.*, 2006). Therefore, texture preservation of high pressure/high temperature treated carrots could be explained by inhibition of the beta-elimination reaction.

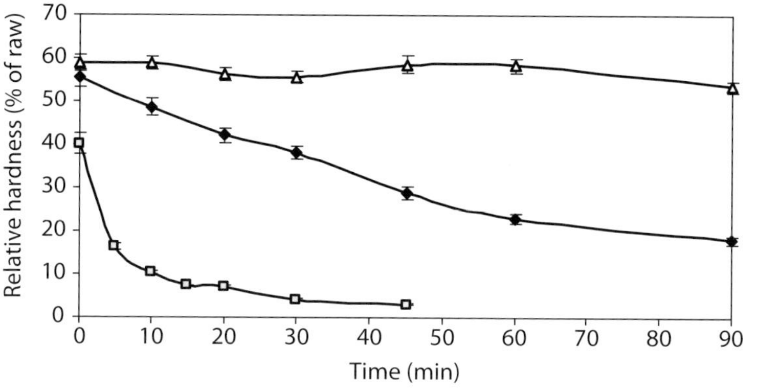

Figure 3. Hardness of the processed carrots relative to the hardness of raw carrots (± standard error) (n=26): (◆) 80 °C – 0.1 MPa, (□) 100 °C – 0.1 MPa, and (△) 80 °C – 600 MPa.

3.2 Influence of processing on the microstructure of carrots

Raw carrot tissue showed well defined, well stained cell walls (Figure 4a). Thermally treated tissue displayed less intense stained, swollen cell walls (Figure 4b, c), probably due to heat induced solubilization of the intercellular cementing pectin facilitating cell wall loosening (Waldron, 2004). In contrast, HP/HT treated tissue bore close similarity to the raw tissue, indicating limited pectin solubilization (Figure 4d).

The way plant tissue fails when an external force is applied, is an indication for the extent of cell-cell adhesion. Raw carrot tissue clearly showed cell breakage (Figure 5a). This means that the forces holding the cells together are stronger than the cell walls and corresponds to strong intercellular adhesion. One can state that the intercellular cementing pectin is doing its job. In contrast, heat treated tissue showed cell separation (Figure 5b), indicating weak intercellular adhesion. Apparently, the pectin is degraded or solubilized during the heat treatment. The HP/HT treated tissue showed both cell separation and cell breakage (Figure 5c). Thus, the microscopic observations suggest there are different phenomena occurring during high temperature versus high pressure/high temperature treatment of carrots. Analyses on molecular scale provided deeper insight.

3.3 Influence of processing on pectin structure

3.3.1 Degree of methyl esterification (DM) of pectin

Raw carrots showed a DM of 64%. During the three different treatments, the DM decreased (Table 1). However, HP/HT showed a more pronounced reduction. The decline at high temperatures may be attributed only to chemical demethoxylation as PME is inactivated at these temperatures at 0.1 MPa. In case of the HP/HT treated

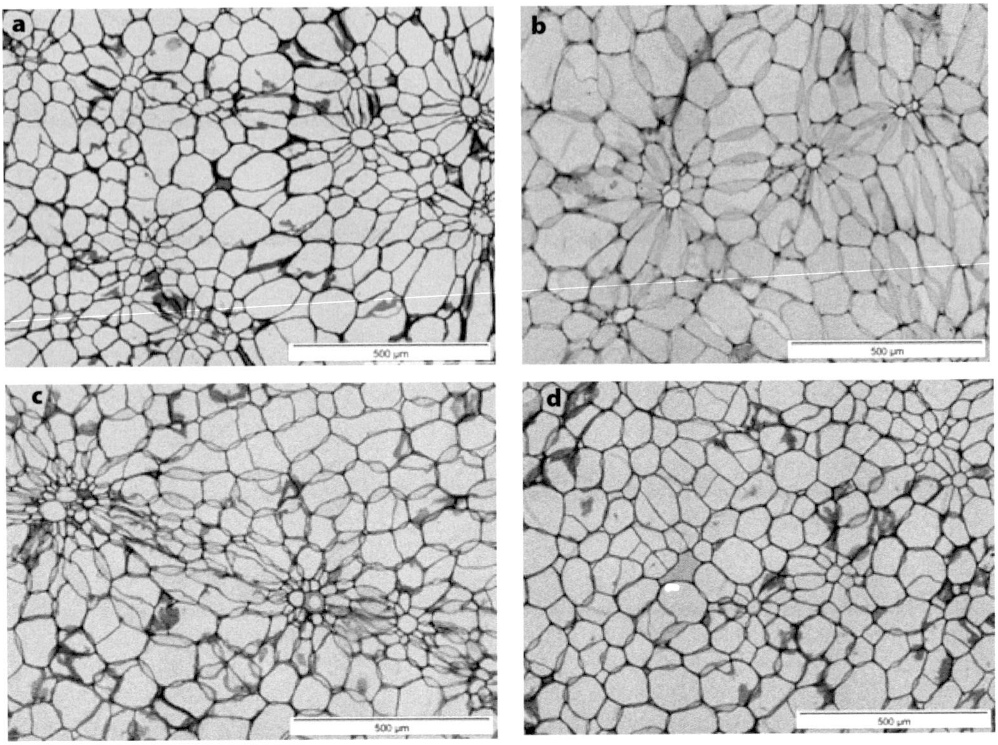

Figure 4. Micrographs displaying carrot tissue: a) raw, b) 100 °C-0.1 MPa-45 min, c) 80 °C-0.1 MPa-90 min, and d) 80 °C-600 MPa-90 min.

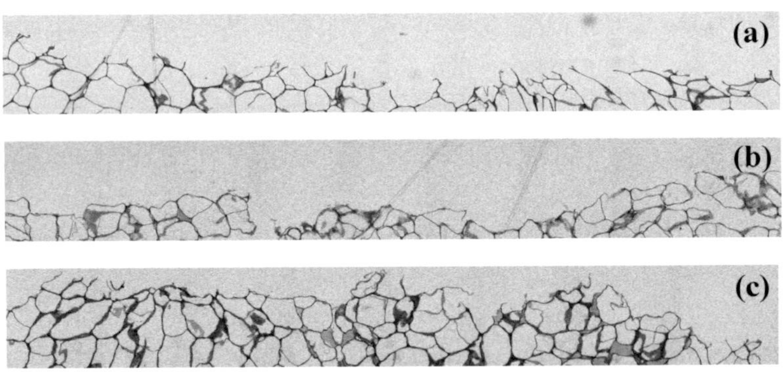

Figure 5. Micrographs showing the cutting edge of carrot tissue: a) raw, b) 80 °C-0.1 MPa-45 min, and c) 80 °C-600 MPa-45 min.

samples, the high reduction may be explained by the combined effect of chemical demethoxylation (which is enhanced by pressurization (Verlent *et al.*, 2004)) and enzymatic demethoxylation. PME is a pressure stable enzyme and it is established

Table 1. Degree of methyl esterification of carrot pectin.

	Degree of methyl esterification (%) ± standard deviation (n=6)				
	0 min	5 min	20 min	45 min	90 min
80 °C - 0.1 MPa	56.8 ± 2.1		52.0 ± 3.0	50.3 ± 2.3	52.7 ± 2.9
100 °C - 0.1 MPa	57.7 ± 3.5	58.0 ± 2.5	55.0 ± 3.2	52.9 ± 1.9	
80 °C - 600 MPa	42.6 ± 3.5		37.4 ± 3.2	31.1 ± 2.0	34.2 ± 1.0

that high pressure has a stabilizing effect on PME and even enhances PME activity (Sila *et al.*, 2007). As methyl esters are one of the main driving forces of the beta-elimination reaction, it is likely that the high reduction in DM of the HP/HT treated carrots triggered inhibition of the beta-elimination reaction, consequently preventing softening.

3.3.2 Changes in pectin fractions

Raw carrots contained predominantly water soluble pectin (~62%), a substantial amount of chelator soluble pectin (~26%), and a low amount of sodium carbonate soluble pectin (~12%). During thermal processing (at 0.1 MPa), changes in pectin fraction ratios were observed (Table 2). The amount of WSP increased, whereas the amount of CSP and NSP decreased. This indicates that a substantial degradation and/or solubilization of pectin occurred. HP/HT treated carrots showed a totally different behaviour. At time zero (after pressure build-up and an equilibration time of 5 min), a low proportion of WSP was noticed while the other fractions were proportionately higher. During processing, the pectin fractions hardly changed, corresponding with the unaltered hardness of these carrots. It is in particular the water soluble, highly methoxylated pectin which is susceptible to beta-elimination. Hence, a decrease of the WSP is believed to entail a decrease in beta-elimination. Less degraded pectin results in a more intact pectin network and, consequently, in a better cell adhesion and higher tissue firmness. How can the different pectin ratios at HP/HT be explained? The low amount of WSP and high amount of CSP could be explained by increased ionic cross-linking as a result of the reduced DM. An explanation for the high amount of NSP could be that certain ester bonds were formed, which are broken by a sodium carbonate treatment. Another explanation could be that pectin with a very low DM is very strongly ionically bound to the cell wall so that it could not or not all be released with the used CDTA solution.

It was remarkable that the extraction yield, defined as the ratio of the galacturonic acid found in the three fractions to the galacturonic acid found in the AIR, was clearly lower for the HP/HT treated carrots (53-70% as opposed to 81%). It is possible there was during fractionation, by accident, just more loss of galacturonic acid with these

Table 2. Changes in pectin fractions during processing.

	Carrot pectin fraction (%)[1]								
	0 min			45 min			90 min		
	WSP	CSP	NSP	WSP	CSP	NSP	WSP	CSP	NSP
80 °C - 0.1 MPa	50.6 ± 2.8	36.7 ± 2.1	12.7 ± 0.6	54.5 ± 3.0	32.8 ± 0.9	12.8 ± 0.5	57.1 ± 1.2	32.6 ± 0.8	10.3 ± 1.0
100 °C - 0.1 MPa	59.0 ± 1.8	29.5 ± 0.6	11.5 ± 0.5	69.7 ± 2.2	26.7 ± 0.7	3.6 ± 0.1			
80 °C - 600 MPa	14.7 ± 0.6	45.1 ± 1.3	40.1 ± 2.1	12.2 ± 0.6	47.0 ± 1.2	40.8 ± 2.0	15.3*	45.6*	39.1*

[1]galacturonic acid in fraction/galacturonic acid in (WSP+CSP+NSP).100.
± standard deviation (n=2; *: n=1).

samples. However, analyzing the galacturonic acid content of the residue confirmed that in the HP/HT treated carrots a high amount of galacturonic acid was still attached to the fractionation residue. This indicates that a part of the pectin was strongly bound to the other cell wall polymers, which was not solubilized with the procedures used. The differences in pectin solubility between differently treated carrots can be related to the differences in textural characteristics of the samples.

3.3.3 Molar mass (MM) distribution

Finally, changes in molar mass distribution of the pectin fractions during processing were investigated by HPSEC. The CSP and NSP fraction of raw carrots eluted at about 10.5 min. WSP of raw carrots comprised of two unresolved peaks, a small peak coinciding with the elution profile of CSP and NSP and a larger peak having a slightly lower MM (elution time ~ 12 min). Mono-galacturonic acid eluted at about 15 min. During thermal processing, distinct changes were apparent, the most pronounced being at 100 °C (Figure 6a). On the one hand, the peak area of the early eluting peak increased. This represents the proportions of the original CSP and NSP fractions which were thermosolubilized, as can be seen in the corresponding decrease in peak area for NSP (Figure 7a). On the other hand, the second peak shifted to lower MM, illustrating transformations in the WSP fraction due to depolymerization. The

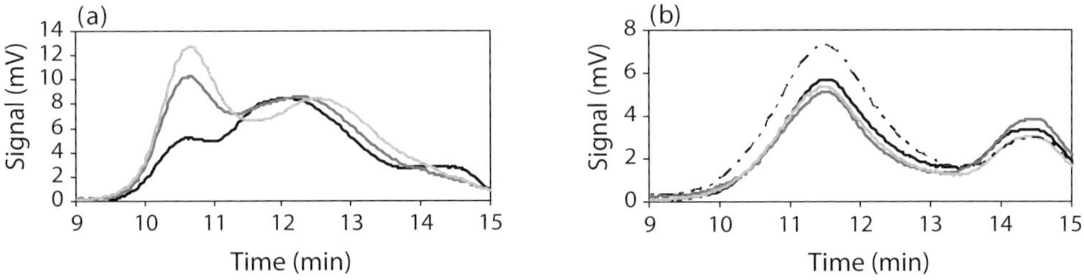

Figure 6. Changes in molar mass distribution of WSP during processing: a) 100 °C – 0.1 MPa; b) 80 °C – 600 MPa. (0 min (—), 20 min (—), 45 min (—), and 90 min (- -)).

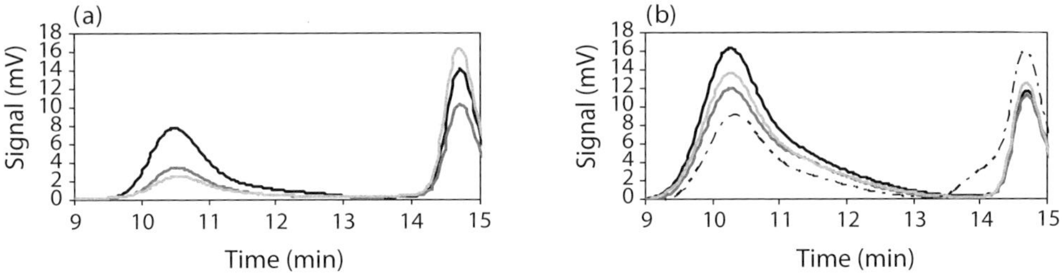

Figure 7. Changes in molar mass distribution of NSP during processing: a) 100 °C – 0.1 MPa; b) 80 °C – 600 MPa. (0 min (—), 20 min (—), 45 min (—), and 90 min (- -)).

results for the HP/HT treated carrots were totally different (Figure 6b en 7b); peaks were neither clearly increasing in area nor shifting towards lower MM, indicating less thermosolubilization. All these observations were in accordance with the results shown in Table 2.

MM distributions of the matrix polymers in the brine (the water surrounding the carrots during treatment) were also analyzed (Figure 8). This pectin can be considered as highly water soluble pectin since it is already solubilized during treatment. Once more, the same results can be observed. Thermal treatment is accompanied with extensive pectin solubilization in contrast to HP/HT treatment.

4. Conclusion

The potential of high pressure sterilization, as alternative for heat sterilization, regarding texture preservation of processed fruits and vegetables was investigated.

After an initial texture loss, the thermally treated carrots underwent further softening whereas the HP/HT treated carrots did not. Microscopic and molecular analyses revealed that differences in intercellular adhesion and pectin solubilization and/or degradation lay at the basis of these macroscopic observations. One explanation for the different pectin fraction ratios at HP/HT treatment could be the lower degree of pectin methyl esterification. Lowly methoxylated pectin is less susceptible to beta-eliminative depolymerization and can form fortifying networks with calcium ions present. Obtaining lowly methoxylated pectin is exactly the aim of the pre-treatments applied before heat treatments to diminish texture loss. During such a pre-treatment step, PME is activated resulting in lowly methoxylated pectin. So it seems that with high pressure sterilization such a step is not necessary, it is implicitly present. However, seen the conversions of WSP into NSP or even sodium carbonate insoluble pectin, other mechanisms than demethoxylation may be playing a role; mechanisms that increase the cell-cell adhesion by enhancing cross-links in the cell wall.

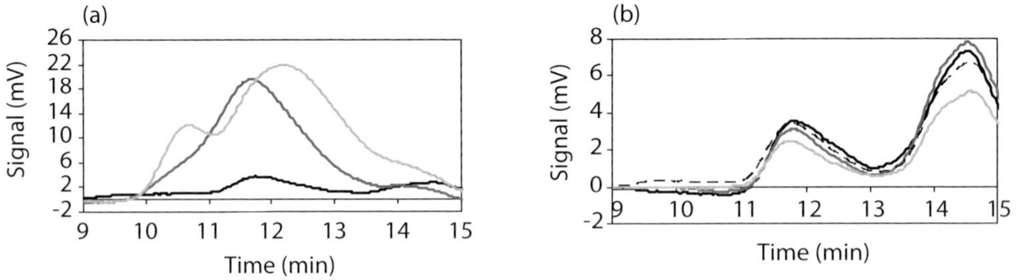

Figure 8. Changes in molar mass distribution of solubilized matrix polymers: a) 100 °C – 0.1 MPa; b) 80 °C – 600 MPa. (0 min (—), 20 min (—), 45 min (—), and 90 min (– –)).

Irrespective of the mechanism, regarding texture, combined high pressure/high temperature treatments offer an interesting alternative for thermal treatments.

Acknowledgements

This study has been carried out with financial support from the Research Fund K.U.Leuven and from the Commission of the European Communities, Framework 6, Priority 5 'Food Quality and Safety', Integrated Project NovelQ FP6-CT-2006-015710.

Figures 3, 4, 6, 7 and 8, and Tables 1 and 2 were reprinted from Food Chemistry, 107, A. De Roeck, D.N. Sila, T. Duvetter, A. Van Loey and M. Hendrickx, Effect of high pressure/high temperature processing on cell wall pectic substances in relation to firmness of carrot tissue, 1225-1235, Copyright (2008), with permission from Elsevier.

References

Ahmed, A.E.R. and J.M. Labavitch, 1977. A simplified method for accurate determination of cell wall uronide content. Journal of Food Biochemistry 1: 361-365.

Basak, S. and H.S. Ramaswamy, 1998. Effect of high pressure processing on the texture of selected fruits and vegetables. Journal of Texture Studies 29(5): 587-601.

Blumenkrantz, N. and G. Asboe-Hansen, 1973. New method for quantitative determination of uronic acids. Analytical Biochemistry 54(2): 484-489.

Greve, L.C., R.N. McArdle, J.R. Gohlke, and J.M. Labavitch, 1994a. Impact of heating on carrot firmness: changes in cell wall components. Journal of Agricultural and Food Chemistry 42: 2900-2906.

Greve, L.C., K.A. Shackel, H. Ahmadi, R.N. McArdle, J.R. Gohlke and J.M. Labavitch, 1994b. Impact of heating on carrot firmness: contribution of cellular turgor. Journal of Agricultural and Food Chemistry 42: 2896-2899.

Hoogland, H., W. de Heij and L. Van Schepdael, 2001. High pressure sterilisation: novel technology, new products, new opportunities. New Food 3: 21-26.

Klavons, J.A. and R.D. Bennett, 1986. Determination of methanol using alcohol oxidase and its application to methyl-ester content of pectins. Journal of Agricultural and Food Chemistry 34(4): 597-599.

Ludikhuyze, L., A. Van Loey, Indrawati and M. Hendrickx, 2001. Combined high pressure thermal treatment of foods. In: P. Richardson (ed.) Thermal technologies in food processing. Woodhead Publishing Limited, Cambridge, United Kingdom, pp. 266-284.

McFeeters, R.F. and S.A. Armstrong, 1984. Measurement of pectin methylation in plant cell walls. Analytical Biochemistry 139: 212-217.

Mertens, B. and D. Knorr, 1992. Developments of nonthermal processes for food preservation. Food Technology 46(5): 124-133.

Ng, A. and K.W. Waldron, 1997. Effect of cooking and pre-cooking on cell-wall chemistry in relation to firmness of carrot tissue. Journal of the Science of Food and Agriculture 73(4): 503-512.

Ridley, B.L., M.A. O'Neill and D.A. Mohnen, 2001. Pectins: structure, biosynthesis, and oligogalacturonide-related signaling. Phytochemistry 57(6): 929-967.

Sila, D.N., C. Smout, F. Elliot, A. Van Loey and M. Hendrickx, 2006. Non-enzymatic depolymerization of carrot pectin: toward a better understanding of carrot texture during thermal processing. Journal of Food Science 71(1): E1-E9.

Sila, D.N., C. Smout, Y. Satara, V. Truong, A. Van Loey and M. Hendrickx, 2007. Combined thermal and high pressure effect on carrot pectinmethylesterase stability and catalytic activity. Journal of Food Engineering 78(3): 755-764.

Smout, C., D.N. Sila, T.S. Vu, A.M.L. Van Loey and M.E.G. Hendrickx, 2005. Effect of preheating and calcium pre-treatment on pectin structure and thermal texture degradation: a case study on carrots. Journal of Food Engineering 67(4): 419-425.

Van Buren, J.P., 1979. The chemistry of texture in fruits and vegetables. Journal of Texture Studies 10: 1-23.

Verlent, I., A. Van Loey, C. Smout, T. Duvetter, B.L. Nguyen en M.E. Hendrickx, 2004. Changes in purified tomato pectinmethylesterase activity during thermal and high pressure treatment. Journal of the Science of Food and Agriculture 84(14): 1839-1847.

Waldron, K.W., 2004. Plant structure and fruit and vegetable texture. In: D. Kilcast (ed.) Texture in food – volume 2: solid foods. Woodhead Publishing Limited, Cambridge, United Kingdom, pp. 241-258.

Waldron, K.W., A.C. Smith, A.J. Parr, A. Ng and M.L. Parker, 1997. New approaches to understanding and controlling cell separation in relation to fruit and vegetable texture. Trends in Food Science & Technology 8: 213-221.

Floatation: scourge of fruit processors?

H.-U. Endress and A. Pirker
Herbstreith & Fox KG Pektin-Fabriken, Turnstraße 37, 75305 Neuenbürg, Germany

Abstract

An increasing demand for a very high quality standard in fruit preparations challenges the pectin producers to develop new pectins that are able to fulfil the high requirements on the production process and the texture of the final product. Beside the gelling behaviour and the texture of the fruit preparation itself an important criterion for high quality is the fruit distribution. The objective of this paper is to show how pectin influences the distribution of fruit pieces by producing a particular rheological behaviour during the production process. It is shown that not only a certain viscosity but also a critical yield stress is needed to prevent floatation in fruit preparations with different contents of soluble solids.

Keywords: pectin, floatation, critical yield stress, fruit preparation

1. Introduction

The use of pectin as gelling and thickening agent in the production of fruit spreads and fruit preparations for industrial processing (e.g. yoghurt fruit preparations) has a long tradition. In addition to a pleasant texture and a naturally fruit typical flavour the preservation of fruits as well as the even fruit distribution in both glassware and containers are decisive quality criteria in the industrial production of fruit preparations containing whole fruits or fruit pieces (Figure 1).

In this context, texture and gelling properties of the fruit preparations essentially depend on recipe parameters such as soluble solids content, type of sugar, pH-value, addition of buffer salts, pH-value of the final product as well as on the used pectin type and the filling temperature.

In order to prevent floatation, that means the undesired rising of the fruits, the recipe parameters can be modified in a way that, at a defined filling temperature, the gelation process has already started in an extend that the fruits are bound in the emerging gel structure. This effect can result in the so called pre-gelation which often has a negative influence on texture and syneresis behaviour of the final products.

Objective of the tests is to develop a method which is able to provide information if a fruit preparation shows floatation or not. At the same time the influence of measures for prevention of floatation on the texture of the final products is studied.

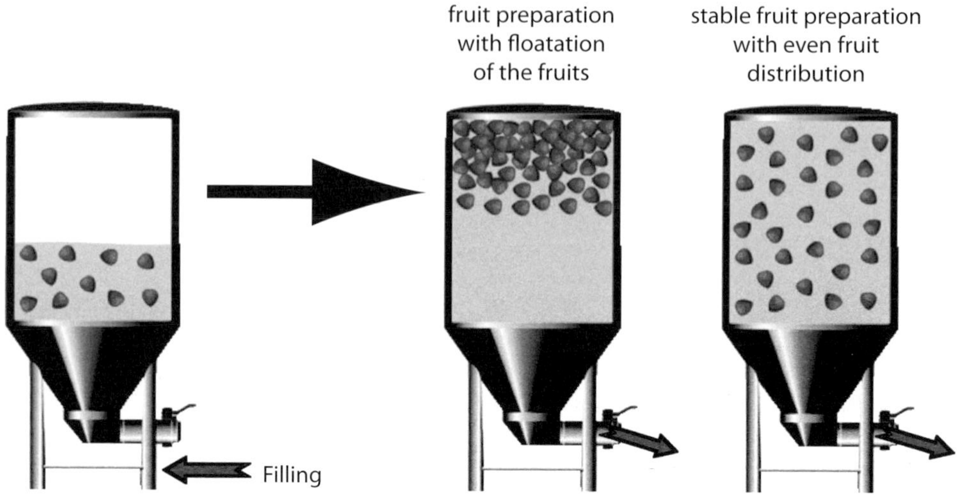

Figure 1. Fruit preparations.

Beside the sensory and visual determination of the floatation behaviour and texture particularly the determination of the rheological parameters yield stress, viscosity and breaking strength are of special importance.

It is the objective to show that at a defined filling temperature not only a defined viscosity but also the presence of a certain yield stress is necessary to prevent floatation.

Furthermore special pectins are investigated which are able to prevent floatation thus resulting in products with an attractive texture and appearance.

2. Physics of floatation

As shown in Figure 2 there are three forces acting on the particle: buoyancy F_B, frictional (drag) force F_D and weight force F_W. Both, buoyancy F_B and frictional (drag) force F_D are acting upwards. Buoyancy tends to float the particles and drag force resists the acceleration of gravity. The weight force F_W is the only force acting downwards and results from the gravitational force (Shearer and Hudson, 2008).

An even distribution of the particles in the fluid is a function of its immobility and results if the described forces are able to counterbalance each other. We obtain

$$F_W - F_B = F_D \qquad (1)$$

F_W = weight force [N]
F_B = buoyancy [N]

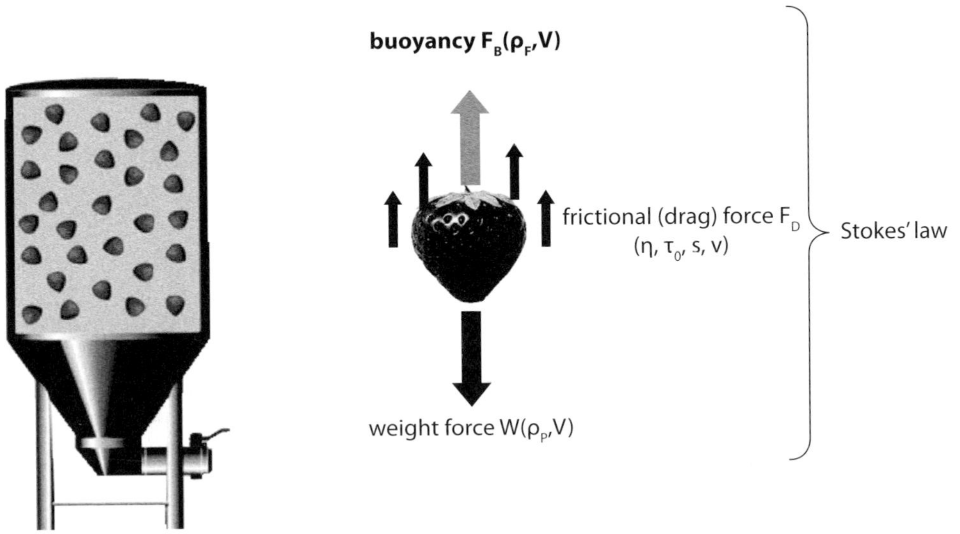

Figure 2. Forces acting on a particle in a fluid.

F_D = frictional (drag) force [N]

Acting on the assumption that the particles have a spherical shape and F_D describes the flow resistance of a sphere in an indefinite yield we can replace F_W - F_B by

$$F_W - F_B = \Delta m \cdot g = V_p \cdot \Delta\varphi \cdot g = V_p \cdot (\varphi_P - \varphi_F) \cdot g \qquad (2)$$

Δm = mass of particle – mass of fluid [kg]
g = gravitational acceleration [m/s^2]
V_P = volume of the particle [m^3]
$\Delta\varphi$ = density of particle - density of fluid [kg/m^3]
φ_P = density of particle [kg/m^3]
φ_F = density of fluid [kg/m^3]

Using Stokes' law F_D can be substituted as follows: for $3 \cdot \pi \cdot d_P \cdot \eta \cdot v$ where d_P stands for the average diameter of the particle, η for the viscosity of the fluid and v for the particles velocity.

$$F_D = 3 \cdot \pi \cdot d_P \cdot \eta \cdot v \qquad (3)$$

d_P = average diameter of particle [m]
η = viscosity of fluid [Pas]
v = velocity of particle [m/s]

The volume of the particle (assumed as a sphere) can be described as

$$V_p = \frac{1}{6} \cdot \pi \cdot d_P^3 \tag{4}$$

Regrouping and rearranging the terms in Equation 4 we amount to the following relationship for the particles velocity:

$$v = \frac{d^2 \cdot (\varphi_P - \varphi_F) \cdot g}{18\eta} \tag{5}$$

If the density of the particle is higher than the fluids density ($\varphi_P > \varphi_F$) the velocity direction is downwards, in contrary if $\varphi_P < \varphi_F$ the velocity direction is upwards. Only in the case that both densities are equal ($\varphi_P = \varphi_F$) the particles' velocity becomes zero.

With the knowledge of Stokes' law the goal of the processor of fruit preparations is to get a velocity as low as possible in order to minimize floatation. Table 1 shows that for this purpose there are different approaches resulting in benefits but also in possible disadvantages.

Regarding the Stokes' equation in the face of viscosity, the particles' velocity can be considered as zero, if the viscosity tends to infinite. This special case occurs if a yield stress τ_0 exists. The yield stress which is needed to prevent floatation is called the critical yield stress $\tau_{0,crit}$.

Table 2 shows that in practice the yield stress must have a defined value to prevent reliably floatation of the fruits or fruit pieces. Above the critical yield stress $\tau_{0,crit.}$ at the filling temperature, the fruit pieces are immobilized in the fruit preparation. The critical yield stress depends on the size and shape of the fruit pieces and the brix of the fluid (fruit preparation syrup).

In order to reach the requested critical yield stress $\tau_{0,crit.}$ either the calcium dosage or the calcium reactivity of the pectin used can be increased in the given recipe and filling temperature.

Table 1. How to minimize floatation.

Measure	Benefits	Possible disadvantages
maximize heat viscosity	increase of η	will destroy fruit pieces
reduce filling temperature	increase of η	pre-gelation, microbiological problems
adapt density of particles to density of fluid, e.g. increase cooking time	reduction of $\Delta\varphi$	off-taste caused by long cooking time, long production time
reduce size of fruit pieces	reduction of buoyancy F_B	fruit pieces too small
maximize agitation	even distribution of fruit pieces	will destroy fruit pieces

Table 2. Critical yield stress.

Fruit	Piece size	Tendency to floatation at filling temperature		
		< 40°Bx	40°Bx	> 40°Bx
Raspberry	broken		low	
Peach	6mm dices		medium	
Cherry	whole fruit	lower	medium	higher
Strawberry	16 mm dices		high	
	10 mm dices		high	
	whole fruit		very high	

3. Materials and methods

3.1 Reactivity of low methylester, amidated pectins

For fruit preparations mainly low methylester, amidated pectins are used. Low methylester, amidated pectins gel with calcium ions according to the egg-box model. Depending on TSS diverse calcium reactivities are necessary.

To find the right pectin for the different TSS ranges in our investigations it is useful to know that low methylester, amidated pectins have different reactivities. The reactivity is a result of different degrees of amidation and esterification (Table 3).

On the basis of previous tests the pectin with the calcium reactivity suitable for the different soluble solids contents was selected. The most suitable pectins are those which are, due to their reactivity, able to form homogeneously firm set gels independent from the calcium dosage.

Figure 3a shows that the most suitable pectin for a very low soluble solids content of e.g. 20% and pH approx. 3.2 is the low methylester, amidated Pectin Amid AF 020 with high calcium reactivity.

Table 3. Reactivity of low methylester, amidated pectins.

Pectin type		Calcium reactivity	Setting rate	Typical degree of esterification	Typical degree of amidation
Apple pectin	Citrus pectin				
Amid AF 005	Amid CF 005	low	slow	35%	15%
Amid AF 010	Amid CF 010	medium	medium	32%	18%
Amid AF 020	Amid CF 020	high	rapid	30%	20%

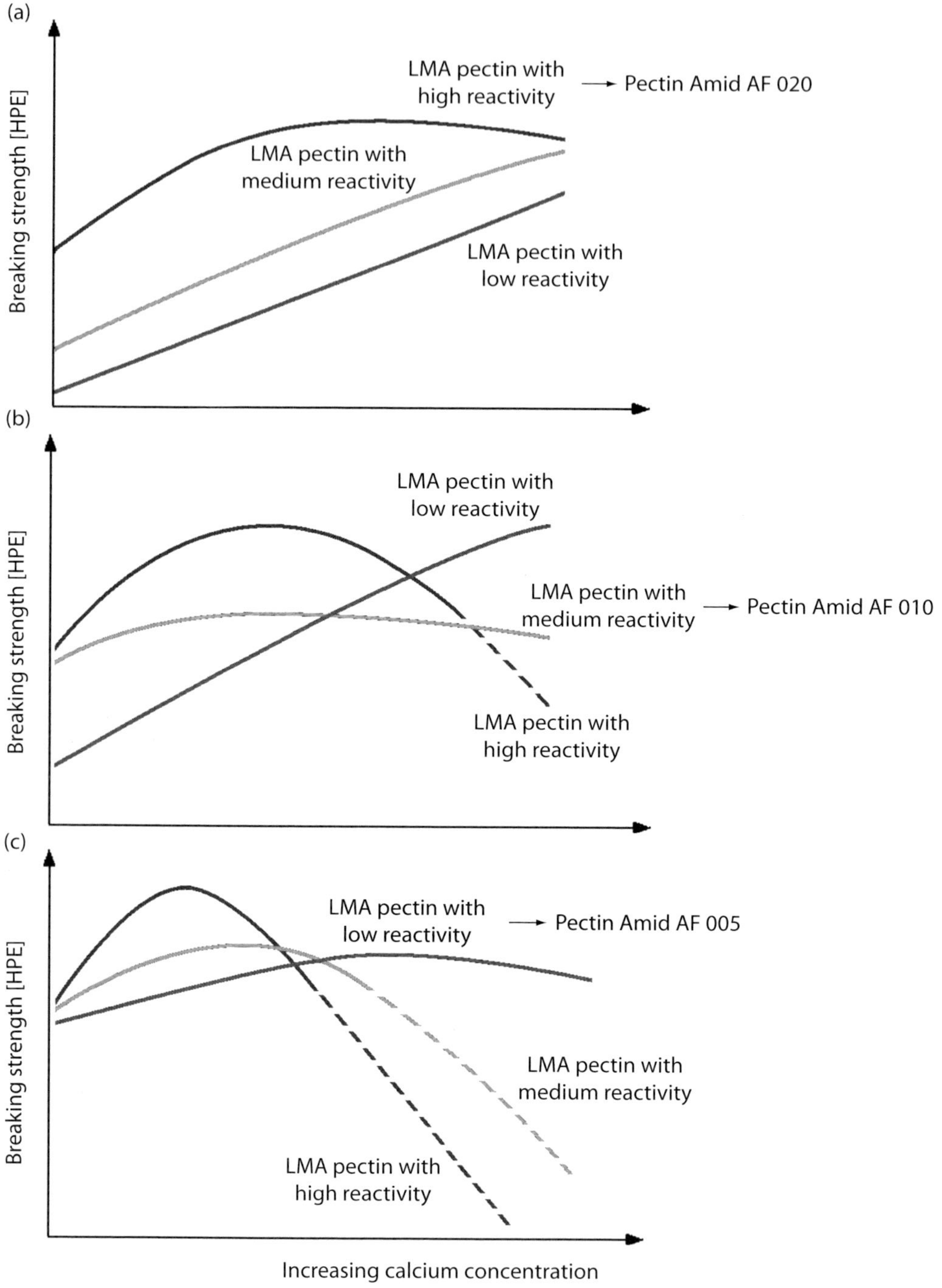

Figure 3. Reactivity of low methylester, amidated pectins (pH 3.2) at a) 20% TSS, b) 40% TSS and c) 60% TSS.

As seen in Figure 3b, the best pectin for a low soluble solids content as e.g. 40% and pH approx. 3.2 is the low methylester, amidated Pectin Amid AF 010 with medium calcium reactivity.

For a high soluble solids content as e.g. 60% and pH approx. 3.2 the low methylester, amidated Pectin Amid AF 005 with low calcium reactivity can be used (see Figure 3c).

3.2 Determination of the critical yield stress and viscosity

The determination of the critical yield stress and viscosity at 70 °C was done in a cherry fruit preparation which was produced in dependence on the calcium dosage with different soluble solids contents. Prior to determining the yield stress and the viscosity using a rheometer, the fruit preparation was sieved in order to avoid measuring errors caused by fibers and fruit pieces. The yield stress being observed at a calcium dosage at which for the first time there was no floatation of the fruits, is defined as critical yield stress.

The determination of yield stress and viscosity at a specified filling temperature of 70 °C was effected using the Rheometer Physica MCR 301. To determine the yield stress the shear stress was reduced logarithmically over time. The assessment can be done by e.g. using the tangent method in the log gamma/log tau diagram.

Measuring system:	Z3 DIN
Shear stress:	0.05-50 Pa
Sweep type:	log ramp, down
Time:	120 s
Temperature:	70 °C

For determining the viscosity the shear rate was increased over the time. The viscosity was determined at a defined shear rate.

Measuring system:	Z3 DIN
Shear rate:	0-120 s^{-1}
Sweep type:	linear ramp, up
Time:	120 s
Temperature:	70 °C

3.3 Evaluation of floatation

Floatation of the fruit preparation at the specified filling temperature has been assessed sensorily (visually) by a sensory research team.

3.4 Determination of the setting temperature

For determination of the setting temperature the oscillating rheometer Bohlin CS 10 was used. For the measurement the preparation was placed in the test chamber and then cooled under defined conditions.

Measurement system:	PP 40 (plate-plate, Ø 40 mm)
Frequency:	1 Hz
Shear stress:	0.06 Pa
Sweep type:	linear ramp
Start temperature:	95 °C
Stop temperature:	20 °C
Cooling rate	2 °C/min.
Gap:	1 mm

The setting temperature of the formula was taken as the temperature when, due to the beginning gel formation, the elastic shares increase very strongly resulting in an immensely increased storage modulus G' (tangent method).

3.5 Determination of breaking strength

As, due to the fruit pieces present, it is not possible to determine the breaking strength of cherry fruit preparations, the same products were produced using strawberry pulp. The determination of breaking strength (20 °C) was done with the Herbstreith Pektinometer Mark IV.

Table 4 illustrates again the recipe parameters used as well as the methods.

Table 4. Recipe parameters and methodology used.

Material	Methods
production of cherry fruit preparations with maintained cherries • increase of Ca^{2+} dosage • variation of TSS (30-60%) • T_{fill} = 70 °C	• sensory evaluation of floatation at T_{fill} and texture of the final product => sensory research team • rheological determination of yield stress and viscosity at T_{fill} (70 °C) • rheological determination of setting temperature
production of strawberry fruit preparations with strawberry pulp • increase of Ca^{2+} dosage • variation of TSS (30-60%) • T_{fill} = 70 °C	• rheological determination of yield stress and viscosity at T_{fill} (70 °C) • rheological determination of breaking strength (20 °C)

4. Results

For each tested soluble solids content (30-60%) it could be seen that the typical curve flow of yield stress and viscosity recur in dependence on calcium dosage.

Regarding the yield stress curve in dependence on calcium dosage (Figure 4), three ranges with different increase can be seen. If a tangent is applied to the curve in each range, two intersections result (A and B). A yield stress is formed as soon as a gel net in the gel preparation is formed. The more calcium ions available, the more bonding points and the higher the yield stress determined (range 1). Starting at a defined calcium dosage (point A) the gel formation starts already at a temperature above the measuring temperature.

This can be shown with the strawberry fruit preparation as the breaking strength curve has its maximum exactly in point A (see Figure 6a). If the maximum breaking strength exceeds due to a higher calcium dosage, the setting temperature is above the filling temperature and the gels become again weaker due to pregelation. Pregelation means that an over-reaction between the pectin molecules and the calcium ions occurs. Fine gel particles are formed, the gel arrangement looses its elastic character and the texture becomes pasty resulting in a reduction of gel strength. At mechanical treatment the gel looses water, syneresis occurs.

The critical yield stress was always observed at higher Ca^{2+} dosages as point A.

The velocity of gel formation gets higher as more calcium ions are available (decrease of setting time). By overlapping with this gelling process the yield stress in range 2 between A and B increases intensely. If the calcium dosage is further increased, the products with calcium dosages above B are pregelled to an extent that the gel looses

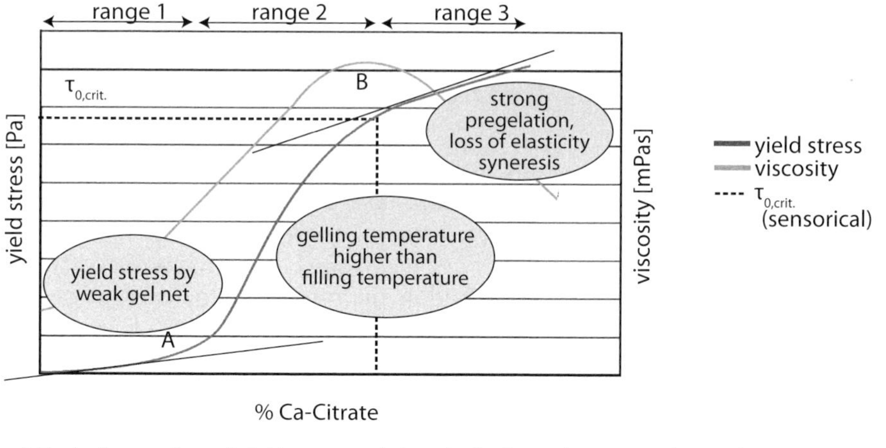

Figure 4. Typical curve flow of yield stress and viscosity in dependence on calcium dosage.

water already when reaching the measuring temperature (70 °C). This results in a strong decrease of both viscosity and also the elastic shares of the fruit preparation. Thus the yield stress is no longer able to increase intensely.

It is conspicuous that the critical yield stress determined lies in the range of calcium dosages above A and often in the range of maximum viscosity. In practice this means that with the addition of calcium ions the critical yield stress can be reached. The texture of the resulting fruit preparation, however, is pre-gelled with high tendency to syneresis.

Figure 5a shows viscosity and yield stress of the sieved cherry fruit preparation in dependence from the calcium dosage. The curve of the yield stress shows a s-shaped flow in dependence on the calcium dosage.

At first the values increase slowly. From a defined concentration of calcium ions the curve shows a much higher rise and, at a very high calcium content, it increases again more slowly.

The viscosity curve of the tested fruit preparations however, increases slowly at the beginning, then more intensely and decreases again when a defined calcium concentration is reached. By means of sensory assessment of the floatation behaviour a critical yield stress of 0.53Pa could be determined which is in the range of the viscosity maximum when calcium is added.

Figure 6a shows breaking strength and yield stress of the fruit preparation produced with strawberry pulp in dependence on the calcium dosage. In dependence on calcium dosage, the curve of yield stress displays a flow comparable to that of the cherry fruit preparation.

The breaking strength increases up to reaching a defined calcium dosage and then decreases again. The texture of the fruit preparation is spreadable and gets a slightly elastic-viscous character with increasing calcium dosage. With calcium dosages above the maximum value of the breaking strength the products become softer and show an increasingly pre-gelled texture.

Figures 5b, c, d illustrate the viscosity and yield stress of cherry fruit preparations in dependence on calcium dosage at 40, 50 and 60% TSS. The fruit preparations have been manufactured using the pectin appropriate for this TSS range. All TSS ranges investigated show comparable curve flows of yield stress and also viscosity. The higher the critical yield stress determined, the higher the soluble solids content of the fruit preparation. With increasing soluble solids content the viscosity of the fruit preparation increases implicating the necessary critical yield stress moving closer to A.

Figures 6b, c, d show yield stress and breaking strength of strawberry fruit preparations in dependence on calcium dosage at 40, 50 and 60% TSS. For the TSS range the

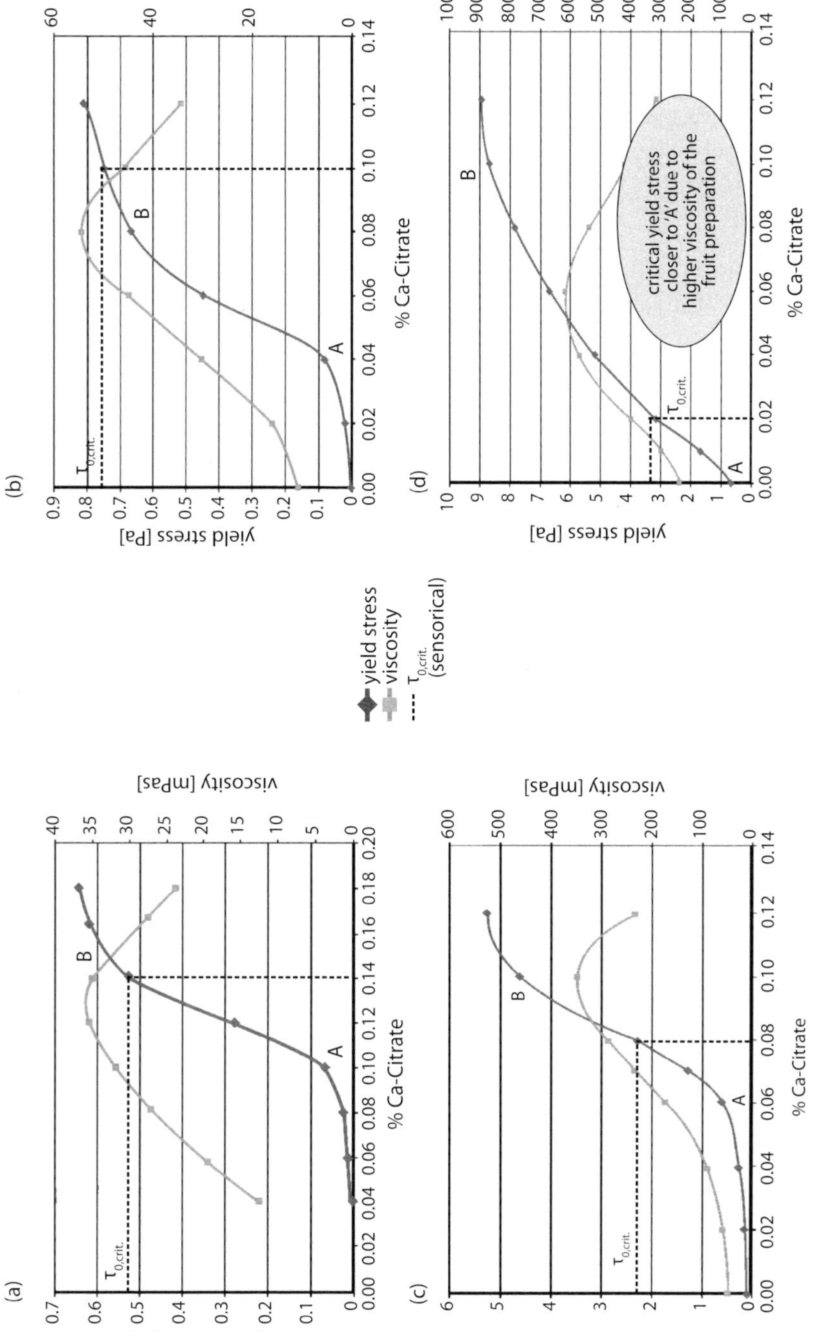

Figure 5. Yield stress and viscosity at 70 °C of a cherry fruit preparation, 30% TSS, pH 3.2, Pectin Amid AF 020, with a) 30% TSS, b) 40% TSS, c) 50% TSS and d) 60% TSS.

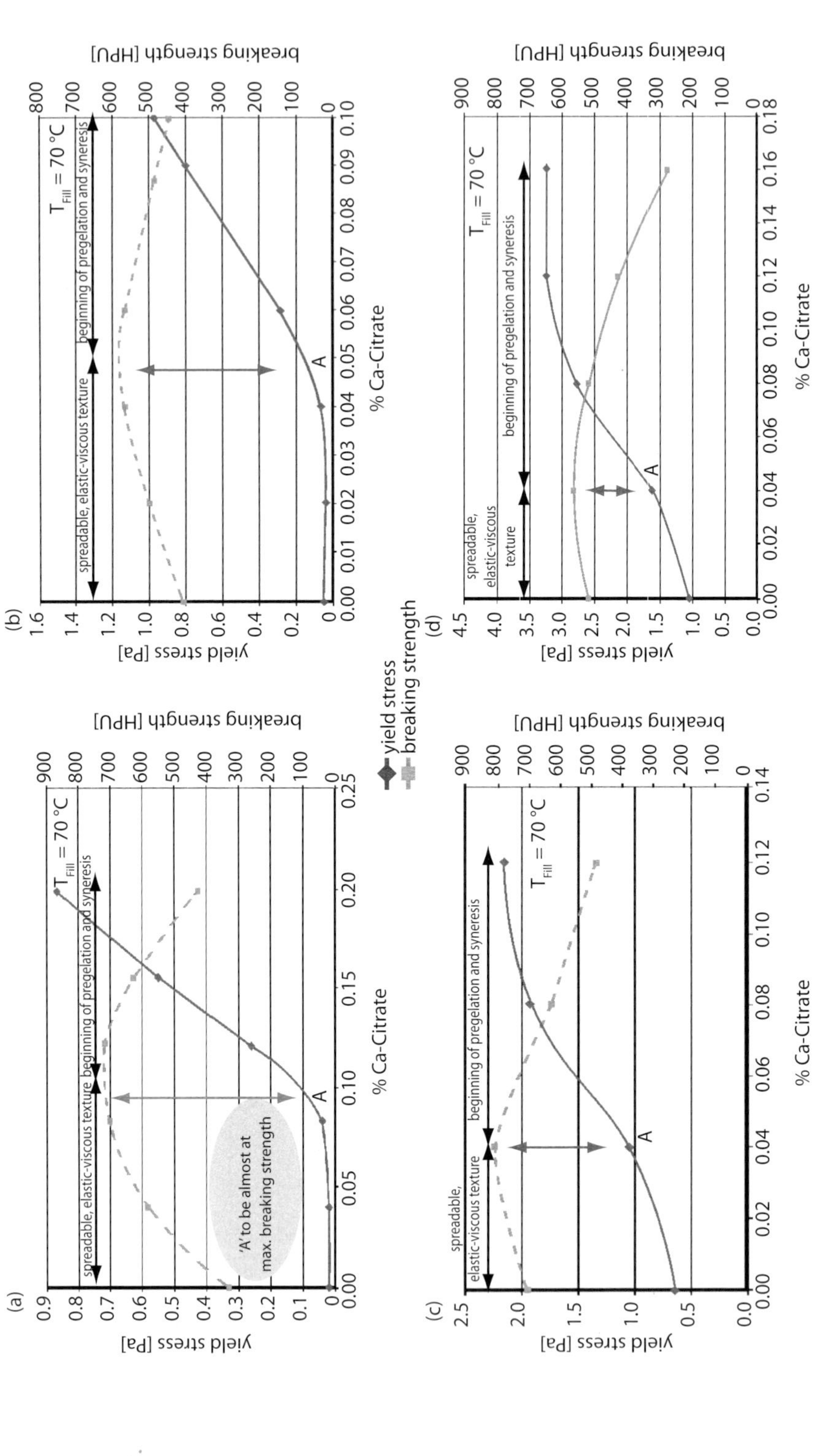

Figure 6. Yield stress (70 °C) and breaking strength (20 °C) of a strawberry fruit preparation, 30% TSS, pH 3.2, Pectin Amid AF 020, with a) 30% TSS, b) 40% TSS, c) 50% TSS and d) 60% TSS.

appropriate pectin was used. The maximum breaking strength of all investigated fruit preparations is reached in A, at higher calcium dosages pre-gelation occurs resulting in a loss of gel strength.

As a summary, Figurc 7 illustrates the yield stress of cherry fruit preparations with different soluble solids contents in dependence on the calcium dosage, also the sensorily determined critical yield stress for each soluble solids content is shown.

Figure 8 shows the critical yield stress in dependence on soluble solids content for the tested cherry fruit preparations (filling temperature 70 °C). It becomes obvious that with increasing soluble solids content the necessary critical yield stress has to be higher to prevent the fruits from floating. It has to be considered that these values cannot be seen as absolute values, they are valid only for the present conditions.

Back to the problem illustrated in the previous figures: floatation can only be surely prevented by presence of a yield stress which can be reached by altering the recipe parameters (in this case the calcium dosage). The resulting products, however, do not meet the requirements of producers of fruit preparations regarding texture.

Herbstreith & Fox has developed special pectins which are, due to their manufacturing process, able to form a yield stress lying above the critical yield stress already without separate calcium addition.

Figure 9 shows yield stress and breaking strength of a strawberry fruit preparation with a soluble solids content of 40% manufactured with pectins I and II. The critical yield stress determined with pectin I (Pectin Amid AF 010) in this system is approx. 0.75Pa and lies in B (see Figure 4). The fruit preparation being manufactured with

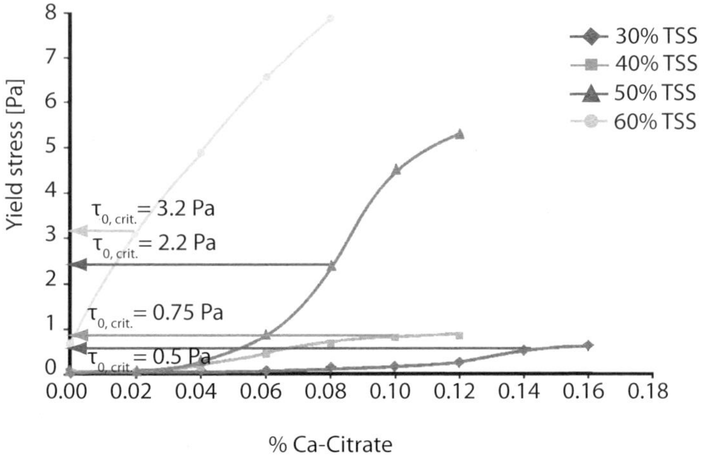

Figure 7. Yield stress and viscosity at 70 °C of cherry fruit preparations, 30-60% TSS, pH 3.2.

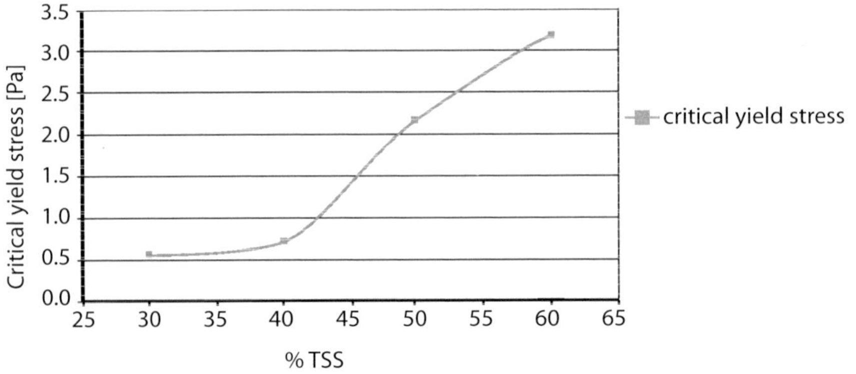

Figure 8. Critical Yield stress at 70 °C of cherry fruit preparations, 30-60% TSS, pH 3.2.

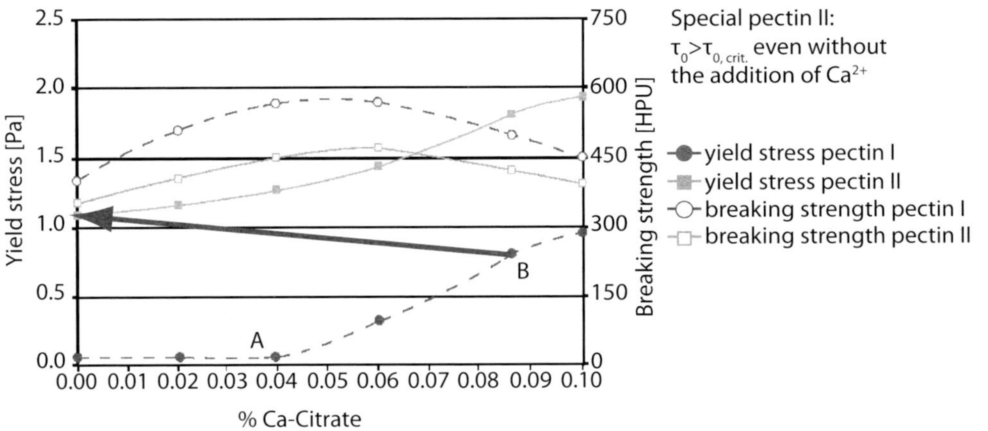

Figure 9. Yield stress (70 °C) and breaking strength (20 °C) of a strawberry fruit preparation, 40% TSS, pH 3.2.

the special pectin II shows, already without any addition of calcium, a yield stress which lies with 1.1Pa clearly above the necessary critical yield stress. The maximum of the breaking strength curve of the fruit preparations manufactured with Pectin II is reached at a calcium dosage of approx. 0.05%. At calcium dosages below this maximum the resulting products have an elastic-viscous and spreadable texture without pre-gelation.

Thus with the special pectins of Herbstreith & Fox fruit preparations can be produced for which floatation of the fruits is surely prevented over a wide soluble solids range and at the same time – depending on the recipe – spreadable-viscous up to elastic-viscous texture without pregelation and syneresis are formed.

Figure 10 illustrates the critical yield stress and the yield stress, which is reached with H&F special pectins without a separate calcium addition, in dependence on the soluble solids content. The values apply for the tested cherry fruit preparation and a filling temperature of 70 °C. It is obvious that H&F special pectins always reach a yield stress which is higher than the critical yield stress thus preventing floatation reliably.

By determining the elastic and viscous shares it can be shown that H&F special pectins for floatation prevention have mainly elastic shares already at very high temperatures, but the gelation process did not start yet. Measuring conditions correspond to the determination of the setting temperature of a jelly fruit mass using the Rheometer Bohlin CS 10 (oscillating measurement: temperature gradient at constant frequency and amplitude). Measuring curve G' and G" resp. the phase shift angle delta provide information on the ratio of elastic and viscous shares. If the elastic shares of a gel preparation overbalance, then a yield stress is available.

Figure 11 shows that Herbstreith & Fox special pectins are able to form a yield stress over a wide temperature range and already at very high temperatures without the gelling process having started yet.

Due to the yield stress, fruit preparations manufactured with Pectin II have more elastic shares (G') already at high temperatures, the phase shift angle is < 45°. Fruit preparations without yield stress have more viscous shares (G" > G'), the phase shift angle delta is then > 45°.

The real gelation, however, does not start until reaching the temperature at which the elastic shares increase disproportionately. After assessing the G' curve using the

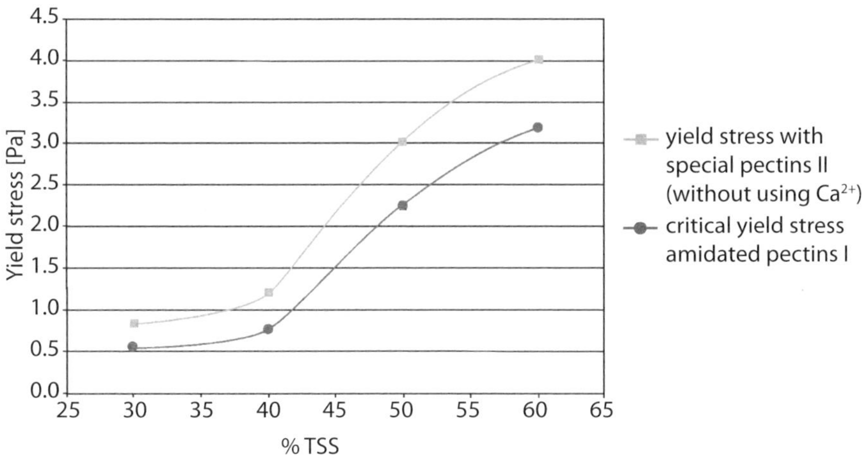

Figure 10. Critical yield stress and yield stress of fruit preparations manufactured with special pectins in dependence on TSS (cherry fruit preparation pH 3.2, 70 °C).

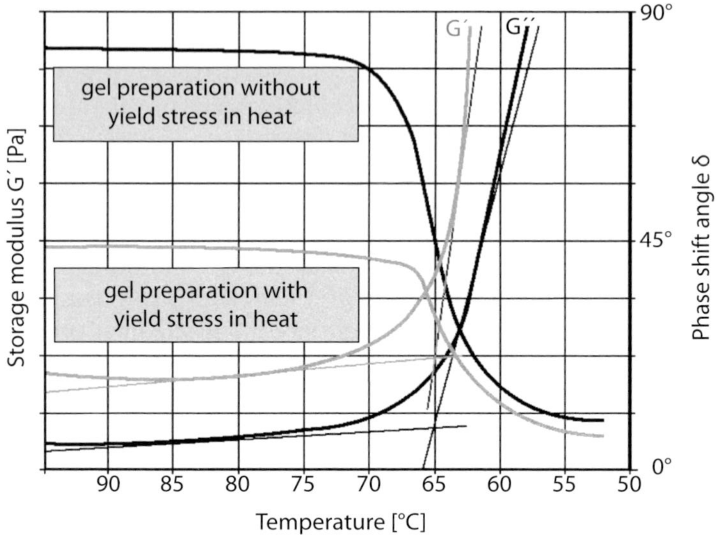

Figure 11. Determination of setting temperature.

tangent method, for both pectins I and II the same setting temperature of 65 °C is determined.

5. Conclusions

For the prevention of floatation, fruit preparations need to have a defined yield stress (critical yield stress $\tau_{0,crit.}$) when reaching the filling temperature. If this yield stress is reached by altering the recipe parameters, e.g. by increasing the calcium dosage, indeed products with even fruit distribution are obtained, however the texture is pre-gelled and the products are liable to syneresis. With specially developed pectins it is possible to manufacture fruit preparations which form a yield stress already during the heating process and, without addition of calcium, provide a pleasant texture at the same time.

Pectins are effective hydrocolloids to prevent the floatation of fruits and fruit pieces and to create a texture suitable for the industrial production process of fruit preparations.

References

Shearer, S.A. and J.R. Hudson, 2008. Fluid Mechanics: Stokes' Law and Viscosity. Measurement Laboratory. Investigation No. 3. www.engr.uky.edu/~egr101/ml/ML3.pdf.

Part 5

Health aspects of pectins

Recent studies on structures and intestinal immunity modulating activities of pectins and pectic polysaccharides from medicinal herbs

H. Yamada, H. Kiyohara and T. Matsumoto
Kitasato Institute for Life Sciences, Kitasato University, 5-9-1, Shirokane, Minato-ku, Tokyo 108-8641 Japan

Abstract

Many plants have been exploited since ancient times for health care of humans, and several pharmacological activities could be attributed to pectins and pectic polysaccharides present in the medicinal plants and plant derived foods. It can therefore be expected that also isolated carbohydrate structures will be biological active and find application in health care. Food polysaccharides are, similar to herbal medicines, generally taken orally, therefore their biological effects are expected to be expressed through the intestinal immune system. Peyer's patches play important roles as an inductive lymphoid organ of the intestinal immune system. The intestinal immune system contributes not only to the defense system of the mucosa, but also regulates the systemic immune system. Therefore these regulative molecules of the intestinal immune system have potential as new immuno-modulators of both systems. Polysaccharide fractions from medicinal herbs were screened for enhancement of Peyer's patch mediated intestinal immune response. Potent activity was observed in extracts containing pectic polysaccharides, and arabino-3,6-galactans from rhizomes of *Atractylodes lancea*, leaves of *Astragalus mongholicus* and roots of *Glycyrrhiza uralensis*. Active carbohydrate structures were characterized to be mainly arabino 3,6-galactans, rhamnogalacturonan (RG) I and RGII. The inner surface of the intestinal tract possesses a large area of mucosal membranes, and the intestinal epithelial cells exist at the interface between an antigen-rich lumen and a lymphocyte-rich lamina propria. When MCE 301 cells from the murine normal colonic epithelial cell line, were stimulated with a pectic polysaccharide, bupleuran 2IIc, from the roots of *Bupleurum falcatum*, only the content of granulocyte colony-stimulating factor (G-CSF) was significantly increased in the conditioned medium. In the carbohydrates structure of RGI, arabinan moiety was suggested to involve in the activity.

Keywords: intestinal immunity, medicinal herb, Peyer's patch, intestinal epithelial cells, G-CSF

1. Introduction

Traditional herbal medicines, and plant derived medicines and functional foods have generally been used as decocted extracts, which consists of numerous kinds of low molecular weight and macromolecular components. Polysaccharides represent 5-10% of the macromolecules in addition to lignin/lignin-carbohydrate complexes and tannins. Although most low molecular weight ingredients show pharmacological activities through direct actions, macromolecular constituents show their activities through host mediated actions. Because the clinical efficacy of most medicinal herbs cannot be explained by only low molecular weight ingredients, elucidation of pharmacological activity of the macromolecules is very important.

Plant polysaccharides have been recognized to bring benefit to human health because several pharmacological activities relating to the immune system, the digestive system and other systems have been reported to be triggered by polysaccharides originating from plants.(Table 1) (Yamada and Kiyohara, 2007). Among them several immunomodulating activities and anti-ulcer activities could be attributed to pectic polysaccharides by our research group (Yamada and Kiyohara, 2007). To understand the role of carbohydrate structures in these pharmacological activities, it is required to elucidate the structural requirements of the pectic polysaccharides necessary for the expression of the activity, and to explain their mechanism of action.

2. Intestinal immune system and its modulating polysaccharides in medicinal herbs

The mucosal immune system in each local site such as gastrointestinal, pulmonary and genitourinary tracts and the exocrine glands are connected to each other through lymphocyte homing, and this inter-tissue communication is named as common mucosal immune system (CMIS) (McGhee *et al.*, 1999). The inner surface of the intestinal tract possesses a large area of mucosal membranes, and they are continuously exposed to various substances in the intestinal lumen. The gut-associated lymphoreticular tissues (GALT) are present on the intestinal mucosal sites, and play an important role in host defense including IgA response in the gastric mucosal immune system (McNabb, 1981). GALT are composed of lamina propria aggregates, Peyer's patches, and mesenteric lymph nodes, which contain lymphocytes (McGhee *et al.*, 1999). Peyer's patch is the important lymphoid organ in the intestine, and is known as inductive site for IgA production (McGhee *et al.*, 1999).

Because herbal medicines and foods containing plant polysaccharides are generally administered or consumed orally, the gastric mucosal immune system may be considered to act as one of the major targets for the expression of pharmacological activity of the active ingredients. If the herbal medicines or foods, which contain intestinal immune system activating substances, are administrated orally, it has a possibility to enhance IgA response in intestinal mucosa.

Table 1. Pharmacological activities of polysaccharides from plant origins.

1. Immune system
   Enhancement
   - Complement activation
   - Antibody production (adjuvant activity)
   - Host-mediated anti-tumor activity
   - B lymphocyte proliferation
   - Phagocytosis
   - Tumocidal activity of macrophage/NK cell
   - Activated oxygen/NO production
   - Protective activity against infection (*Candida albicans*, Pneumococci, Cytomegarovirus)
   - Apoptosis of tumor cells

   Suppression/Improvement
   - Erythropoietin-resistant anemia
   - Metastasis of prostate/lung cancer cells
   - Anti-inflammatory activity
   - Anti-nephritic effect
2. Digestive system
   - Anti-emetic acitity
   - Anti-ulcer activity
   - Suppresion of colon cancer
   - Inhibition of heavy metal absorption
   - Suppression of colitis induced by methotrexate
3. Others
   - Enhancement of long learning memory
   - Lowering effect of blood glucose level
   - Anti-glaucoma effect
   - Cholesterol lowering effect

Previously we have reported that when the immunostimulating pectic polysaccharide fraction from *Bupleurum falcatum* was administrated orally to the mice, ramified (hairy/rhamnogalacturonan I) regions could be detected in the T cell area of follicles in Peyer's patches by immunohistochemical staining with anti-ramified region antibody (Sakurai *et al.*, 1996). This result suggests that oral administration of certain pectic polysaccharides affects the gastric mucosal immune system through Peyer's patches without systemic absorption from the intestine. Hong *et al.* (1998) developed an *in vitro* and *in vivo* assay system to evaluate the function of Peyer's patch cells and the intestinal immune system modulating activity. The Peyer's patch cells, obtained from the small intestine of lipopolysaccharide (LPS) low responder C3H/HeJ mice, were cultured with test polysaccharides, and the proliferation of Peyer's patch cells were assayed by Alamar Blue reduction methods. To obtain information on the intestinal immune system modulating activity, the resulting conditioned medium was also tested for the proliferation of bone marrow cells by the same assay method. We have screened intestinal immune system modulating polysaccharides present in medicinal herbs used clinically in Japanese traditional herbal medicines, and we found significant activities in polysaccharides from the rhizomes of *Atractylodis lancea*, the roots of *Glycyrrhiza uralensis*, and aerial part of *Astragalus mongholicus* (Yu *et al.*, 1998, 2001; Taguchi *et al.*, 2004; Kiyohara *et al.*, 2008) These active polysaccharides showed different extents of activity, and were characterized to be arabinogalactans, rhamnogalacturonan II (RGII) like polysaccharides, pectic arabinogalactans, heteroglycans and pectins. Active sites were characterized to be galactan, RGII like region and RGI. When the polysaccharide fractions from these three active medicinal herbs having intestinal immune system modulating activity were each administrated orally to the different types of mice, different cytokine production patterns in Peyer's patch cells were observed depending on the type of polysaccharide fraction and the test mouse strains. This result suggests that the intestinal immune system modulating activity is altered depending on the structure of the active structures in the polysaccharides.

3. Structural requirements for intestinal immune system modulating activity

Pectins

When the active pectin fraction, GR-PS-15 from *G. uralensis*, which showed potent activity for proliferation of Peyer's patch cells and intestinal immune system modulating activity, was digested with endo-polygalacturonase, the resulting digest had a decreased activity indicating that the whole pectin molecule is required for the expression of activity. When the endo-polygalacturonase digest was fractionated by gel filtration, RGI, RGII and oligogalacturonic acids were obtained from the pectin fraction from *G. uralensis* and the pectic arabinogalactan from *A. lancea*, respectively. Of these carbohydrate fragments, RGI from *G. uralensis* and RGII from *A. lancea* showed significant activity. Several modified hairy regions (MHR/RGI, obtained from Professor Voragen and Dr. Schols group at Wageningen University) were also tested for biological activity (Kiyohara *et al.*, unpublished). Beet MHR and Potato MHR showed weak but significant activity. When the side chains in RGI from the active pectin fraction from *A. mongholics*, Potato MHR and Beet MHR were digested with exo-β-(1→3)-galactanase or endo-β-(1→6)-galactanase, the enzymatic digestion products partially or completely lost their activity. β-D-(1→3,6) galactan structures in these RGI fractions possessed only oligomeric galactosyl chains having an average d.p. < 2 (Kiyohara *et al.*, unpublished). These results suggest that short β-(1→3,6) galactan chains in the hairy region/ RGI region are involved in the activity of certain RGI (Kiyohara *et al.*, unpublished).

Arabinogalactans

Several intestinal immune system modulating arabinogalactans have also been purified from the hot water extracts of *A. lancea*, *G. uralensis* and *A. morgholicus* (Table 2). These arabinogalactans mainly consist of arabinose and galactose, and most of them strongly react with β-D-glucosyl Yariv antigen. Among them, the potently active arabinogalactan, ALR-5IIa-1-1 was isolated from *A. lancea*, and its intestinal immune system modulating activity was compared with those of other known arabinogalactans. The arabino-3,6-galactan from *A. lancea*, ALR-5IIa-1-1, showed a potent activity, while arabinogalactans from Larch wood, Gum arabic, and Soy bean did not show this activity. This result suggests that a particular arabino-3,6-galactan structure is required for activity. The structure of ALR-5IIa-1-1 was analyzed using enzymic digestions and oligosaccharide analysis by MALDI-TOF-MS and NMR spectroscopy. It was established that ALR-5IIa-1-1 was composed of a β-D-(1→3)-linked galactosyl backbone with β-D-(1→6) linked mono- to octameric galactosyl

Table 2. Intestinal immune system modulating arabinogalactans found in Atractylodes lancea, Astragalus mongholics *and* Glycyrrhiza uralensis.

	Atractylodes lancea	*Glycyrrhiza uralensis*			*Astragalus mongholics*	
	ALR-5IIa-1-1	GR-PS-1	GR-PS-6	GR-PS-10	AM-PS-6	AM-PS-10
MW	120,000	67,000	56,000	230,000	87,000	72,000
Carbohydrate content (%)	91.9	83.4	79.5	77.9	91.5	91
Uronic acid content (%)	0.9	7.1	7.3	9.6	6.2	6.9
Protein content (%)	1.1	0.1	0.1	0.2	2.3	2.1
Component sugar (mol %)						
Rha	0.6	1.9	2.3	6.4	2	3.9
Ara	40.6	29.4	59.9	55.2	36.4	35.3
Fuc		1.1		2	0.4	0.4
Xyl	2.3	2.4	1.2	3	4.7	2.4
GlcA	0.5	0.8	0.9	1.4	2	3.4
GalA	0.5	1.2	2.2	2.1	0.4	2
Man	3.3	6.2	0.9	4.2	5.1	4.1
Gal	49.4	52.2	31.8	20.6	45.2	44.9
Glc	3.1	5	0.6	5.1	3.6	3.7
Relative reactivity with β-D-glucosyl-Yariv antigen	+++	+++	+++	+	+++	+++
Intestinal immune system modulating activity	+	+++	+	+	+	+

chains attached to O-6 of the backbone. (Figure 1) (Taguchi *et al.*, 2004). It was suggested that the chains longer than trisaccharide are involved in the activity. When ALR-5IIa-1-1 was also digested with exo-β-D-(1→3) galactanase, which can hydrolyze the β(1→3) galactan backbone chain, the resulting digest had lost its activity. When the ALR-5IIa-1-1 was digested with endo-β-D-(1→6) galactanase, the resulting digest also lost its activity. These results suggest that the side chains are attached to β-D-(1→3) galactan backbone in the form of glyco-cluster like structures. The length of of the β (1→6)-galactosyl side chains determines expression of the activity. The results also suggest that certain cellular receptor molecules may recognize the oligosaccharide glyco-cluster structure, and modulate the immunofunction of Peyer's patch cells (Figure 2).

Synthetic glyco-clusters

Because it was suggested that the glyco-cluster structure is important for the expression of activity, we collaborated with Professor Takeda and Dr. Hada group, and they synthesized trivalent phenyl-glyco-clusters based on β-D-(1→6)-galacto mono- to di-saccharides bearing 4-OMe-glucuronic acid (GlcA) as non-reducing terminal moiety, and tetravalent peptidyl-glyco-cluster and dendrimer of β-D-(1→6)-galacto-disaccharides bearing 4-O-Me GlcA as non-reducing terminal moity. (Hada *et al.*, 2001; Jin *et al.*, 2006). Among them, trivalent phenyl-glyco-custers based on di and tri-saccharides showed significant intestinal immune system enhancing activity on Peyer's patch cells obtained from either young or aged mice. Therefore this result

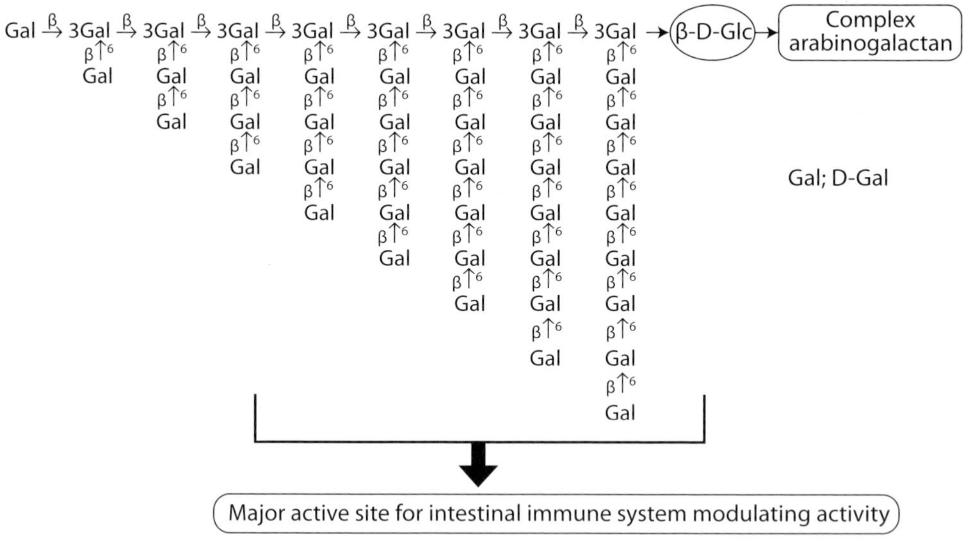

Figure 1. Major structure of galactan moiety in non-reducing terminal side of intestinal immune system modulating arabinogalactan (ALR-5IIa-1-1) from Atractycodes lancea.

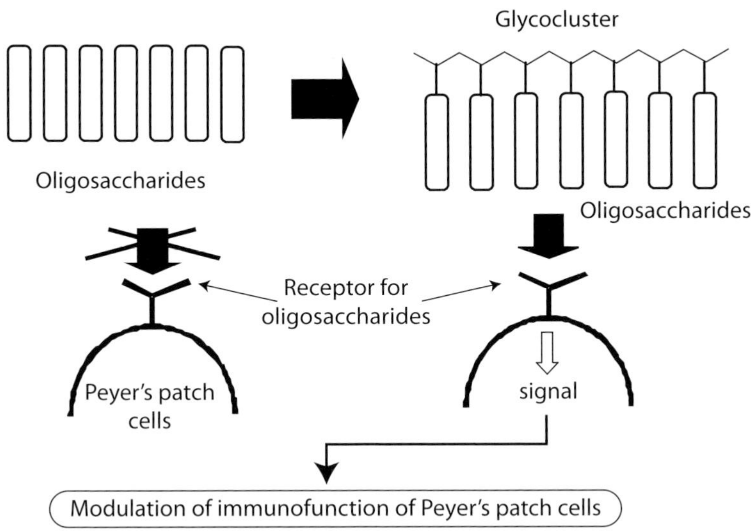

Figure 2. Requirement of oligosaccharide-based cluster (glyco-cluster) formation for expression of modulating activity against Peyer's patch cells.

partly confirmed the importance of the glycol-cluster structure of the side chain oligosaccharide for the expression of the activity.

Rhamnogalacturonan IIs (RGIIs)

We have also isolated active RGII from *A. lancea*. The proliferating activity of Peyer's patch cells and intestinal immune system modulating activity of this RGII were compared with those of three other RGIIs (GL-RI, GL-RII and GL-RIII) isolated from the leaves of *Panax ginseng* (Shin *et al.*, 1998). Among these RGIIs, GL-RI and GL-RII from *P. ginseng* showed significant proliferating activity of Peyer's patch cells. RGII from *A. lancea* and GL-RI from *P. ginseng* showed significant intestinal immune system modulating activity. However, other RGII showed no such activity suggesting that the activity is also dependent on the structural diversity of RGII. It is known the presence of a borate-RGII dimer in plant cell walls including medicinal herbs (O'Neill *et al.*, 1996; Kobayashi *et al.*, 1996; Shin *et al.*, 1998) Among three RGIIs from *P. ginseng*, GL-RI was characterized to be a mainly RGII monomer, GL-RIII was a mainly RGII dimer and GL-RII was almost equal mixture of RGII monomer and dimer. Therefore this result suggests dimerization of RGII are not necessary for the activity.

Structure-activity relationship

Present results suggest that glyco-cluster like long β(1→6)-linked galactosyl side chains attached to β(1→3)-linked galactan backbone in arabino 3,6-galactan, and β(1→3,6)-linked short galactosyl side chains attached to rhamnogalacturonan backbone in RGI are responsible for the expression of the activity (Figure 3). Some β(1→4)-galactan chains were also suggested to be involve in the activity based on results of activity measurements of β(1→4)-galactanase digests (Kiyohara *et al.*, unpublished) (Figure 3). In addition, certain RGII structures were also suggested to be involve in the activity (Figure 3).

4. Stimulatory effect of medicinal herb and its polysaccharide fraction on G-CSF secretion from intestinal epithelial cells

Because herbal medicines are taken orally, intestinal epithelial cells may act as one of the major targets in the gastric mucosal immune system for the expression of immunopharmacological activity. Therefore the effects of herbal medicines on cytokine secretion of intestinal epithelial cells are also studied using the murine colonic epithelial cell line, MCE 301 (Tabuchi *et al.*, 2000). MCE 301 has been cultured from a primary culture of gastric mucosal cells of transgenic mice harboring a temperature-

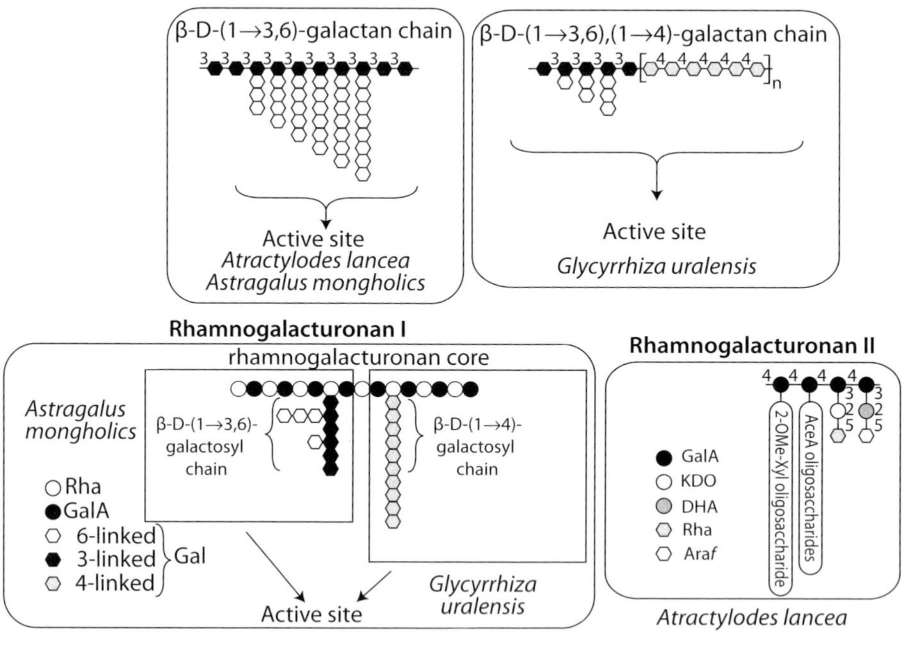

Figure 3. Intestinal immune system modulating polysaccharides from medicinal herbs.

sensitive simian virus 40 large T-antigen gene. MCE301 cells have been shown to retain many of the characteristics of normal colonic epithelial cells such as presence of microvilli-like structure and tight junction, and expression of Muc2 mucin (Tabuchi *et al.*, 2000). The cytokine network plays an important role in the inflammation and immune responses. Therefore, we have investigated the effect of a Kampo (Japanese traditional herbal) medicine, Hochu-ekki-to (HET) on cytokine secretion of intestinal epithelial cells. HET is a herbal formulation consisting of ten component herbs, and its hot water extract has been used clinically as an effective drug to improve the function of digestive system and to strengthen the defense system against various infections. When MCE301 cells were stimulated with HET, and the resulting conditioned medium was assayed for 22 different cytokines by the murine cytokine protein array, only the contents of granulocyte colony-stimulating factor (G-CSF) was significantly increased in comparison with the control (Matsumoto *et al.*, 2008a). Enhancement of G-CSF by HET was detected in a dose dependent manner by ELISA and RT-PCR. G-CSF is a preiotropic cytokine, and was identified initially as a growth factor for neutrophiles. G-CSF modulates NKT cells and macrophage function, and induces regulatory T cells and regulatory dendritic cell-like cells. G-CSF modulates the balance between Th1 and Th2 immune response by affecting cytokine production. Therefore, the enhanced G-CSF secretion from intestinal epithelial cells by HET may contribute to the expression of various clinically observed immunopharmacological activities by this herbal medicine. When the extract of the herbal medicine, HET was fractionated, only the polysaccharide fraction, F-5, enhanced the G-CSF secretion from MCE301 cells. F-5 was rich in carbohydrates, and was composed mainly of arabinose, galacturonic acid, galactose and glucose. The activity of F-5 was lost after its carbohydrate had been submitted to peridate oxidation. After α-amylase treatment the activity was higher than that of the original F-5 fraction, suggesting that pectin-like structures in this herbal medicine may involved in the G-CSF secretion enhancing activity (Matsumoto *et al.*, 2008a).

5. Stimulatory effect of a pectic polysaccharide, bupleuran 2IIc on G-CSF secretion from intestinal epithelial cells.

We also found the potent G-CSF secreting activity in the extract of one of component herbs of HET, the roots of *B. falcatum* (Matsumoto *et al.*, 2008b). We have previously reported that *B. falcatum* contains interesting bioactive pectin fractions, bupleuran 2IIb and 2IIc, which have several immunomodulating activities (Yamada and Kiyohara, 2007). One of these fractions, bupleuran 2IIc enhanced only G-CSF secretion among the 22 cytokines from MCE301 cells in a dose dependent and time dependent manners which were measured by ELISA method and RT-PCR (Figure 4). Because this was observed with the MCE301 cell line, it was further examined whether the G-CSF secretion is also possible in the primary cultured colonic epithelial cells. We have established primary cell culture of colon epithelial cells from C3H/HeJ mouse. When the primary cultured cells were treated with bupleuran 2IIc for 24hr, significantly enhanced G-CSF secretion was observed (Matsumoto *et al.*, 2008b). So far to our

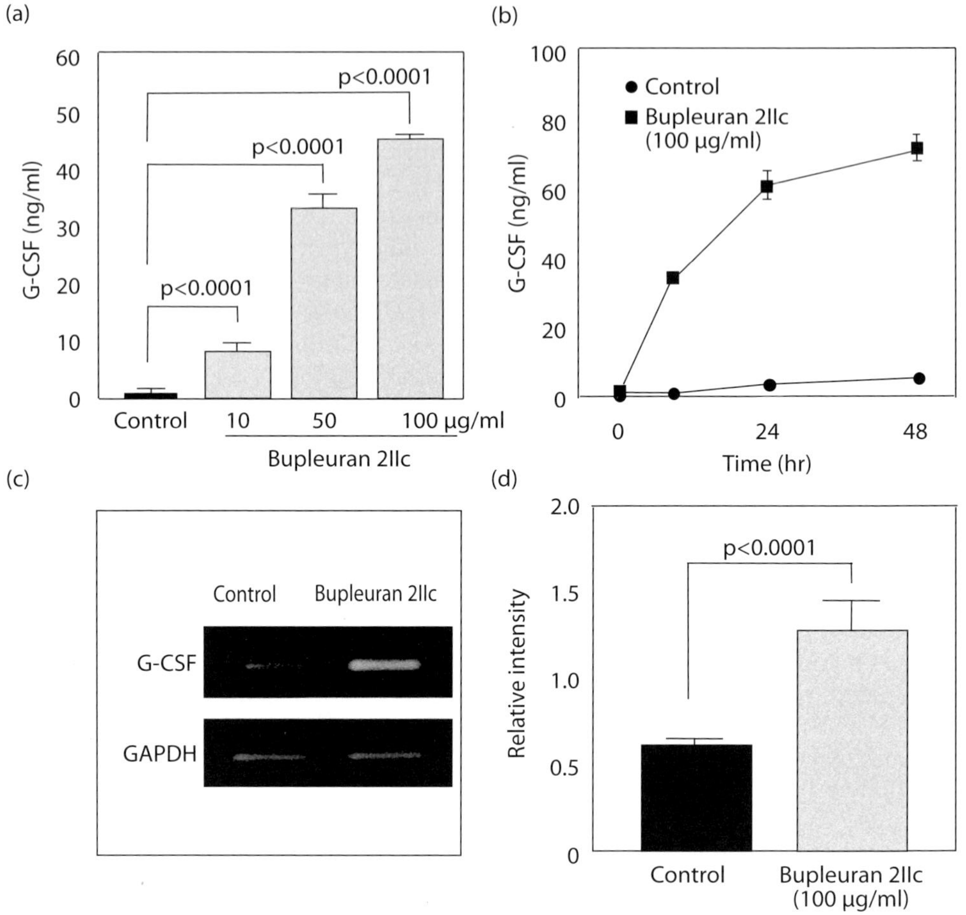

Figure 4. Enhanced secretion of G-CSF from MCE301 cells by bupleuran 2IIc stimulation. a) dose-dependent increases in production of G-CSF induced by bupleuran 2IIc (ELISA method); b) time-dependent increases in production of G-CSF induced by bupleuran 2IIc (ELISA method); c) PCR product on a 1.5% agarose gel; d) The relative level of G-CSF mRNA (by NIH Image-J).

knowledge, the finding that pectic polysaccharides enhances G-CSF secretion by intestinal epithelial cells is novel. To elucidate the active carbohydrate moiety of bupleuran 2IIc responsible for G-CSF production, bupleuran 2IIc was digested with endo-α-(1→4)-polygalacturonase. Only the ramified region (PG-I) showed potent G-CSF secretion enhancing activity (Matsumoto *et al.*, 2008b). This ramified region of bupleuran 2IIc consists of a rhamnogalacturonan backbone with several galactosyl and arabinofuranosyl side chains attached(Yamada *et al.*, 1991; Hirano *et al.*, 1994; Sakurai *et al.*, 1998, 1999). We have previously reported that the ramified region of bupleuran 2IIc showed mitogenic activity of B lymphocytes and immunoglobulin production, and that B cells recognize β(1→3) galactan chains with β(1→6) galactosyl residue containing side chains terminated with 4-O-Me-glucuronic acid or glucuronic acid residues as the active site (Sakurai *et al.*, 1999). When the ramified region, PG-1 was digested with endo (1→5)-α-L-arabinanase, the resulting enzyme-resistant moiety, EA-1 lost its G-CSF secretion enhancing activity (Matsumoto *et al.*, 2008b). This result suggests that the arabinan moiety of the ramified region plays an important role in expression of the G-CSF secretion enhancing activity. Because the present result suggests that the active site for G-CSF secreting activity from colonic epithelial cells was different from that for B-cell mitogenic activity, it can be concluded that different carbohydrate receptor molecules which can recognize particular carbohydrate structures are present on particular cell surfaces in order to express the different immunopharmacological activity.

6. Conclusion

Present results indicate that some bioactive pectins and pectic polysaccharides modulate the mucosal immune system through the Peyer's patches and intestinal epithelial cells resulting in various immunopharmacological activities in a carbohydrate structure-depending manner.

References

Hada, N., T. Ogino, H. Yamada and T. Takeda, 2001. Synthesis of model compounds related to an antigenic epitopes in pectic polysaccharides from *Bupleurum falcatum* L. Carbohydr. Res. 334: 7-17.

Hirano, M., H. Kiyohara, T. Matsumoto and H. Yamada, 1994. Structural studies of endo-polygalacturonase-resistant fragments in anti-ulcer pectin from the roots of *Bupleurum falcatum* L. Carbohydr. Res. 251: 145-162.

Hong, T., T. Matsumoto, H. Kiyohara and H. Yamada, 1998. Enhanced production of hematopoietic growth factors through T cell activation in Peyer's patches by oral administration of Kampo (Japanese herbas) medicines, 'Juzen-taiho-to'. Phytomedicine 5: 353-360.

Jin, Y., N. Hada, J. Oka, O. Kanie, S. Daikoku, Y. Kanie, H. Yamada and T. Takeda, 2006. Synthesis of model compounds related to an antigenic epitope in pectic polysaccharides from *Bupleurum falcatum* L. (II). Chem. Pharm. Bull. 54: 485-492.

Kiyohara, H., T. Matsuzaki, T. Matsumoto, T. Nagai and H. Yamada, 2008. Elucidation of structures and functions through Peyer's patches of responsible carbohydrate chains in intestinal immune system modulating polysaccharides from Japanese medicinal herbs. YAKUGAKU ZASSHI 128(5): 709-716.

Kobayashi, M., T. Matoh and J. Azuma, 1996. Two chains of rhamnogalacturonan II are cross-linked by borate-diol ester bonds in higher plant cell walls. Plant Physiol. 110: 1017-1020.

Matsumoto, T., M. Moriya, H. Kiyohara, Y. Tabuchi and H. Yamada, 2008a. (in press) Hochuekkito, a Kampo (Traditional Japanese herbal) Medicine, and its polysaccharide portion stimulate G-CSF secretion from intestinal epithelial cells. Evid. Based Complement. Alternat. Med. (doi: 10.1093/ecam/nen007).

Matsumoto, T., M. Moriya, M. H. Sakurai, H. Kiyohara, Y. Tabuchi and H. Yamada, 2008b. Stimulatory effect of a pectic polysaccharide from a medicinal herb, the roots of *Bupleurum falcatum* L., on G-CSF secretion from intestinal epitherial cells. Int. Immunopharmacol. 8: 581-588.

McGhee, J.R., M.E. Lamm, W. Strober, 1999. Mucosal immune responses, An overview. In: P.L. Ogra, J. Mestecky, M.E. Lamm, W. Strober, J. Brienenstock and J.R. McGhee (ed.) Mucosal Immunology, 2nd ed., Academic Press, New York, USA, pp. 485-506.

McNabb, P.C., 1981 Host defense mechanisms at mucosal surfaces. Annual Rev. Microbiol. 35:477-496.

O'Neill, M.A., D. Warrenfeltz, K. Kates, P. Pellerin, T. Doco, A.G. Darvill and P. Albersheim, 1996. Rhamnogalacturonan-II, a pectic polysaccharide in the walls of growing plant cell, forms a dimer that is covalently cross-linked by a borate ester. J. Biol. Chem. 271: 22923-22930.

Sakurai, M.H., T. Matsumoto, H. Kiyohara and H. Yamada, 1996. Detection and tissue distribution of anti-ulcer polysaccharides from *Bupleurum falcatum* L. by polyclonal antibody. Planta Med. 62: 341-346.

Sakurai, M.H., H. Kiyohara, T. Matsumoto, Y. Tsumuraya, Y. Hashimoto and H. Yamada, 1998. Characterization of antigenic epitopes in anti-ulcer pectic polysaccharides from *Bupleurum falcatum* L. using several carbohydrases. Carbohydr. Res. 311: 218-229.

Sakurai, M.H., T. Matsumoto, H. Kiyohara and H. Yamada, 1999. B cell proliferation activity of pectic polysaccharide from a medicinal herb, the roots of *Bupleurum falcatum* L. and its structural requirement. Immunology 97: 540-547.

Shin, K.-S., H. Kiyohara, T. Matsumoto and H. Yamada, 1998. Rhamnogalacturonan II dimers cross-linked by borate diesters from the leaves of *Panax ginseng* C. A. Meyer are responsible for expression of their IL-6 production enhancing activities. Carbohydr. Res. 307: 97-106.

Tabuchi, Y., S. Ohta, Y. Arai, M. Kawahara, K. Ishibashi, N. Sugiyama, T. Horiuchi, M. Furusawa, M. Obinata, H. Fuse, N. Takeguchi and S. Asano, 2000. Establishment and characterization of a colonic epithelial cell line MCE301 from transgenic mice harboring temperature-sensitive simian virus 40 large T-antigen gens. Cell Struct. Funct. 25:297-307.

Taguchi I., H. Kiyohara, T. Matsumoto and H. Yamada, 2004. Structure of oligosaccharide side chains in intestinal immune system modulating arabinogalactan from rhizomes of *Atractylodes lancea* DC. Carbohydr. Res. 339: 763-770.

Yamada, H., and H. Kiyohara, 2007. Immunomodulating Activity of Plant Polysaccharides Structures. In: A.G.J. Voragen (ed.) Complehensive Glycoscience Vol. 4. Elsevier, Amsterdam, pp.663-694.

Yamada, H., X.-B. Sun, T. Matsumoto, K.-S. Ra, M. Hirano and H. Kiyohara, 1991. Purification of anti-ulcer polysaccahrides from the roots of *Bupleurum falcatum* L.. Planta Medica 57: 555-559.

Yu, K.-W., H. Kiyohara, T. Matsumoto, H.-C. Yang and H. Yamada, 1998. Intestinal immune system modulating polysaccharides from rhizomes of *Atractylodes lancea*. Planta Med. 64: 714-719.

Yu, K.-W., H. Kiyohara, T. Matsumoto, H.-C. Yang and H. Yamada, 2001. Characterization of pectic polysaccharides having intestinal immune system modulating activity from rhizomes of *Atractylodes lancea* DC. Carbohydr. Polymer 46: 125-134.

New bioactive and biobased product applications of pectin

A. Hotchkiss[1], R. Rastall[2], G. Gibson[2], I. Eliaz[3], L.S. Liu[1] and M. Fishman[1]
[1]Eastern Regional Research Center, Agricultural Research Service, U.S. Department of Agriculture, 600 E. Mermaid Lane, Wyndmoor, PA 19038, USA
[2]School of Food Biosciences, The University of Reading, P.O. Box 226, Whiteknights, Reading RG6 6AP, United Kingdom
[3]Amitabha Medical Clinic and Healing Center, 7064 Corline Court, Suite A, Sebastapol, CA 95472, USA

Abstract

Pectin is well known for its bioactive health-promoting properties and use in biobased products. Recent reports have demonstrated that pectin and pectic fractions have potential as prebiotics, prevent pathogenic bacterial adhesion, increase prostate specific antigen doubling time in patients with recurrent prostate cancer, induce apoptosis in prostate and colon cancer cells, and systemically chelate heavy metals which are eliminated by urinary excretion. New biobased product applications include bioplastics and films derived from pectin or pectin-rich aricultural residues such as sugar beet pulp. Pectin has also been used in scaffolds for bone and connective tissue repair and controlled-release matrices for colon-specific drug delivery. While the biobased product applications of pectin require high molecular weight, the new bioactive properties of pectin are enhanced or only observed following reduction of the polysaccharide's molecular weight.

Keywords: pectic oligosaccharides, bioactive, biobased product

1. Introduction

Pectins are plant polysaccharides typically utilized for their gelling or viscosity modifying properties (May, 1990). However, pectin or fractions derived from pectin, are also well known for various bioactive health-promoting properties including lowering LDL-cholesterol, post-parandial blood glucose and insulin plus body weight reduction as well as immunostimulation and anti-peptic ulcer activity (Endress, 1991; Yamada, 1996). Pectin's use in biobased products or non-food, non-feed products composed of biological or renewable agricultural materials has also been reported (Endress, 1991). These multiple uses for pectin make it a valuable component of biomass and agricultural processing residues.

2. Prebiotics

Pectin has been investigated for its prebiotic potential. A prebiotic is able to selectively enhance the growth and/or activity of one or a limited number of gut bacteria that are beneficial to host health (Gibson and Roberford, 1995), To date, only *in vitro* evaluations have been carried out and prebiotic activity remains to be confirmed in vivo. Oligosaccharides derived from either low-methoxy (8%) apple pectin or high-methoxy (66%) citrus pectin had more prebiotic activity in vitro than the parent pectin (Olano-Martin *et al.*, 2002). The pectic oligosaccharides (POS) derived from the low-methoxy pectin were more prebiotic than those derived from the high-methoxy pectin (Olano-Martin *et al.*, 2002). Prebiotic POS was derived from the soluble supernatant fraction following hot acid extraction of pectin from orange peel (Manderson *et al.*, 2005). This is the fraction typically discarded during commercial pectin production. The POS was bifidogenic, increased levels of *Eubacterium rectale* (Manderson *et al.*, 2005) and was used in synbiotic matrices where it enhanced the growth of lactobacilli (Hotchkiss *et al.*, 2008). These pectic oligosaccharides were not initially as selective as fructo-oligosaccharides, however, the POS prebiotic index matched the FOS prebiotic index by 24 h of fecal culture, while the fructo-oligosaccharide prebiotic index declined after an initially high value and then recovered. The prebiotic index of orange albedo also consistently increased during fecal culture and was the highest prebiotic index observed at 24 h (Manderson *et al.*, 2005). Orange albedo is the white part of the peel and the source of orange peel POS. Another POS was extracted from Bergamot citrus peel and had an even higher prebiotic index compared to FOS (Mandalari *et al.*, 2007). The prebiotic index of the Bergamot POS rose quickly, like FOS, but it maintained a higher prebiotic index while the FOS prebiotic index declined by 24 h of fecal culture (Mandalari *et al.*, 2007). The Bergamot POS also enhanced growth of bifidobacteria, lactobacilli and eubacteria (Mandalari *et al.*, 2007). It is not possible to directly compare the prebiotic index values between different studies, even performed by the same lab, since different fecal donors are used. However, in all reports POS was consistently prebiotic equal to or better than FOS. The pattern of POS prebiotic index change in fecal culture suggests that POS may persist into more distal regions of the colon.

Pectin includes many polysaccharides including homogalacturonan, xylogalacturonan, rhamnogalacturonan I with arabinan, galactan and arabinogalactan side chains, and rhamnogalacturonan II. Therefore, which structure in pectin is responsible for its potential prebiotic activity? Using oligosaccharides prepared from apple homogalacturonan, apple rhamnogalacturonan I with single galactose side chains attached to rhamnose, sugar beet arabinan and soybean arabinogalactan, only the later two oligosaccharide preparations supported fermentation by individual intestinal bacteria (Van Laere *et al.*, 2000). Based on degradation of the substrates and acidification of the medium following fermentation, arabinogalactan I structure in pectin was proposed to be responsible for its prebiotic activity (Voragen *et al.*, 2005). However, sugar beet arabino-oligosaccharides with lower degree of polymerization were bifidogenic in vitro (Al-Tamimi *et al.*, 2005). Additionally, we will present evidence elsewhere that orange peel POS is rich in arabino-oligosacchrides and the structure of these oligosaccharides is responsible for in vitro prebiotic activity. At least it appears that the structures responsible for prebiotic activity are pectin's neutral sugar side chains rather than the acidic homogalacturonan- or rhamnogalacturonan-rich regions. The structure of these pectic side chains may select for different species of prebiotic bacteria.

3. Bacterial adhesion

The acidic homogalacturonan-rich structure in pectin was reported to block adhesion of food pathogenic *E. coli* to uroepithelial cells (Guggenbichler *et al.*, 1997a, 1997b). In order for *E. coli* to cause diarrhea or other pathogenic conditions, it must first adhere to intestinal epithelial cells and colonize the surface in order to produce effective concentrations of toxins that bind to specific receptors. Adhesion is mediated by lectin-carbohydrate interactions between bacterial pili or fimbriae and epithelial cell surface receptors. The minimum receptor for p-fimbriated *E. coli* is an α-(1-4)-linked galactose disaccharide. Oligogalacturonic acid with a degree of polymerization (DP) of 2 and3 had the highest activity to block the adhesion of *E. coli* to uroepithelial cells (Guggenbichler *et al.*, 1997a). However, orange peel POS inhibited the adhesion of veritoxigenic and enteropathogenic strains of *E. coli* to HT29 cells (Rhoades *et al.*, 2008). Adhesion of the probiotic *Lactobacillus acidophilus* was not inhibited but adhesion of *L. gasseri* was inhibited by POS. Adhesion of *Desulfovibrio desulfuricans* NCIMB 12833, was also inhibited by POS. Additionally, POS inhibited the toxicity of *E. coli* O157:H7 verocytotoxins toward HT29 cells (Olano-Martin *et al.*, 2003b; Rhoades *et al.*, 2008). POS does not contain DP 2 or 3 oligogalacturonic acids or the galactose disaccharide that terminates the *E. coli* p-fimbriae receptor. Therefore, the mechanism for pectic oligosaccharide inhibition of bacterial adhesion does not involve receptor mimicry (Rhoades *et al.*, 2008). POS may contain an oligogalacturonic acid tetrasaccharide methyl-esterified at one carboxyl group (Rhoades *et al.*, 2008) and POS does contain small amounts of higher molecular weight partially methyl-esterified oligogalacturonic acids detectable only after saponification (Manderson *et al.*, 2005). Further investigation is necessary to determine the mechanism of action for acidic pectic oligosaccharides in bacterial adhesion to epithelial cells.

4. Cancer

Pectin fractions are known to have anti-cancer properties. A pH-modified citrus pectin inhibited melanoma and prostate cell metastasis to the lung (Platt and Raz, 1992; Pienta *et al.*, 1995). The authors proposed that the pH-modified citrus pectin was a reduced molecular weight, partially debranched form of commercial citrus pectin and that galactose-rich side-chain binding to galectin-3 inhibited tumor cell metastasis. Following intravenous administration, the untreated commercial citrus pectin bound to the tumor cell surface, induced hemotypic tumor cell aggregation and increased metastasis of melanoma cells to the lung (Platt and Raz, 1992). Therefore, reduction of pectin molecular weight and disruption of its global tertiary structure was critical for its anti-cancer properties. It should be emphasized that the molecular weight of commercial citrus pectin is too large for intestinal absorption and that pectin as a fiber component passes through the gastrointestinal tract to the colon largely intact (Saito *et al.*, 2005). Therefore, a pectin-rich diet would not induce tumor cell metastasis.

POS induced apoptosis of colon cancer cells (Olano-Martin *et al.*, 2003a). A reduced molecular weight fraction of citrus pectin produced by autoclave hydrolysis or a combination of heating and enzymatic hydrolysis (Mitchell and Mitchell, 1996) also induced apoptosis of both androgen-responsive and androgen-independent prostate cancer cells (Jackson *et al.*, 2007). Saponification with base (NaOH) destroyed the apoptotic activity of this fraction which lead the authors to suggest that a base-sensitive linkage was responsible for activity (Jackson *et al.*, 2007). The nature of this linkage is unknown but not likely to be a methyl ester linkage since fungal pectin methyl esterase treatment had no effect on apoptotic activity (Jackson *et al.*, 2007). An enzymatically-hydrolyzed citrus pectin increased the prostate specific antigen (PSA) doubling time in patients following surgery, radiation and hormone therapies used to treat prostate cancer (Strum *et al.*, 1999; Guess *et al.*, 2003). This fraction is rich in homogalacturonan with relatively lower amounts of rhamnogalacturonan I (galactose- and arabinose- rich side chains) and approximately 10% rhamnoglacturonan II (Eliaz *et al.*, 2006; Zhao *et al.*, 2008). The small phase I and phase II clinical trials provided the first human evidence that a pectin fraction can be a potential therapy for recurrence of prostate cancer (Guess *et al.*, 2003). Another hydrolyzed citrus pectin increased clinical benefits and quality of life and decreased pain in 29 patients with advanced sold tumors representing many types of cancer (Azemar *et al.*, 2007). This study confirmed the previous report that an enzymatically hydrolyzed citrus pectin fraction could alter the progression of recurrent prostate cancer. One patient had a 50% decrease in serum PSA following 14 months of treatment (Azemar *et al.*, 2007).

5. Heavy metal chelation and detoxification

Oral administration of the enzymatically-hydrolyzed citrus pectin fraction in small clinical trials removed arsenic and cadmium from healthy adults and lead from children hospitalized with toxic lead levels (Eliaz *et al.*, 2006; Zhao *et al.*, 2008). Significantly, this pectin fraction chelated only heavy metals without altering levels of essential mineral nutrients. The authors attributed this chelation selectivity to the presence of rhamnogalacturonan II. It is well known that rhamnoglacturonan II dimers bind to Pb, Ba, Sr, La, Eu, Ce, Pr and Nd, while Ca, Mg, Fe, Zn and Cu are not affected (O'Neill *et al.*, 1996; Tahiri *et al.*, 2000). Further research is needed to determine how this pectic fraction is absorbed leading to systemic chelation of heavy metals in the blood stream followed by urinary excretion. However, rhamnogalacturonan I side chains have been immunolocalized in Payer's patches and the liver following oral administration in mice (Sakurai *et al.*, 1996). Therefore, this could be how rhamnogalacturonan II in the enzymatically-hydrolyzed citrus pectin is absorbed into the blood stream.

6. Biobased products

Pectin has been used in various film and bioplastic applications. Extrusion of pectin with starch and glycerol formed biodegradable films (Fishman *et al.*, 2000; Fishman *et al.*, 2004). Pectin was also blended with poly(vinyl alcohol) in various ratios to form biodegradable films (Fishman and Coffin, 1998). Edible films and antimicrobial packaging bioplastics were formed from composites of pectin and polylactic acid (Liu *et al.*, 2007a; Liu *et al.*, 2007b). Pectin-rich sugar beet pulp and polylactic acid (PLA) composites were formed that had the tensile properties of PLA alone but at a fraction of the cost (Liu *et al.*, 2005; Finkenstadt *et al.*, 2007; Liu *et al.*, 2007c). These composites could be used for construction or automotive materials. Biomedical products containing pectin included scaffolds for bone and connective tissue repair (Liu *et al.*, 2003a) and controlled-release matrices for colon-specific drug delivery (Liu *et al.*, 2003b; Liu *et al.*, 2006). It is anticipated that many pectin-rich residues from agricultural processing will be utilized for biobased as well as biofuel applications.

7. Conclusions

The structural diversity of pectins extracted from different sources, and the hydrolysis fractions derived from them, present a myriad of carbohydrate structures with valuable health-promoting bioactive properties. Hydrolyzed citrus pectin with anti-cancer properties has reached the clinical trial stage of new bioactive product development. More fundamental research is needed to determine how various bioactive pectin fractions are absorbed following oral administration. Human feeding studies are needed to prove that POS has prebiotic activity in humans. Pectin-rich residues of agricultural processing have recently been used in bioplastic composites. It is anticipated that these biobased applications of pectin will continue to expand as we seek less reliance on petrochemical products.

References

Al-Tamimi, M.A.H.M., R.J. Palframan, J.M. Cooper, G.R. Gibson and R.A. Rastall, 2005. In vitro fermentation of sugar beet arabinan and arabino-oligosaccharides by the human gut microflora. J. Appl. Microbiol. 100: 407-414.

Azemar, M., B. Hildenbrand, B. Haering, M.E. Heim and C. Unger, 2007. Clinical benefit in patients with advanced solid tumors treated with modified citrus pectin: A prospective pilot study. Clinical Medicine: Oncology 1: 73-80.

Eliaz, I., A.T. Hotchkiss, M.L. Fishman and D. Rode, 2006. The effect of modified citrus pectin on urinary excretion of toxic elements. Phytother. Res. 20: 859-864.

Endress, H.U., 1991. Nonfood uses of pectin. In: R.H. Walter (ed.) The Chemistry and Technology of Pectin. Academic Press, pp. 251-268.

Finkenstadt, V.L., L. Liu and J.L. Willett, 2007. Evaluation of sugar beet pulp and poly(lactic acid) green composites. J. Polym. Environ. 15: 1-6.

Fishman, M.L and D.R. Coffin, 1998. Mechanical, microstructural and solubility properties of pectin/poly(vinyl alcohol blends. Carbohydrate Polymers. 35: 195-203.

Fishman, M.L., D.R. Coffin, R.P. Konstance and C.I. Onwulata, 2000. Extrusion of pectin/starch blends plasticized with glycerol. Carbohydr. Polym. 41: 317-325.

Fishman, M.L., D.R. Coffin, C.I. Onwulata and R.P. Konstance, 2004. Extrusion of pectin and glycerol with various combinations of orange albedo and starch. Carbohydr. Polym. 57: 401-413.

Gibson, G.R. and M.B. Roberford, 1995. Dietary modulation of the human colonic microbiota: introducing the concept of prebiotics. J. Nutr. 125: 1401-1412.

Guess, B.W., M.C. Scholz, S.B. Strum, R.Y. Lam, H.J. Johnson and R.J. Jenrich, 2003. Modified citrus pectin (MCP) increases the prostate-specific antigen doubling time in men with prostate cancer: a phase II pilot study. Prostate Cancer and Prostatic Diseases 6: 301-304.

Guggenbichler, J.P., A. De Bettignies-Dutz, P. Meissner, S. Schellmoser and J. Jurenitsch, 1997a. Acidic oligosaccharides from natural sources block adherence of Escherichia coli on uroepithelial cells. Pharm. Pharmacol. Lett. 7: 35-38.

Guggenbichler, J.P., P. Meissner, J. Jurenitsch and A. De Bettignies-Dutz, 1997b. Blocking the attachment of germs to human cells. US Patent 5,683,991.

Hotchkiss, A.T, L. Liu, J. Call, P. Cooke, J.B. Luchansky and R.A. Rastall, 2008. Synbiotic matrices derived from plant oligosaccharides and polysaccharides. In: N. Parris, L. Liu, C. Song and V.P. Shastri (eds.) New Delivery Systems for Controlled Drug Release from Naturally Occurring Materials. ACS Symposium Series 992, American Chemical Society, Washington, DC, pp. 69-77.

Jackson, C.L., T.M. Dreaden, L.K. Theobald, N.M. Tran, T.L. Beal, M. Eid, M.Y. Gao, R.B. Shirley, M.T. Stoffel, M.V. Kumar and D. Mohnen, 2007. Pectin induces apoptosis in human prostate cancer cells: correlation pf apoptotic function with pectin structure. Glycobiology 17: 805-819.

Liu, L., Y.J. Won, P.H. Cooke, D.R. Coffin, M.L. Fishamn, K.B. Hicks and P.X. Ma, 2003a. Pectin/poly(lactide-co-glycolide) composite matrices for biomedical applications. Biomaterials 25: 3021-3210.

Liu, L., M.L. Fishman, J. Kost and K.B. Hicks, 2003b. Pectin-based systems for colon-specific drug delivery via oral route. Biomaterials 24: 3333-3343.

Liu, L., M.L. Fishman, K.B. Hicks and C.-K. Liu, 2005. Biodegradable composites from sugar beet pulp and poly(lactic acid). J. Agric. Food Chem. 53: 9017-9022.

Liu, L., M.L. Fishman, K.B. Hicks, M. Kende and G. Ruthel, 2006. Pectin/zein beads for potential colon-specific drug delivery: synthesis and in vivo evaluation. Drug Delivery 13: 417-423.

Liu, L., V.L. Finkenstadt, C.-K. Liu, T. Jin, M.L. Fishman and K.B. Hicks, 2007a. Preparation of poly(lactic acid) and pectin composite films intended for applications in antimicrobial packaging. J. Appl. Polym. Sci. 106: 801-810.

Liu, L., C.-K. Liu, M.L. Fishman and K.B. Hicks, 2007b. Composite films from pectin and fish skin gelatin or soybean flour protein. J. Agric. Food Chem. 55: 349-355.

Liu, L., V.L. Finkenstadt, C.-K. Liu, D.R. Coffin, J.L. Willett, M.L. Fishman and K.B. Hicks, 2007c. Green composites from sugar beet pulp and poly(lactic acid): structural and mechanical characterization. J. Biobased Materials and Bioenergy 1: 1-8.

Mandalari, G., C. Nueno Palop, K. Tuohy, G.R. Gibson, R.N. Bennett, K.W. Waldron, G. Bisignano, A. Narbad and C.B. Faulds, 2007. *In vitro* evaluation of the prebiotic activity of a pectic oligosaccharide-rich extract enzymatically derived from bergamot peel. Appl. Microbiol. Biotechnol. 73: 1173-1179.

Manderson, K., M. Pinart, K.M. Tuohy, W.E. Grace, A.T. Hotchkiss, W. Widmer, M.P. Yadav, G.R. Gibson and R.A. Rastall, 2005. *In vitro* determination of prebiotic properties of oligosaccharides derived from an orange juice manufacturing by-product stream. Appl. Environ. Microbiol. 71: 8383-8389.

May, C.D., 1990. Industrial pectins: sources, production and applications. Carbohydr. Polym. 12: 79-99.

Mitchell, C.R. and P.R. Mitchell, 1996. Treated pectinic acid process and product. U.S. Patent 5,498,702.

Olano-Martin, E., G.R. Gibson and R.A. Rastall, 2002. Comparison of the in vitro bifidogenic properties of pectins and pectic-oligosaccharides. J. Appl. Microbiol. 93: 505-511.

Olano-Martin, E., G.H. Rimbach, G.R. Gibson and R.A. Rastall, 2003a. Pectin and pectic-oligosaccharides induce apoptosis in in vitro human colonic adenocarcinoma cells. Anticancer Res. 23: 341-346.

Olano-Martin, E., M.R. Williams, G.R. Gibson and R.A. Rastall, 2003b. Pectins and pectic-oligosaccharides inhibit *Escherichia coli* O157:H7 Shiga toxin as directed towards the human colonic cell line HT29. FEMS Microbiol. Lett. 218: 101-105.

O'Neill, M.A., D. Warrenfeltz and K. Kates, 1996. Rhamnogalacturonan-II, a pectic polysaccharide in the walls of growing plant cells, forms a dimer that us covalently cross-linked by a borate ester. In vitro conditions for the formation and hydrolysis of the dimer. J. Biol. Chem. 271: 22923-22930.

Pienta, K.J., H. Naik, A. Akhtar, K. Yamazaki, T.S. Repogle, J. Lehr, T.L. Donat, L. Tait, V. Hogan and A. Raz, 1995. Inhibition of spontaneous metastasis in a rat prostate cancer model by oral administration of modified citrus pectin. J. Natl. Cancer Inst. 87: 348-353.

Platt, D. and A. Raz, 1992. Modulation of the lung colonization of B16-F1 melanoma cells by citrus pectin. J. Natl. Cancer Inst. 84:438-442.

Rhoades, J., K. Manderson, A. Wells, A.T. Hotchkiss, G.R. Gibson, K. Formentin, M. Beer and R.A. Rastall, 2008. Oligosaccharide-mediated inhibition of the adhesion of pathogenic *Escherichia coli* strains to human gut epithelial cells *in vitro*. J. Food Protect. 71: 2272-2277.

Saito, D., S. Nakaji, S. Fukuda, T. Shimoyama, J. Sakamoto and K. Sugawara, 2005. Comparison of the amount of pectin in the human terminal ileum with the amount of orally administered pectin. Nutrition 21: 914-919.

Sakurai M.H., T. Matsumoto, H. Kiyohara and H. Yamada. 1996. Detection and tissue distribution of anti-ulcer pectic polysaccharide from Bupleurum falcatum by polyclonal antibody. Planta Med. 62: 341-346.

Strum, S., M. Scholz, J. McDermed, M. McCulloch and I. Eliaz, 1999. Modified citrus pectin slows PSA doubling time: A pilot clinical trial. International Conference on Diet and Prevention of Cancer, Tampere, Finland.

Tahiri, M., P. Pellerin and J.C. Tressol, 2000. The rhamnogalacturonan-II dimer decreases intestinal absorption and tissue accumulation of lead in rats. J. Nutr. 130: 249-253.

Van Laere, K.M.J., R. Hartemink, M. Bosveld, H.A. Schols and A.G.J. Voragen, 2000. Fermentation of plant cell wall derived polysaccharides and their corresponding oligosaccharides by intestinal bacteria. J. Agric. Food Chem. 48: 1644-1652.

Voragen, A.G.J., S.E. Guilotin, G.J. Coenen, J.S. Zandleven, S.W.A. Hinz, J.-P. Vincken, G. Beldman and H.A. Schols, 2005. Potential of pectins as structurally complex polymers. 229th American Chemical Society Meeting, San Diego, CA, CELL 93.

Yamada, H., 1996. Contribution of pectins on health care. In: J. Visser and A.G.J. Voragen (eds.) Pectins and Pectinases. Elsevier Science B. V., Amsterdam, pp. 173-190.

Zhao, Z.Y., L. Liang, X. Fan, Z. Yu, A.T. Hotchkiss, B.J. Wilk and I. Eliaz, 2008. The role of modified citrus pectin as an effective chelator of lead in children hospitalized with toxic lead levels. Alternative Therapies 14: 34-38.

PectiCoat: immobilized enzymatically-tailored pectins to improve the biocompatibility of medical devices

M. Vayssade[1], R. Verhoef[2], A. Haeger[3], G. Cascardo[4], C. Bussy[1], M. Gallet[1], M. Morra[4], J.P. Knox[3], H.A. Schols[2] and M.-D. Nagel[1]
[1]CNRS UMR 6600, Université de Technologie de Compiègne, BP 20529, 60205 Compiègne cedex, France
[2]Wageningen University, Department of Agrotechnology and Food Sciences, Laboratory of Food Chemistry, P.O. Box 8129, 6700 EV Wageningen, the Netherlands
[3]Centre for Plant Sciences, University of Leeds, LS2 9JT Leeds, United Kingdom
[4]Nobil Bio Ricerche srl, V. Valcastellana 26, 14037 Portomaro d'Asti (AT), Italy

Abstract

During the last decade, biomimetic strategies have been developed to improve the biocompatibility of artificial surfaces by promoting tissue/implant integration. For several cell types, it has been suggested that control of cell behaviour on substrata is, at least partly, sugar-mediated. Our multidisciplinary effort aimed at endowing biomaterial surfaces with appropriate properties to control cell behaviour by using immobilization of enzymatically tailored pectins. Techniques for tailoring Rhamnogalacturonan-I (RGI) side-chain structure by specific enzymes were used. Modified RGI hairy regions (MHRs) decorated with or without side chains were obtained from apple, carrot or potato pectins. Polystyrene (PS) Petri dishes, aminated by a plasma deposition, were surface modified by the covalent linking of MHRs. Here, we report some biological results which illustrate that MHRs differing in charge, molecular weight, degree of branching and acetylation differently control cell behaviour. Such studies pave the way for the design of intelligently-tailored biomaterial coatings, able to induce specific cell functions.

Keywords: biomaterials, cell adhesion, cell-surface interactions, rhamnogalacturonan-I

1. Introduction

The extracellular matrix (ECM) regulates cell behaviour by influencing cell shape, migration, proliferation, differentiation and survival. Hence, artificial surfaces are specially designed to improve the biocompatibility of medical devices by ensuring appropriate cell behaviour.

Extensive efforts have concentrated on surface treatments and coatings to improve host tissue/implant integration (Morra, 2007). For instance, various surface treatments and biomimetic strategies have been developed to enhance bone tissue healing and implant osseointegration (Le Guéhennec *et al.*, 2007; Reyes *et al.*, 2007; Kalinina *et al.*, 2008; Petrie *et al.*, 2008).

Recent approaches have shown that the control of cell behaviour on substrata is, for several cell types, at least partly sugar-mediated and some adhesive peptide sequences can be recognized by polysaccharide molecules on the cell membrane (Manton *et al.*, 2006; Wu *et al.*, 2006; Manton *et al.*, 2007a; Manton *et al.*, 2007b).

Rhamnogalacturonan-I (RGI) is an abundant subunit of pectin. RGI is built from a backbone of alternating rhamnose and galacturonic acid residues (Voragen *et al.*, 1995; Rydley *et al.*, 2001) while arabinan and (arabino-)galactan side chains may be present on the rhamnosyl residues, as well as acetyl groups on the galacturonic acid residues. Techniques for tailoring side chain structures *in vitro* by specific enzymes are well-established (Schols *et al.*, 1994). It is therefore possible to obtain modified pectic polymers enriched in RGI decorated with or without side chains. A simple procedure to isolate RGI rich fractions on semi-large scale is to use commercial pectinase preparations to liquefy the tissue of fruit and vegetable (Schols *et al.*,1990; Schols and Voragen, 2003). The enzymes degrade the homogalacturonan part (smooth regions) of the pectin leaving the so-called hairy regions (HRs).

Polystyrene (PS) Petri dishes, aminated by a plasma deposition process, were surface modified by the covalent linking of different enzymatically modified hairy regions (MHR) from apple, carrot or potato pectin (Morra *et al.*, 2004; Bussy *et al.*, 2008; Nagel *et al.*, 2008).

Two MHRs from apple pectin, differing in charge, molecular weight, degree of branching and acetylation, induced opposite effects on fibroblast and bone cell behaviour. MHR-α which caused cells to aggregate, decreased proliferation and cell survival, in contrast to MHR-B which induced cell adhesion, spreading and proliferation. Here, we report biological investigations which attest that MHR-coatings are potential innovative candidates for controlling cell reactions and improving the biocompatibility of medical devices, in various applications.

2. Materials and methods

2.1 Pectin MHR substrates

Pectin MHR substrates were prepared as described by Schols *et al.*, (1990, 1994) by treatment of apple (MHR-A, MHR-α, MHR-B), carrot (MHR-C) or potato (MHR-P) pulp with commercial enzyme preparations. MHR-A, -α, -C and -P were prepared by using the experimental pectolytic enzyme preparation Rapidase C600, while MHR-B was made by the use of commercial pectinase/hemicellulase preparation Rapidase Liq+. The resulting suspension was centrifuged, the juice ultrafiltrated, and the ultrafiltration retentate was lyophilized to yield the MHRs. Characterization of the polymers was performed as previously described (Morra *et al.*, 2004). In general the difference between MHRs is reflected by the sugar composition and the molecular weight distribution which are presented in Table 1.

2.2 Surface modification of polystyrene Petri dishes (PS)

PS was first surface-functionalized by the introduction of amino groups *via* deposition from allylamine plasma, as described by Morra *et al.* (2004). As a second step, MHRs were covalently coupled to the surface amino groups on PS. In brief, 5 ml of an unbuffered 0.5% solution of MHR was placed into the aminated PS and coupled by carbodiimide mediated condensation between carboxyl groups present in the MHR and amino groups of the surface.

2.3 Biological studies

All methods used are detailed in Bussy *et al.*, 2008 (bone explant studies, statistical analysis) and Nagel *et al.*, 2008 (fibroblast culture, harvesting, observation and counting, cell cycle analysis, apoptosis quantification, characterization of fibronectin adsorption and conformation, statistical analysis).

Table 1. Sugar and substituent composition (in mol%, rhamnose/galacturonic acid ratio and total sugar in w/w%) of MHRs obtained by enzymatic liquefaction of apple (MHR-A, -α, - B), carrot (MHR-C) or potato (MHR-P).

MHR	-A	-α	-B	-C	-P
Fuc	0	1	0	2	1
Rha	9	15	11	22	8
Ara	51	25	11	8	14
Gal	10	15	20	32	28
Glc	0	1	3	2	4
Xyl	8	12	18	2	nd
GalA	22	31	37	30	42
GlcA	2	nd	nd	nd	nd
O-Me	42	37	34	16	57
O-Ac	60	45	11	46	59
Rha/ GalA	0.41	0.49	0.30	0.75	0.24
Total w/w%	86	86	78	82	70

O-Me: moles MeOH per 100 moles GalA (max 100%).
O-Ac: moles Ac per 100 moles GalA (max 200%).
nd: non determined.

3. Results and discussion

3.1 Enzymatically-tailored pectins differentially influence the morphology, adhesion, cell cycle progression and survival of fibroblasts

Cell shape was found to govern whether individual cells grow or die (Re *et al.*, 1994; Chen *et al.*, 1997; Chen *et al.*, 1998). Thus, we first examined morphology of cells cultured on the different MHRs. These observations were completed by cell counts and cell cycle analysis to address cell proliferation. Cell survival was also investigated through programmed cell death (apoptosis) measurement. Cells were seeded at 10000 cells/cm^2 and all experiments carried out 48 hour post-seeding. Cultures performed on tissue culture polystyrene (TCPS) were used as controls. Swiss 3T3 fibroblasts are adherent cells commonly used for biomaterial surface cytocompatibility studies. They were grown in DMEM medium supplemented with 10% foetal bovine serum, L-glutamine and antibiotics at 37 °C under humidified atmosphere (10% CO_2, 90% air).

Five different pectic polysaccharides (modified hairy regions (MHRs)) obtained by enzymatic liquefaction of fruit/ vegetables were covalently linked to polystyrene Petri dishes, previously aminated by a plasma deposition process. MHR-A (apple), MHR-C (carrot), MHR-P (potato) and MHR-α (a new batch of MHR-A), were prepared by using the experimental enzyme preparation Rapidase C600, while MHR-B (apple) was made by the use of Rapidase Liq+. The MHRs differed in their monosaccharide composition, molecular architecture and length of side chains (Table 1).

The impact of the pectic polymers on Swiss 3T3 fibroblast morphology, adhesion, proliferation, cell cycle progression and survival was addressed, through cell culture on immobilized MHRs (MHR-grafted PS Petri dishes). Results gained from fibroblasts grown on MHR-coated Petri dishes are summarized in Table 2. Cell behaviours on MHR-B and TCPS control were similar: cells spread and proliferated well, percentage of apoptotic cells was low. It can be concluded that MHR-B favours cell proliferation and survival. On the contrary, in a decreasing order, MHR-P, MHR-C, MHR-A and MHR-α appeared to be resistant to cell adhesion. Only few cells could be collected for cell cycle analysis on MHR-α, 48 hour post-seeding. Percentage of cells in DNA synthesis was dramatically decreased, and that of early apoptotic cells (cells entering in programmed cell death) was increased. Figure 1 illustrates fibroblast morphology on TCPS, MHR-B and MHR-α.

Thus, results clearly demonstrate that MHR-B induces cell adhesion, proliferation and survival, in contrast to the other MHRs. Moreover, MHR-α causes cells to aggregate, and decreases cell proliferation and survival.

The expected cell-biomaterial surface interactions depend on the application of biomedical devices. For instance, bone implant or vascular prosthesis integration needs colonization by cells *in situ*. On the contrary, intraocular lens or mammary

Table 2. Behaviour of Swiss 3T3 fibroblasts cultured on MHR-coated Petri dishes for 48 hours.

	TCPS[1]	MHR-B[1]	MHR-P	MHR-C	MHR-A	MHR-α[2]
Cell spreading	+++++	+++++	+++	++	+	-
Cell adhesion	+++++	+++++	+++	++	+	-
DNA synthesis	+++++	+++++	+++	++	+	-
Cell survival	+++++	+++++	++	++	++	+

[1] > 99% spread and adherent cells, > 17.5% cells synthesizing DNA, > 90% cell survival.
[2] all cells round, aggregated, poorly adherent, <6% cells synthesizing DNA, > 80% cell survival.

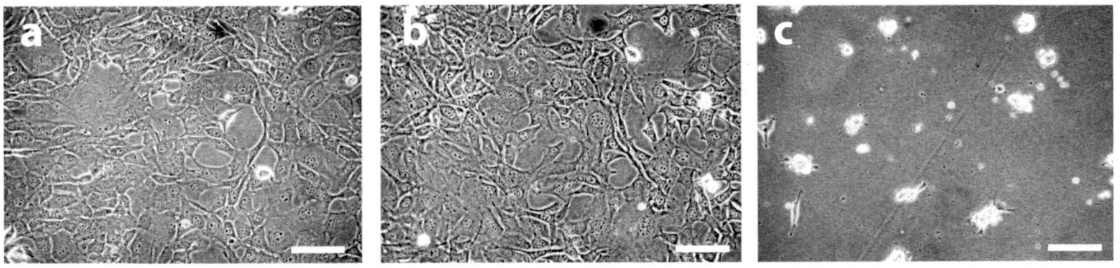

Figure 1. Swiss 3T3 fibroblast morphology 48 hour post-seeding on TCPS (a) MHR-B (b), and MHR-α (c). Cells had spread well on TCPS and MHR-B, on MHR- α, all cells were round and aggregated. Observations were performed using an inverted microscope Olympus CKX 41. Scale bar = 100μm.

implant biocompatibility involves a surface resistant to cell adhesion. In order to verify whether fibroblast results could be reproduced with other cell types, we made the decision to explore bone cell reactions in contact with MHR-B and MHR-α.

3.2 Enzymatically-tailored pectins differentially influence the migration, growth and differentiation of bone cells

Bone explants from 14-day-old chick embryos were cultured for 14 days on Thermanox® (a reference material for cell culture, as TCPS is) or MHR-B and MHR-α-grafted Thermanox®. Cell-layers developed around explants were observed by Scanning Electron Microscopy (SEM). They were stained with neutral red and measured, using a stereo-microscope fitted with a camera lucida and digitizing tablet connected to a microcomputer. Cell migration area was calculated using image analysis software. Alkaline phosphatase (ALP) activity which is recognized as an early indicator of bone cell differentiation was assessed. Cells forming cell-layers were trypsinized and counted using a Coulter Multisizer cell counter.

SEM observation of cultures (results not shown) evidenced cell layers around explants on Thermanox® and MHR-B. On the contrary, cells did not appear to escape from explants cultured in contact with MHR-α. Measurements of bone cell migration area around explants and mean number of cells per explant are presented in Table 3. They indicate that bone cell migration was increased on Thermanox® compared to MHR-B. On MHR-α, cell migration was negligible. Mean cell number of cells per explant confirm that cell layers were more developed on Thermanox® control than on MHR-B, and not detectable on MHR-α. ALP staining showed that on Thermanox® as well as on MHR-B, cell layers contained ALP positive cells.

Thus, the whole of the results clearly demonstrates that MHR-B promotes bone cell migration, growth and differentiation, while MHR-α does not. Our results agree with previous observations of Kokkonen *et al.* (2007) who have examined primary rat bone cells and MC 3T3 E1 pre-osteoblasts on MHR-A and MHR-B, with fluorescence and confocal microscopy.

Studies presented here, from fibroblasts as well as bone cells, demonstrate that in contact with MHR-grafted PS, cell behaviour can be affected by the type of grafted MHR. Thus, coatings of tailored pectins induce different biological activities *in vitro*. MHR-α contains less Ara than MHR-A. MHR-α therefore is suggested to have shorter arbinan side chains than the chains in MHR-B. Nevertheless, MHR-α exerts a stronger interaction than MHR-A does. MHR-B and MHR-C, in spite of their roughly similar Ara content, induce different cell behaviours. This suggests that Ara content is not responsible for the observed differences in fibroblast behaviour. Compared with that of other MHRs, the degree of MHR-B acetyl esterification is dramatically low. Further studies targeted to changes in MHR acetyl esterification should be performed to verify a possible involvement of acetyl esterification in the control of fibroblast behaviour.

3.3 Adhesive proteins appear to be involved in MHR grafted surface-cell interactions

It is well known that serum adhesive proteins, for example fibronectin (Fn), depending on their adsorption onto biomaterial surfaces, can play a major role in the control of adherent cell behaviour.

In order to address a possible MHR-protein interaction, we compared Fn adsorption and conformation onto MHR-B and MHR-α coated multi-well plates, using a modified ELISA approach. Both polyclonal antibodies to measure Fn global recognition, and

Table 3. Bone cell migration from tibia explant cultures of 14-day-old-chick embryo, 14 day post-seeding on Thermanox® control, MHR-B and MHR-α.

	Cell migration (mm^2)	Number of cells per explant
Thermanox®	37.94 ± 1.01	34888 ± 667.51
MHR-B	16.16 ± 6.44	7064.13 ± 3696.74
MHR-α	0.07 ± 0.05	1247.63 ± 818.74

monoclonal antibody to explore accessibility of Fn RGD cell binding site were used. Secondary antibodies were horse radish peroxydase-conjugated. Results presented in Table 4 indicate that Fn adsorbed similarly on MHR-B and TCPS control, but poorly on MHR-α. RGD site was the most accessible on MHR-B and very poorly on MHR-α.

Fn adsorption level and RGD site accessibility on MHR-B and TCPS can explain the similarity of cell behaviour observed on both substrata. Moreover, results strongly suggest that at least in part, proteins mediate pectin-cell interactions. Complementary experiments have demonstrated that echistatin, a disintegrin which prevents cell receptor binding to RGD site of adhesive proteins, inhibits fibroblast adhesion and spreading on MHR-B-grafted PS. (Nagel *et al.*, 2008). Thus, integrin receptors appear to be involved in cell spreading on MHR-B *via* adsorbed serum adhesive proteins which act as mediators between the MHR-B coating and cell membrane.

3.4 MHRs in solution and MHR coated-surfaces control differently cell behaviour

To determine whether MHRs in solution could induce the same effects as MHR grafted-PS on fibroblasts, we performed standard cultures on TCPS with or without 1% MHR-B or MHR-α in the culture medium. As shown in Table 5, MHRs in solution and MHR coatings induced opposite behaviours of fibroblasts 48 hour post-seeding.

Table 4. Fibronectin adsorption/conformation on TCPS, MHR-B and MHR-α coated multi-well plates assessed by an ELISA modified method (results are absorbance, λ=490 nm).

	TCPS	MHR-B	MHR-α
Polyclonal antibody (Fn global recognition)	0.119 ± 0.021	0.125 ± 0.044	0.029 ± 0.026
Monoclonal antibody (RGD cell receptor binding site specific recognition)	0.107 ± 0.013	0.143 ± 0.03	0.003 ± 0.009

Table 5. Behaviour of Swiss 3T3 fibroblasts cultured in standard conditions on TCPS with or without MHR-B or MHR-α, 48 hour post-seeding.

	TCPS (control)[1]	MHR-B[2]	MHR-α[1]
Cell spreading	+++++	-	+++++
Cell adhesion	+++++	-	+++++
DNA synthesis	+++++	-	+++++
Cell survival	+++++	+	+++++

[1] > 99% spread and adherent cells, > 22% cells synthesizing DNA, > 90% cell survival.

[2] all cells round, aggregated, poorly adherent, <6% cells synthesizing DNA, > 70% cell survival.

Figure 2 illustrates fibroblast morphology when cultured on TCPS in standard conditions, with or without MHR-B and MHR-α.

When cultured on TCPS with or without MHR-α, cells spread and proliferated well. On the contrary, they poorly adhered, rounded and aggregated with MHR-B. MHR-B addition to the culture medium decreased cell proliferation, percentages of cells synthesising DNA and cell survival.

Other studies have demonstrated *in vitro* the influence of polysaccharides on the behaviour of various cell types. For example, Olano-Martin *et al.* (2003) observed an activation of apoptosis in human colon adenocarcinoma cells treated with 10mg/ml of low-methylated apple pectin and Jackson *et al.* (2007) correlated pectin apoptotic function to human prostate cancer cells with pectin structure. Further experiments to address which apoptotic pathways in cell signalling are activated by MHR-B in solution and MHR-α coating should be carried out.

3.5 MHRs differently adhere to fibroblast cell surface

An intriguing question was related to the existence of interactions between MHRs in solution and cell membrane. We therefore explored the possible adherence of MHR-B and MHR-α to fibroblast surface by immunolabelling with fluorescent (FITC-conjugated) monoclonal antibodies to MHR epitopes and confocal microscopy observation.

Fibroblasts that had been incubated for 1 hour in complete culture medium without and with MHR-B or MHR-α in solution, washed twice and cytospinned on glass slides, were immunolabelled in order to determine the extent of pectin adherence to cell surfaces. Antibodies JIM13 (anti-AGP) and LM6 (anti RG-1 arabinan) were used. Contrasting patterns of staining were observed between MHR-B and MHR-α (Figure 3). MHR-α was detected at the cell surface by JIM13 and LM6. JIM13 was unable to detect MHR-B on the surfaces of the cells despite recognition of MHR-B

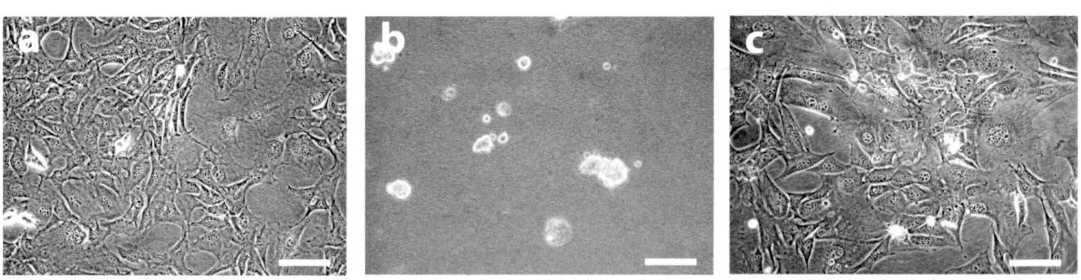

Figure 2. Swiss 3T3 fibroblast morphology 48 hour post-seeding on TCPS without MHR (a), with MHR-B (b), and MHR-α (c). Cells had spread well on TCPS without MHR and with MHR-α, while with MHR-B all cells were round and aggregated. Observations were performed using an inverted microscope Olympus CKX 41. Scale bar = 100μm.

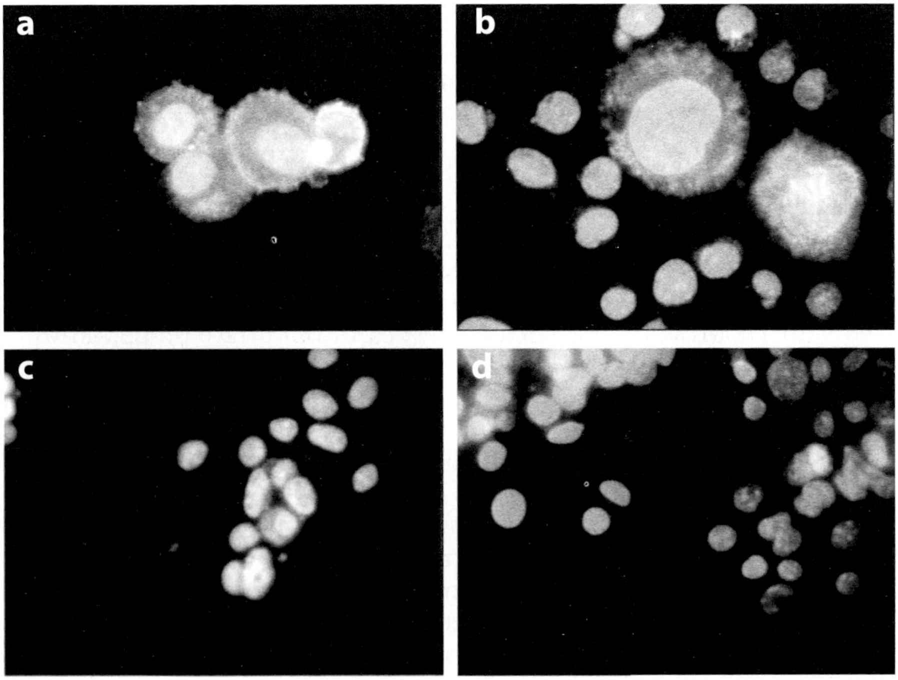

Figure 3. Swiss 3T3 fibroblasts previously incubated with MHR-B and MHR-α for 1 hour at 37 °C, immunolabelled with fluorescent (FITC-conjugated) monoclonal antibodies. Figure 3 (a) and (c) JIM13 (anti AGP) and Figure 3 (b) and (d) LM6 (anti RGI arabinan). Cell nuclei were stained with DAPI. Images are overlays. MHR-α (a, b) was detected at the cell surface by both JIM13 and LM6, MHR-B (c, d) was detected neither by JIM13 nor LM6. 400x.

epitopes in previous immunoassays, suggesting that the act of adhesion to the plasma membrane may obscure specific polysaccharide structural features. LM6 immunolabelling in cytoplasm suggests that MHR-α might be internalized by cells. Further work using a range of monoclonal antibodies in order to determine whether a specific pectin structural domain is responsible for adherence to the cell membrane should be performed.

4. Conclusions

It is well known that biomaterial surface properties (chemistry, hydrophilicity, roughness…) influence protein adsorption and elicit diverse cellular responses (Keselowsky *et al.*, 2003; Tajima *et al.*, 2008; Takamori *et al.*, 2008). This work is part of a multidisciplinary effort which aims at providing the framework for the future design of 'intelligent' tailored biomaterial coatings able to induce specific cell functions. Previously published MHR-grafted PS characterization data by Morra *et al.*,

(2004) confirmed the successful modification of the PS surface. X-ray photoelectron spectroscopy, contact angle measurement, and atomic force microscope force-separation curves were used to characterize the effects of the surface modification process on surface chemistry and interfacial forces.

Our results indicate that grafted and soluble MHRs, though they express the same epitopes, induce opposite cell behaviour. Cells spread and proliferate on MHR-B grafted PS, they are aggregated and cannot grow when cultured on TCPS in a medium enriched with 1% MHR-B. On the contrary, cells which are aggregated and apoptotic on MHR-α grafted PS, spread and proliferate when cultured on TCPS in a medium enriched with 1% MHR-α. This strongly suggests that surface properties of MHR coatings play a major role in the control of cell behaviour. Moreover, we have shown that adhesive proteins are involved in the control of fibroblast behaviour by MHR-grafted surfaces. MHR-α grafted-PS yields very hydrophilic surface (Cassinelli *et al.*, 2006; Ceccone *et al.*, 2008), on which adhesive proteins poorly adsorb. That might explain, at least in part, cell behaviour.

Our studies have evidenced that coatings of tailored pectins induce different biological activities *in vitro* and are therefore potential candidates for improving the biocompatibility of various medical devices. The presented data encourage further studies of the molecular trigger mechanisms which are initiated by MHR coatings. 'Sugar language' decoding, in other words: MHR structure/function understanding, is in progress.

Acknowledgements

Mrs Pascale Vigneron and Jean-Luc Duval are acknowledged for their participation to the work and technical collaboration.

Contract grant sponsor: European Union, Sixth Framework Programme for Research and Technological development (EU, PectiCoat – contract grant number 517036, NMP3).

References

Bussy, C., R. Verhoef, A. Haeger, M. Morra, J-L. Duval, P. Vigneron, A. Bensoussan, E. Velzenberger, G. Cascardo, C. Cassinelli, H. Schols, J.P. Knox and M-D. Nagel, 2008. Modulating in vitro bone cell and macrophage behavior by immobilized enzymatically tailored pectins. J Biomed Mater Res A. 86(3): 597-606.

Cassinelli, C., M. Morra, G. Cascardo, C. Della Volpe, S. Siboni, M.A. Rodriguez, R. Verhoef, H. Schols, B. Jorgensen, P. Ulvskov, C. Bussy, M-D. Nagel, I. Liakos, D. Gilliland and G. Ceccone, 2006. Surface characterization and effects on SaOS-2 cell behavior of enzymatically tailored, surface-linked pectic polysaccharides. ESB 20th European Conference on Biomaterials, 27 Sept -1 Oct Nantes, France.

Ceccone, G., D. Gilliland, I. Liakos, G. Cascardo, C.Cassinelli, M. Morra, C. Della Volpe, S. Siboni, M.A. Rodriguez-Valverde, G. Ischia, G. Segatta, R. Verhoef, H. Schols, B. Jorgensen, P. Ulvskov and M-D. Nagel, 2008. Physico-chemical characterization and cell behavior influence of surfaces modified by enzymatically tailored pectic hairy polysaccharides. 8th World Biomaterials Congress, 28 May -1 June Amsterdam, the Netherlands.

Chen, C.S., M. Mrksich, S. Huang, G.M. Whitesides and D.E. Ingber, 1997. Geometric control of cell life and death. Science 276(5317): 1425-1428.

Chen, C.S., M. Mrksich, S. Huang, G.M. Whitesides and D.E. Ingber, 1998. Micropatterned surfaces for control of cell shape, position, and function. Biotechnol Prog 14(3): 356-363.

Jackson, C.L., T.M. Dreaden, L.K. Theobald, N.M. Tran, T.L. Beal, M. Eid, M.Y. Gao, R.B. Shirley, M.T. Stoffel, M.V. Kumar and D. Mohnen, 2007. Pectin induces apoptosis in human prostate cancer cells: correlation of apoptotic function with pectin structure. Glycobiology 17(8): 805-819.

Kalinina, S., H. Gliemann, M. López-García, A. Petershans, J. Auernheimer, T. Schimmel, M. Bruns, A. Schambony, H. Kessler and D. Wedlich, 2008. Isothiocyanate-functionalized RGD peptides for tailoring cell-adhesive surface patterns. Biomaterials 29(20): 3004-3013.

Keselowsky, B.G., D.M. Collard and A.J. Garcia, 2003. Surface chemistry modulates fibronectin conformation and directs integrin binding and specificity to control cell adhesion. J Biomed Mater Res A 66: 247-259.

Kokkonen, H.E., J.M. Ilvesaro, M. Morra, H.A. Schols and J. Tuukkanen, 2007. Effect of modified pectin molecules on the growth of bone cells. Biomacromolecules 8(2): 509-515.

Le Guéhennec, L., A. Soueidan, P. Layrolle and Y. Amouriq, 2007. Surface treatments of titanium dental implants for rapid osseointegration. Dent Mater 23(7): 844-54.

Manton, K.J., M. Sadasivam, S.M. Cool and V. Nurcombe, 2006. Bone-specific heparan sulfates induce osteoblast growth arrest and downregulation of retinoblastoma protein. J Cell Physiol 209: 219-229.

Manton, K.J., L.M. Haupt, K. Vengadasalam, V. Nurcombe and S.M. Cool, 2007a. Glycosaminoglycan and growth factor mediated murine calvarial cell proliferation. J Mol Histol 38: 415-424.

Manton, K.J., F.M. Leong, S.M. Cool and V. Nurcombe, 2007b. Disruption of heparan and chondroitin sulfate signaling enhances mesenchymal stem cell derived osteogenic differentiation via BMP signaling pathways. Stem Cells 25(11): 2845-54.

Morra, M., 2007. Biomolecular modification of implant surfaces. Expert Rev med Devices, 4(3): 361-372.

Morra, M., C. Cassinelli, G. Cascardo, M-D. Nagel, C. Della Volpe, S. Siboni, D. Maniglio, M. Brugnara, G. Ceccone, H.A. Schols and P. Ulvskov, 2004. Effects on interfacial properties and cell adhesion of surface modification by pectic hairy regions. Biomacromolecules 5: 2094-2104.

Nagel, M-D., R. Verhoef, H. Schols, M. Morra, J.P. Knox, G. Ceccone, C. Della Volpe, P. Vigneron, C. Bussy, M. Gallet, E. Velzenberger, M. Vayssade, G. Cascardo, C. Cassinelli, A. Haeger, D. Gilliland, I. Liakos, M. Rodriguez-Valverde and S. Siboni, 2008. Enzymatically-tailored pectins differentially influence the morphology, adhesion, cell cycle progression and survival of fibroblasts. Biochim Biophys Acta. 1780(7-8): 995-1003.

Olano-Martin, E., G.H. Rimbach, G.R. Gibson and R.A. Rastall, 2003. Pectin and pectic-oligosaccharides induce apoptosis in *in vitro* human colonic adenocarcinoma cells. Anticancer Res 23(1A): 341-346.

Petrie, T.A., J.E. Raynor, C.D. Reyes, K.L. Burns, D.M. Collard and A.J. Garcia, 2008. The effect of integrin-specific bioactive coatings on tissue healing and implant osseointegration. Biomaterials 29(19): 2849-2857.

Re, F., A. Zanetti, M. Sironi, N. Polentarutti, L. Lanfrancone, E. Dejana and F. Colotta, 1994. Inhibition of anchorage-dependent cell spreading triggers apoptosis in cultured human endothelial cells. J Cell Biol 127(2): 537-546.

Reyes, C.D., T.A. Petrie, K.L. Burns, Z. Schwartz and A.J. García, 2007. Biomolecular surface coating to enhance orthopaedic tissue healing and integration. Biomaterials 28(21): 3228-3235.

Ridley, B.L., M.A. O'Neill and D. Mohnen, 2001. Pectins: structure, biosynthesis, and oligogalacturonide-related signaling. Phytochemistry 57(6): 929-967.

Schols, H.A. and A.G.J. Voragen, 2003. The chemical structure of pectins. In: G.B. Seymour and J.P. Knox (ed.) Pectins and their Manipulation. Blackwell Publishing, Oxford, United Kingdom, pp 1-29.

Schols, H.A., M.A. Posthumus and A.G.J. Voragen, 1990. Structural features of hairy regions of pectins isolated from apple juice produced by the liquefaction process. Carbohydrate Res 206(1): 117-129.

Schols, H.A., A.G. Voragen and I.J. Colquhoun, 1994. Isolation and characterization of rhamnogalacturonan oligomers, liberated during degradation of pectic hairy regions by rhamnogalacturonase. Carbohydrate Res 256(1): 97-111.

Tajima, S., J.S. Chu, S. Li and K. Komvopoulos, 2008. Differential regulation of endothelial cell adhesion, spreading, and cytoskeleton on low-density polyethylene by nanotopography and surface chemistry modification induced by argon plasma treatment. J Biomed Mater Res A. 84(3): 828-36.

Takamori, E.R., R. Cruz, F. Gonçalvez, R.V. Zanetti, A. Zanetti and J.M. Granjeiro, 2008. Effect of roughness of zirconia and titanium on fibroblast adhesion. Artif Organs 32(4): 305-309.

Voragen, A.G.J., W. Pilnik, J-F. Thibault, M.A.V. Axelos and C.M.G.C. Renard, 1995. In: A.M. Stephen (ed.) Food polysaccharides and their applications. Marcel Dekker Inc., New York-Basel-Hong Kong, pp 287-339.

Wu, S.C., J.C. Yu, S.H. Hsu and D.C. Chen, 2006. Artificial extracellular matrix proteins contain heparin-binding and RGD-containing domains to improve osteoblast-like cell attachment and growth. J Biomed Mater Res A 79: 557-565.

Index

A

B

C

F

G

H

N

O

P

Q

R

S